T0100623

Medial Representations

Computational Imaging and Vision

Managing Editor

MAX VIERGEVER
Utrecht University, The Netherlands

Series Editors

GUNILLA BORGEFORS, *Centre for Image Analysis, SLU, Uppsala, Sweden*
RACHID DERICHE, *INRIA, France*
THOMAS S. HUANG, *University of Illinois, Urbana, USA*
KATSUSHI IKEUCHI, *Tokyo University, Japan*
TIANZI JIANG, *Institute of Automation, CAS, Beijing*
REINHARD KLETTE, *University of Auckland, New Zealand*
ALES LEONARDIS, *ViCoS, University of Ljubljana, Slovenia*
HEINZ-OTTO PEITGEN, *CeVis, Bremen, Germany*
JOHN K. TSOTSOS, *York University, Canada*

This comprehensive book series embraces state-of-the-art expository works and advanced research monographs on any aspect of this interdisciplinary field.

Topics covered by the series fall in the following four main categories:
- Imaging Systems and Image Processing
- Computer Vision and Image Understanding
- Visualization
- Applications of Imaging Technologies

Only monographs or multi-authored books that have a distinct subject area, that is where each chapter has been invited in order to fulfill this purpose, will be considered for the series.

Volume 37

Medial Representations

Mathematics, Algorithms and Applications

Edited by

Kaleem Siddiqi and Stephen M. Pizer

Authors

Kaleem Siddiqi and Stephen M. Pizer – Principal Authors

James N. Damon and Peter J. Giblin – Principal Mathematics Authors

Nina Amenta   Diego Macrini

Gunilla Borgefors   Ingela Nyström

Sylvain Bouix   Jayant Shah

Robert Broadhurst   Gabriella Sanniti di Baja

Edward Chaney   Ali Shokoufandeh

Sunghee Choi   Gábor Székely

Sven Dickinson   Martin Styner

P. Thomas Fletcher   Timothy Terriberry

Qiong Han   Andrew Thall

Sarang Joshi   Paul Yushkevich

Benjamin B. Kimia   Juan Zhang

Fréderic F. Leymarie

Springer

Editors
Kaleem Siddiqi
McGill University
School of Computer Science
3480 University St.
Montreal QC H3A 2A7
Canada
siddiqi@cim.mcgill.ca

Stephen M. Pizer
Department of Computer Science
University of North Carolina
Campus Box 3175
Chapel Hill NC 27599-3175
Sitterson Hall
USA
pizer@cs.unc.edu

ISBN: 978-1-4020-8657-1   e-ISBN: 978-1-4020-8658-8

Library of Congress Control Number: 2008930851

© 2008 Springer Science + Business Media B.V.
No part of this work may be reproduced, stored in a retrieval system, or transmitted
in any form or by any means, electronic, mechanical, photocopying, microfilming, recording
or otherwise, without written permission from the Publisher, with the exception
of any material supplied specifically for the purpose of being entered
and executed on a computer system, for exclusive use by the purchaser of the work.

Cover illustration: WMXDesign Gmbh

Printed on acid-free paper.

9 8 7 6 5 4 3 2 1

springer.com

This book is dedicated to Harry Blum, who started it all.

Preface

The last half century has seen the development of many biological or physical theories that have explicitly or implicitly involved medial descriptions of objects and other spatial entities in our world. Simultaneously mathematicians have studied the properties of these skeletal descriptions of shape, and, stimulated by the many areas where medial models are useful, computer scientists and engineers have developed numerous algorithms for computing and using these models. We bring this knowledge and experience together into this book in order to make medial technology more widely understood and used.

The book consists of an introductory chapter, two chapters on the major mathematical results on medial representations, five chapters on algorithms for extracting medial models from boundary or binary image descriptions of objects, and three chapters on applications in image analysis and other areas of study and design. We hope that this book will serve the science and engineering communities using medial models and will provide learning material for students entering this field.

We are fortunate to have recruited many of the world leaders in medial theory, algorithms, and applications to write chapters in this book. We thank them for their significant effort in preparing their contributions. We have edited these chapters and have combined them with the five chapters that we have written to produce an integrated whole.

We are very grateful to Vrinda Narain and Lyn Pizer, both for their generous support and for hosting each of us as we met on several occasions in Montréal and Chapel Hill to write this book. We are grateful to one another for the many ideas we have shared, discussed and refined since this project began. We have learned much from this collaboration. We are indebted to the UNC graduate students and visitors who took the course on Medial Representations in the fall of 2006; their input in identifying needed improvements in a draft of this book was invaluable. We thank Delphine Bull and Kim Jones for their technical help with preparing this manuscript.

Montréal and Chapel Hill   *Kaleem Siddiqi*
July 2007   *Stephen M. Pizer*

Contributors

Nina Amenta
Department of Computer Science, University of California at Davis, USA.

Gunilla Borgefors
Centre for Image Analysis, Swedish University of Agricultural Sciences, Sweden.

Sylvain Bouix
Psychiatry Neuroimaging Laboratory, Harvard University Medical School, USA.

Robert Broadhurst
MIDAG group, University of North Carolina at Chapel Hill, USA.

Edward Chaney
MIDAG group, University of North Carolina at Chapel Hill, USA.

Sunghee Choi
Computer Science Division of EECS, Korea Advanced Institute of Science and
Technology, Korea.

James Damon
Department of Mathematics, University of North Carolina at Chapel Hill, USA.

Sven Dickinson
Department of Computer Science, University of Toronto, Canada.

P. Thomas Fletcher
Department of Computer Science, University of Utah, USA.

Peter Giblin
Department of Mathematical Sciences, University of Liverpool, UK.

Qiong Han
MIDAG group, University of North Carolina at Chapel Hill, USA.

Sarang Joshi
MIDAG group, University of North Carolina at Chapel Hill, USA.

Benjamin B. Kimia
Division of Engineering, Brown University, USA.

Frederic F. Leymarie
Department of Computer Science, Goldsmiths College, University of London, UK.

Diego Macrini
Department of Computer Science, University of Toronto, Canada.

Ingela Nyström
Centre for Image Analysis, Uppsala University, Sweden.

Stephen M. Pizer
MIDAG group, University of North Carolina at Chapel Hill, USA.

Gabriella Sanniti di Baja
Institute of Cybernetics "E. Caianiello," C. N. R., Pozzuoli, Naples, Italy.

Jayant Shah
Department of Mathematics, Northeastern University, USA.

Ali Shokoufandeh
Department of Computer Science, Drexel University, USA.

Kaleem Siddiqi
School of Computer Science and the Centre for Intelligent Machines, McGill
University, Canada.

Martin Styner
MIDAG group, University of North Carolina at Chapel Hill, USA.

Gábor Székely
Computer Vision Laboratory, ETH Zurich, Switzerland.

Timothy Terriberry
MIDAG group, University of North Carolina at Chapel Hill, USA.

Andrew Thall
Department of Computer Science, Allegheny College, USA.

Paul A. Yushkevich
Department of Radiology, University of Pennsylvania, USA.

Juan Zhang
School of Computer Science and the Centre for Intelligent Machines, McGill
University, Canada.

Contents

Part III Applications

9 Statistical Applications with Deformable M-Reps 269
Stephen Pizer, Martin Styner, Timothy Terriberry,
Robert Broadhurst, Sarang Joshi, Edward Chaney,
and P. Thomas Fletcher

Chapter 1
Introduction

Stephen Pizer, Kaleem Siddiqi, and Paul Yushkevich

Abstract In the late 1960s Blum (1967) first suggested that medial loci, later generalized and called symmetry sets and central sets by mathematicians (Yomdin, 1981; Mather, 1983; Millman, 1980), would provide an effective means of representing objects appearing in 2D images. Soon thereafter Blum suggested the extension of medial loci to objects in 3D images. These object representations have since come to have an important role in the description of shape. In this introductory chapter we explore the forms the medial representation of objects can take, and we give some history of its development and use. To start, Section 1.1 places the medial representation of objects into the context of alternative representations.

This chapter sets the context for the subsequent material covered in this book, which is organized in three parts. Part I of the book describes in detail the the mathematical properties of the medial representation and the relation between it and the corresponding object boundary. Section 1.2 of this chapter introduces this material by giving the basic mathematical definition and properties of the medial representation. Section 1.3 cites evidence for medial representations as models of human vision. Part II of the book presents algorithms that have been proposed for going from a boundary to a medial representation or from a medial representation to a boundary representation, and it discusses their performance. Section 1.4 of this chapter overviews such methods, including some that are not detailed later in the book. Finally, Part III of the book covers selected applications of the medial representation in image analysis. These include segmentation, shape characterization,

S. Pizer
Medical Image Display & Analysis Group, University of North Carolina at Chapel Hill, USA,
e-mail: pizer@cs.unc.edu

K. Siddiqi
School of Computer Science & Centre for Intelligent Machines, McGill University, Canada,
e-mail: siddiqi@cim.mcgill.ca

P. Yushkevich
Department of Radiology, University of Pennsylvania, USA,
e-mail: pauly2@grasp.upenn.edu

K. Siddiqi and S. Pizer (eds.) *Medial Representations – Mathematics, Algorithms and Applications*.
© Springer Science + Business Media B.V. 2008

recognition, object labeling, and registration. Section 1.5 of this chapter introduces these applications.

1.1 Object Representations

A variety of alternative means of representing objects or multi-object ensembles in 3D or 2D have appeared. The main alternatives to the medial representation fall into three categories. In *landmark representations* (Fig. 1.1e) an object or object ensemble is described by an ordered set of geometrically recognizable and salient locations on the object(s). Typically this is a sparse set, due to the difficulty of manually or automatically extracting a large set of landmarks from the 2D or 3D image data. In *boundary representations* (b-reps) the object or objects are represented by a relatively dense set of points sampling its (their) boundary (Fig. 1.1a), by a mesh of tiles whose vertices form a boundary sampling set but from which normal and curvature information can also be extracted (Fig. 1.1b), or by an orthogonal function decomposition of the boundary surface, e.g., by Fourier (spherical, in 3D) harmonics (Fig. 1.1c). In this decomposition the representation is formed by the collection of coefficients of the basis orthogonal functions. In *displacement by voxel representations* the object is represented by an atlas and a displacement vector field (Fig. 1.1d). The atlas is formed as a template label image giving for each voxel the name of the object class (including background as a class) in which the voxel falls in a template image. The displacement vector field gives a displacement of each voxel and thereby describes a mapping of space and thus of the object labels from the template image to a particular instance of the object or object ensemble. Finally, there is the medial representation (Fig. 1.1f), which, like the displacement by voxels representation, directly represents the object interior.

Landmark representations have a long and rich history. They have the major strength that, by the very choice of the landmarks, there is a strong correspondence between each specified landmark in one image and that in another image in the population. While their sparseness typically limits them to object descriptions of large spatial scale and the statistics has been largely in terms of positional displacements and not local changes of orientation or magnification, there has been an extensive study of their geometry and their statistics. Prolific and early authors on this subject are Bookstein (1991) and Kendall (1989). A very effective summary of statistics of landmark geometry is provided in Dryden and Mardia (1998). Methods to compute diffeomorphic warps from landmark displacements can be found in Joshi and Miller (2000). Medial representations typically capture many of the landmark correspondences but also allow the study, at many spatial scale levels, of such additional transformations as bending, widening, and elongation.

Boundary representations via tile meshes (e.g., Delingette, 1999) provide all of the capabilities of boundary representations via points and share the difficulty of analysis at larger scales. In addition, they allow access to boundary normals and curvatures and thus to the analyses of differential geometry. Nevertheless, they remain shell and not interior representations.

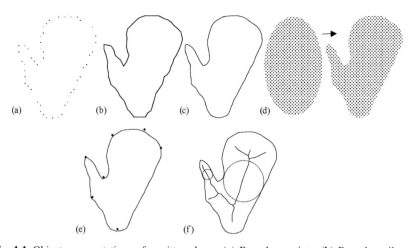

Fig. 1.1 Object representations of a mitten shape. (**a**) Boundary points. (**b**) Boundary tiles. (**c**) Fourier harmonics. This parametrized representation is by coefficients of sinusoids on the parameter. (**d**) Atlas displacements with labels. Shown are the binary labels (inside vs. outside the object) in a base space and the new locations of the labels after the displacements have been applied. (**e**) Landmarks. (**f**) Medial. Shown is the continuous locus of centers of tangent disks (spheres in 3D) and two of these disks, which in the representation are given at every center point. Notice how the natural landmarks, at boundary vertices, correspond to end points of the internal medial axis (shown) or the external medial axis (not shown)

Boundary representation via basis functions includes both spline fits and orthogonal function decomposition. These represent an object of a particular topology (e.g., spherical topology) by a smooth continuous mapping from the surface of a standard object of the same topology (in our example, a sphere). This vector mapping function can be decomposed into orthonormal functions (Kelemen et al., 1999). Being smooth functions, all derivatives of the object surface are available, and hence boundary normals and curvatures can be derived analytically. Nevertheless, interior properties are not directly accessible.

The displacement by voxel representation gives easy access both to boundaries and to interiors. It requires methodology to limit the displacement maps to diffeomorphisms (1-to-1 smooth warps); these typically involve velocity fields integrated via differential equations (Grenander and Miller, 1998). The difficulties of this representation derive from the large computer storage space required by the representation, the correspondingly large computations needed to derive and analyze these representations, and the difficulty of determining large scale properties from this small scale representation. However, both displacement by voxel and displacements of boundary representations provide a useful small scale description when combined with a medial representation describing the larger spatial scales.

The medial representation has special strengths in directly capturing various aspects of shape, in giving direct access to both object interiors and object boundaries, and in providing rich geometrical relationships within objects. In the remainder of this chapter we give a careful definition of this representation, and we introduce some of its basic mathematical properties. In addition we introduce some of the basic algorithms for extraction of the medial representation of an object, and we sketch some of the applications of this representation in computer vision and image analysis that will be covered in this book.

1.2 Medial Representations of Objects

The medial approach to representing an object (Figs. 1.1f, 1.2) is to describe a locus midway between (at the center of a sphere bitangent to) two sections of boundary and to give the distance to the boundary (called the medial radius), yielding the object as the union of overlapping bitangent spheres. One way to think of this representation is a locus of $(\mathbf{p}, r)$, where $\mathbf{p}$ gives the sphere center and r gives the radius of the sphere. In some representations the vectors from the medial point to the two or more corresponding boundary points are included; in others they are derived via first derivatives of the medial locus and of the radius function on that locus. When these vectors are included, the primitive, called a medial atom, is a hub point $\mathbf{p}$ with two equal-length spokes $\mathbf{S}^{\pm 1}$ (Fig. 1.2b). Blum (1967) described the interior medial locus by restricting the bitangent spheres to those entirely contained in the object's interior. The spokes of these medial atoms do not overlap and sweep out the object's interior.

The Blum medial axis is a transformation of an object boundary with the same topology as the object. Thus the boundary can generate the medial locus of $(\mathbf{p}, r)$ or the medial atoms, but equally the medial locus, in either form, can also generate the object boundary. In the first direction the transformation is a function, but in the second direction it is one-to-many, since a medial point describes more than one boundary point. In the boundary-to-medial mapping, low level boundary noise can be transformed into large changes in the medial locus. Therefore, the inverse of the boundary-to-medial mapping, which maps in the medial-to-boundary direction,

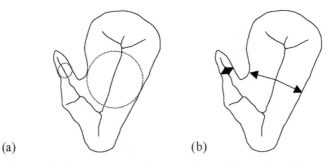

(a)   (b)

Fig. 1.2 (a) The medial spheres representation: a continuous locus (or collection of continuous loci) of medial spheres (disks in 2D). (b) The medial atoms representation, a continuous locus (or collection of continuous loci) of medial atoms, where a medial atom consists of a hub and two spokes

turns certain large changes in the medial locus into small changes in the boundary. We cover methods working in both directions in this book.

One of the strengths of using the medial representation as a primitive is that any unbranching, connected subset of the medial locus generates intrinsic space coordinates for the part of the object interior that the medial atoms in that subset reach. These coordinates can be thought of as follows: coordinate system location in the medial sheet, choice of spoke (left or right) and fraction of the spoke length along that spoke.

Two alternative definitions of the Blum medial locus have appeared. (1) In the so-called "grassfire" or "eikonal flow" definition the locus is the set of quench points along with their times of formation when a fire burning on equally dense grass within the object is lit at time $t = 0$ at all points on the object boundary. This has been shown to be equivalent to the location of shocks in a partial differential equation of motion at fixed speed in a direction initially normal to the object boundary (Kimia et al., 1995). (2) The "Maxwell set" definition of the medial locus is the set of locations internal to the object with more than one corresponding closest boundary point in the sense of Euclidean distance (Mather, 1983). Each point in this set is augmented with its distance to the boundary.

The Blum medial locus generalizes to the locus defined by all bitangent spheres. This locus is called the *symmetry set* and has been analyzed in some detail in the literature (Bruce et al., 1985; Giblin and Brassett, 1985). As an object deforms, the symmetry set can undergo a complex deformation. Likewise, as discussed in Chapter 3, as the medial locus and the associated spokes deform, the correspond-ing boundary can undergo a complex deformation. Other medial loci than the Blum medial locus and the symmetry set have also been derived from the set of spheres bitangent to the object. Asada and Brady (1986) suggested the use of the locus of middles of the chord connecting the two points of bitangency. This representation is also referred to as the *midpoint locus* and its properties in relation to that of the symmetry set are studied in Giblin and Brassett (1985). It is the basis for the skele-tonization technique developed in Zhu (1999). This locus turns out to be of special interest when specialized to the family of tubes, i.e., objects with circular cross-sections, whose skeletons turn out to be curves in three-space. In this case the chord center becomes the center of a disk orthogonal to the skeletal space curve.

Leyton (1992) described another locus derived from the set of spheres bitan-gent to the object boundary. This locus is defined by connecting the two points of bitangency along the shortest-distance geodesic path on the bitangent sphere and taking the associated point to be the center of this geodesic path. Leyton called this locus the Process Induced Symmetric Axis, or PISA. In his theory more complex objects are formed by protrusion or indentation processes acting on the bound-ary of simpler ones (Leyton, 1988, 1989). A formal justification for this view is a symmetry-curvature duality principle whereby the end points of this construction are related to curvature extrema on the object's boundary (Leyton, 1987; Yuille and Leyton, 1990).

Both the locus of chordal centers and the locus of geodesic centers require more information than that required by the Blum medial axis, namely just a point and a distance. Whereas the midpoint locus has been shown to be less singular than the symmetry set (Giblin and Brassett, 1985), reconstruction of the object from it is far from clear. Since neither of these medial loci have found broad acceptance or application, we will not cover them further in this book.

Medial loci based on spheres change topologically as an object deforms. This behavior was motivation to seek a medial axis that is stable under affine deformation. Such an axis, based on a consideration of affine area enclosed by a region was invented by Betelu et al. (2000); Niethammer et al. (2004). This construct is also interesting because it, unlike the Blum locus, is insensitive to noise. However, it is unintuitive and is very new, and as a result it has so far received little study or use. Therefore, we will not cover it in this book.

The major disadvantage of medial representations based on bitangent spheres is the fact that the branching structure of the medial locus is very sensitive to the small scale geometric aspects of the object. In Fig. 1.2 the lower joint of the thumb produces its own longish branch, and though this may be desirable in this case, an even smaller pimple due to noise would also produce a similar branch. Blum noticed this deficiency and coined the term ligature (meaning glue) to describe such portions of the medial axis that describe only small portions of the boundary such as regions of high negative curvature. Such ideas have lead to modifications of the medial axis to identify salient portions of it (August et al., 1999; Katz, 2002) but can also be overcome in other ways, including the marrying of the medial representation to the displacement by voxel representation or by displacement of the boundary representation at small scale.

The medial representation has a variety of strengths and as a result we have written this book. In particular:

1. It is an interior representation of the object and thus is subject to both geometric and mechanical operations applicable on the object's interior, such as bending, widening, elongation, and warping.
2. It provides rich geometric information, giving simultaneously locational, orientational, and metric (size) description in any locality of the interior and near exterior of an object.
3. It provides a basis for description at multiple spatial scales and thus provides efficiency of computation and efficiency in the number of population samples needed to estimate object geometry probabilities.
4. It allows one to distinguish object deformations into along-object deviations, namely elongations and bendings, and across-object deviations, namely bulgings and attachment of protrusions or indentations.
5. Its branches at the larger scales divide objects in a way that makes automatic object recognition effective.
6. It provides descriptions of objects and their geometric transformations that are intuitive to nonmathematical users.
7. It generates object-relative coordinate systems for object interiors and their near neighborhoods that provide useful correspondences across instances of an object.

8. It provides a means for describing the locational, orientational, and size relations between one object and a neighboring region of another object within a complex of objects (see Chapter 8).

It is often the case that the interior of an object is described as a binary image, e.g., a set of pixels in $\mathbb{R}^2$ or a set of voxels in $\mathbb{R}^3$. A frequent algorithmic objective for researchers is to transform this binary image, or equivalently the curve or surface that forms its boundary, to its medial axis. In this book we refer to this operation as the *Medial Axis Transform* or the MAT. If the output of the MAT is discretely represented, i.e., as a set of pixels or voxels, we will refer to the operation as the Discrete MAT.

In the remainder of this section we define medial loci and some of their properties in detail, discuss their history, sketch the basic algorithms for their extraction from boundaries and overview their applications in computer vision. This material is organized as follows. Section 1.2.1 precisely defines the *Blum medial locus*. Section 1.2.2 describes the structural geometry of Blum medial loci and their use for decomposing objects into simple figures. Section 1.2.3 explains the local geometric properties of Blum medial loci. Finally, Section 1.2.4 defines medial atoms and describes their use in skeletonization via deformable models. This section also provides an introduction to m-reps which are used as an object representation, with medial atoms as the building blocks.

1.2.1 The Definition of the Medial Locus

Medial loci enjoy wide use in computer vision and image analysis, as well as in other fields of computer science such as graphics, computer aided design, and human-computer interfaces (Bloomenthal and Shoemake, 1991; Sherstyuk, 1999; Storti et al., 1997; Blanding et al., 1999; Igarashi et al., 1999). The modern interest in medial loci originates with the work of Blum, who defined the medial locus of a two-dimensional object, studied its geometric properties, and noted its usefulness for describing the shape of objects (Blum, 1967; Blum and Nagel, 1978). While the definition itself is deceptively simple, the thorough understanding of the properties of the medial locus requires rigorous mathematical treatment. Only recently have a number of mathematically rigorous studies of the medial loci of higher-dimensional objects been published (Damon, 2003; Giblin and Kimia, 2002, 2004). These studies and their extensions are covered in Chapters 2 and 3.

We begin by defining the medial locus using a formulation that slightly extends Blum's original definition. Later in this section we mention some alternative definitions of the medial locus. The basic element of Blum's definition is the *maximal inscribed ball*.

Definition 1.2.1. Let S be a connected closed set in $\mathbb{R}^n$. A closed ball $B \subset \mathbb{R}^n$ is called a *maximal inscribed ball* in S if $B \subset S$ and there does not exist another ball $B' \neq B$ such that $B \subset B' \subset S$.

Next let us give a formal definition of the term *object* to avoid any ambiguity that the use of this word may bring.

Definition 1.2.2. A set in $\mathbb{R}^n$ is called an *n*-dimensional *object* if it is homeomorphic to the *n*-dimensional closed ball.

Let Ω denote an *n*-dimensional object and let $\partial\Omega$ denote its boundary. We are now ready to define the medial locus of Ω.

Definition 1.2.3. The *internal medial locus* of Ω is the set of centers and radii of all the maximal inscribed balls in Ω.

Definition 1.2.4. The *external medial locus* of Ω is the set of centers and radii of all the maximal inscribed balls in the closure of $\mathbb{R}^n/\Omega$.

Definition 1.2.5. The *Blum medial locus* of Ω is the union of its internal and external medial loci.

Definition 1.2.6. The tuple $\{\mathbf{p}, r\}$ that belongs to the Blum medial locus of an object Ω is called a *medial point* of Ω.

The medial locus is thus a subset of the space $\mathbb{R}^n \times [0, +\infty]$. We will sometimes use the term medial locus to refer just to the set of the centers of the maximal inscribed balls, forgetting about their radii. The terms *skeleton, medial axis*, and *symmetric axis* have been used by other authors to describe both the internal medial locus and the entire medial locus as a whole, with or without the inclusion of the radial component.

It turns out that the medial locus consists of a countable number of manifolds whose codimension in the space $\mathbb{R}^n \times (0, +\infty)$ is no less than 2. Hence, the medial locus of a two-dimensional object consists of a number of curves and isolated points, and the medial locus of a three-dimensional object consists of surface patches, curves, and isolated points. The manifolds composing the internal medial locus lie inside the object and are bounded, while the manifolds in the external medial locus lie outside of the object and extend to infinity. Figure 1.3a shows the internal and external medial loci of a two-dimensional object.

Moreover, it turns out that the inscribed disks whose centers and radii compose the medial locus of an object generically[1] are bitangent to the object's boundary. In fact, the medial locus is a subset of a more general geometric construct called the *symmetry set*, defined as the closure of the locus of centers and radii of all balls bitangent to the boundary of an object (Giblin and Brassett, 1985). The balls that generate the symmetry set are not restricted to lie either inside or outside of the boundary of the object. Hence, in addition to the medial locus, the symmetry set of an object contains connective structures such as the cusps shown on Fig. 1.3b.

The medial locus can also be defined analytically using the following *grassfire analogy*. The object is imagined to be a patch of grass whose boundary is set on

---

[1] The word *generic* is being used in a precise mathematical sense, as defined in the glossary.

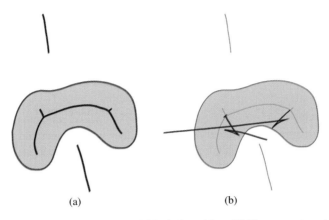

(a)   (b)

Fig. 1.3 (**a**) The internal and external medial loci of an object. (**b**) The symmetry set of the same object, which contains the internal and external medial loci (shown in blue) as well as some cusped structures

fire instantaneously. As the grass burns away, the fire fronts propagate from the boundary. This propagation can be described by the following differential equation:

$$\frac{\partial \mathscr{C}(t,p)}{\partial t} = -\alpha \mathbf{n}(p),$$

(1.1)

where $\mathscr{C}(t,p)$ denotes the fire front at time t, parameterized by p, $\mathbf{n}(p)$ is the unit outward normal to the fire front, and α is a constant, positive for inward propagation and negative for outward propagation. As the propagation progresses, segments of fire fronts that originate from disjoint parts of the boundary begin to meet and quench themselves at points that are called *shocks*. The medial locus is defined as the set of all the shocks, along with associated values of time t at which each shock is formed. This analytic definition of the medial locus is equivalent to the geometric definition given previously; a proof was given by Calabi (1965a,b); Calabi and Hartnett (1968). Kimia et al. (1995) combine the grassfire flow with an additional additive term based on the Euclidean curvature of the evolving front to yield a reaction-diffusion space for shape analysis.

1.2.2 Structural Geometry of Medial Loci

A rigorous description of the structural composition and local geometric properties of Blum medial loci of three-dimensional objects is given in Chapters 2 and 3 (see also Giblin and Kimia, 2000). The classification of types of points on the medial locus was also given in (Yomdin, 1981) and in (Mather, 1983). Their description classifies medial points based on the multiplicity and order of contact that occurs between the boundary of an object and the maximal inscribed ball centered at a medial point.

Each medial point $P = \{\mathbf{p}, r\}$ in the object Ω is assigned a label of form A_k^m. The superscript m indicates the number of distinct points at which a ball of radius r centered at $\mathbf{p}$ has contact with the boundary $\partial \Omega$. The subscript k indicates the order of contact between the ball and the boundary. The order of tangent contact is a number that indicates how tightly a ball B is fitted to a surface S at a point of contact P.

The following theorem specifies all the possible types of contact that can generically occur between the boundary of a three-dimensional object and the maximal inscribed balls that form its medial locus. The theorem also specifies how medial points with different associated type of contact are organized to form surfaces and curves in the medial locus.

Theorem 1.2.1 (Giblin and Kimia). *The internal medial locus of a three-dimensional object Ω generically consists of*

1. *Sheets (manifolds with boundary) of A_1^2 medial points*
2. *Curves of A_1^3 points, along which these sheets join, three at a time*
3. *Curves of A_3 points, which bound the free (unconnected) edges of the sheets and for which the corresponding boundary points fall on a crest*
4. *Points of type A_1^4, which occur when four A_1^3 curves meet*
5. *Points of type A_1A_3 (i.e., A_1 contact and A_3 contact at a distinct pair of points) which occur when an A_3 curve meets an A_1^3 curve*

Proof. See Giblin and Kimia (2000) for a rigorous proof.

In two dimensions, a similar classification of medial points is possible. The internal medial locus of a two-dimensional object generically consists of (i) curves of bitangent A_1^2 points, (ii) points of type A_1^3 at which these curves meet, three at a time, and (iii) points of type A_3 which form the free ends of the curves. The three classes of contact are illustrated in Fig. 1.4a. In two dimensions, A_3 contact means that the inscribed disk and the boundary osculate at a local maximum of boundary curvature.

The geometric properties of the external medial locus are similar to those of the internal locus, with the exception that the sheets and curves are no longer completely bounded and may stretch out to infinity. Less effort has been devoted in the literature to the study of external medial loci.

Theorem 1.2.1 states that each surface composing the internal medial locus of an object joins another two such surfaces or terminates at a point of type A_3^1, which correspond to ridges of curvature on the boundary surface. Similarly, curve segments composing the internal medial loci of two-dimensional objects either connect with pairs of other curve segments or terminate at points corresponding to positive maxima of boundary curvature. Hence, the number of such ridges or maximal points limits the number of surfaces and curves in the internal medial locus. It can be shown by induction that the number of curve segments composing the internal medial locus of an object whose boundary has M positive maxima of curvature may not exceed $2M - 3$.

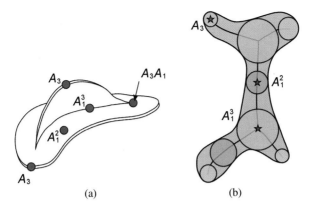

Fig. 1.4 (a) Different classes of points that compose the medial locus of a three-dimensional object, as categorized by Giblin and Kimia. (b) Three possible ways in which maximal inscribed disks can be tangent to the boundary of a two-dimensional object

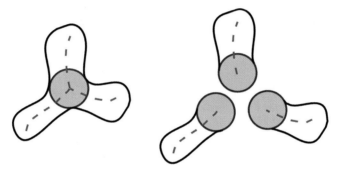

Fig. 1.5 Decomposition of a planar object into figures with joints. Each curve in the medial locus corresponds to a single figure

We will use the term *stratum* to refer collectively to curves in medial loci of two-dimensional objects and to surfaces in medial loci of three dimensional objects. The composition of medial loci into interconnected strata makes it possible to decompose geometrically complex objects into simple components called *figures* (Fig. 1.5). Roughly speaking, a figure is the part of an object that corresponds to a particular stratum in the medial locus. A particularly simple mathematical definition of a figure is the following:

Definition 1.2.7. The union of closed balls whose centers and radii form a single stratum in the medial locus of an object is called a figure with joint.[2]

Figures generated by strata belonging to the internal medial locus of an object are bounded, and the union of all such figures is the object itself. The intersection of a pair of figures with joints is non-empty if the generating strata of the two figures

---

[2] As distinguished from *figure*, which will be discussed later in the context of m-reps.

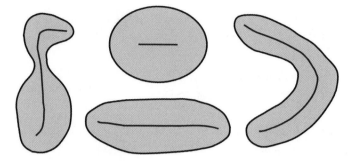

Fig. 1.6 All four of these objects fall into the category of *figures with joints* according to Definition 1.2.7, even though none of them have an actual "joint". Notice that the figure on the right has more than two positive maxima of curvature

are connected. This non-empty intersection is called the *joint,* and it comprises of balls of triple boundary contact.

The internal medial locus of a figure has only a single stratum and figures can be said to be geometrically simple and easier to study than whole objects. Figure 1.6 shows some examples of two-dimensional figures with joints.

The relationship between the structure of symmetry sets and the extrema of boundary curvature of two-dimensional objects are central to Leyton's theory of symmetry (Leyton, 1987). For planar objects, Leyton's symmetry-curvature duality theorem states that

> Any section of curve, that has one and only one curvature extremum, has one and only one symmetry axis. This axis is forced to terminate at the extremum itself.

The extension of this theorem to three dimensions is given by Yuille and Leyton (1990).

Leyton's theory states that the curves composing the symmetry set of an object represent the history of events that have formed the object. According to Leyton's postulates, "memory is always in the form of asymmetry," meaning that asymmetry makes it possible to recover information about the formation of an object, while "symmetry is always the absence of memory." He suggests that the extrema of curvature are the places where the boundary has been pushed in from the outside or pushed out from the inside, indicating a growth or deformation process. The medial curves that terminate at these extrema are in a sense arrows in the direction of the push. Hence, the symmetry set is a diagram of protrusion and indentation operations that have been applied to an object (Leyton, 1992).

1.2.3 Local Geometry of Medial Loci

Prior to describing the local geometry of Blum medial loci, let us introduce a useful notation for referring to the points of contact between a ball placed at a medial point and the boundary of an object.

Definition 1.2.8. If $P = \{\mathbf{p}, r\}$ is a medial point of an object Ω, then the set of points of contact between a ball of radius r centered at $\mathbf{p}$ and $\partial\Omega$ is called the *boundary pre-image* of P.

In other words, a medial point labeled A_k^m has a boundary pre-image that contains m points. Since for most of the medial points $m = 2$, the following definition is quite useful.

Definition 1.2.9. If points A and B form the boundary pre-image of a medial point P, then A is called a *medial involute* of B and vice versa.

Said in another way, medial involutes are pairs of points on the boundary of an object that are symmetric with respect to the medial locus. It is possible for a point to have multiple medial involutes, for example one with respect to the interior medial locus and one with respect to the exterior medial locus.

The major part of Blum's work on the internal medial loci of two-dimensional objects is devoted to the study of the geometric relationships between medial points and their boundary pre-images (Blum, 1967; Blum and Nagel, 1978). Blum showed that the points in the boundary pre-image can be expressed in terms of the position and radius of the medial point and from their derivatives with respect to movement along the medial locus.

For the purposes of studying local geometry of internal two-dimensional medial loci it suffices to focus on medial points that lie on the interior of the curves composing the medial locus and thus have two-point boundary pre-images. The geometric properties of the free end points and connecting end points of medial curves can be derived as the limit cases of the interior point properties.

In addition to using the position $\mathbf{p}$ and the radius r to characterize each point on the medial locus, Blum uses two first order properties. The first property is the slope of the medial curve at the medial point, which can be expressed as a unit-length tangent vector $\mathbf{U}^0$. The second is called the *object angle* and is given by

$$\theta = \arccos\left(-\frac{dr}{ds}\right), \tag{1.2}$$

where s is the arc length along the medial curve. The object angle is indicative of the narrowing rate of the object with respect to movement along the medial curve. When the object angle is equal to $\pi/2$, the radius has a critical point, and as one moves along the medial locus, the object retains its local width.

A medial point can be characterized by $\mathbf{p}, r, \mathbf{U}^0$, and θ, or preferably, by $\mathbf{p}, r, \mathbf{U}^{+1}$, $\mathbf{U}^{-1}$, where $\mathbf{U}^{+1}$ and $\mathbf{U}^{-1}$ are unit-length vectors orthogonal to the boundary of the object at $\mathbf{b}^{\pm 1} = \mathbf{p} + r\mathbf{U}^{\pm 1}$.

Figure 1.7a describes the local geometry of a medial point and associated boundary pre-image points $\mathbf{b}^{-1}$ and $\mathbf{b}^{+1}$. The angle formed by the points $\mathbf{b}^{-1}$, $\mathbf{p}$, and $\mathbf{b}^{+1}$ is bisected by the vector $\mathbf{U}^0$, the unit tangent vector of the medial curve at $\mathbf{p}$. The angle between $\mathbf{U}^0$ and the vectors $\mathbf{b}^{-1} - \mathbf{p}$ and $\mathbf{b}^{+1} - \mathbf{p}$ is the object angle θ.

The quantities $\mathbf{p}, \mathbf{b}^{\pm 1}, r, \mathbf{U}^{\pm 1}, \mathbf{U}^0$, and θ appear frequently in this book. To better remember these quantities, consider an analogy between a medial point and a

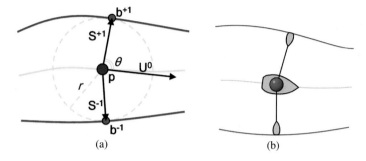

(a)   (b)

Fig. 1.7 Local medial geometry. (**a**) Local geometric properties of a medial point and its boundary pre-image. (**b**) The rowboat analogy for medial points: the oars (spokes) are $\mathbf{S}^{\pm 1} = r\mathbf{U}^{\pm 1}$

one-person rowboat, illustrated in Fig. 1.7b. The position of the rower in the boat corresponds to $\mathbf{p}$, and the length of the oars corresponds to r. The vector $\mathbf{U}^0$ represents the direction in which the boat is moving and θ is the angle that each oar makes with $\mathbf{U}^0$. The points $\mathbf{b}^{-1}$ and $\mathbf{b}^{+1}$ correspond to the tips of the oars, and the directions of the oars are given by the vectors $\mathbf{U}^{+1}$ and $\mathbf{U}^{-1}$. The movement of a point along the medial locus is analogous to the rowboat navigating down the middle of a stream, with the rower adjusting his oars in such a way that their tips always just touch the banks of the stream (of course, the oars are made of a stretchable material, and as the boat moves, their length changes). A similar analogy to a wheel, corresponding to the bitangent disk, is made in m-rep literature, and the term *spoke* is used instead of the term *oar*. In this book we have adopted the term *spoke* for this vector between the medial locus and the boundary.

The values of $\mathbf{p}$, r, and their derivatives can be used to qualitatively describe the local bending and thickness of an object, as first shown by Blum and Nagel (1978). The measurements $\mathbf{p}$ and $\mathbf{U}^0$ along with the curvature of the medial curve describe the local shape of the medial locus, and subsequently describe how a figure bends at $\mathbf{p}$. A figure that has a line for its medial curve is symmetrical under reflection across that line. The measurement r describes how thick the figure is locally, while $\cos \theta$ describes how quickly the object is narrowing with respect to movement along the medial curve. A figure with a constant value of r has the shape of a worm.

Free ends of medial curves, where the maximal inscribed disk and the boundary osculate and the boundary pre-image contains a single point, are a limiting case of the bitangent disk situation. As our imaginary rowboat approaches such a point, its oars, i.e., the spokes, come closer and closer together until they collapse infinitely quickly at the end-point, forming a single vector in the direction $\mathbf{U}^0$. The object angle θ, which is equal to half of the angle between the spokes, is zero at such end points.

The geometry of medial loci of three-dimensional objects is harder to visualize and express than the planar medial geometry. A number of researchers have studied the differential geometry of three dimensional medial loci (Nackman, 1982; Vermeer, 1994; Gelston and Dutta, 1995; Hoffmann and Vermeer, 1996; Teixeira, 1998). However, the basic relations are not too different from the two-dimensional case. Instead of bitangent circles, we have bitangent spheres, with the spokes still forming normal vectors to the object boundary and their vector difference forming a normal $\mathbf{N}$ to the medial locus. Generically the medial locus forms two-dimensional manifolds and there are still two spokes at all but branch points and end points. The edges of the medial locus form space curves, as do the branches. Branch curves typically end; the normal situation is of a branching sheet attached to a parent sheet to form a fin.

In non-generic situations, sections of the sheet can degenerate to a space curve, in which case the object is a tube, i.e., has circular cross section and the medial atom becomes an entire cone of spokes with the $\mathbf{U}^0$-vector as their axis. The relation between the $\mathbf{U}^0$ vector, the object angle θ and derivatives of r on the medial sheet becomes

$$\nabla r = -\mathbf{U}^0 cos(\theta),$$

where the gradient is intrinsic to the medial sheet.

The local reconstruction of the boundary from the medial axis or the symmetry set is discussed in Chapter 2 (Sections 6 and 7) following which the relationship between medial points of three-dimensional objects and the boundary pre-images of these points is presented in detail in Chapter 3. Many of the results reported in Chapter 3 are based on Damon's work on *skeletal structures* (Damon, 2003). This work follows the generative approach to medial geometry, as opposed to the previously described approaches that derive the medial locus from the boundary description of a given object. In the generative approach, the medial locus is defined first, and the object and its boundary are generated by an outward flow from it. Since the medial locus can be defined to contain a fixed number of figures with a specified topology, this approach to medial geometry is often better suited to problems of object modeling and shape description than the approaches deriving the medial axis from the object boundary. In the latter class of methods the derived medial loci can vary greatly and can thus be difficult to compare or analyze. The generative approach is the cornerstone of m-rep methodology, which will be introduced in the following section.

Damon's skeletal set is a stratified set[3] of arbitrary dimension, on which a multi-valued vector field $\mathbf{S}$, called the *radial vector field*, is defined. At most points in the skeletal set a pair of radial vectors is defined; these vectors point in the different directions relative to the tangent space of the skeletal set. At edges of skeletal manifolds that are not shared by more than one manifold, the radial vectors come together to form a single vector that lies in the tangent space of the manifold. At shared edges, more than two radial vectors are defined. The end points of the radial vectors form a locus that is called the boundary of the object described by the skeletal set. Damon describes a number of constraints that must be satisfied by the skeletal set and the radial vector field in order for the boundary to be continuous and differentiable. These constraints are expressed in terms of the *radial shape operator* S_{rad} and a skeletal *edge shape operator* S_E, which measure how the radial vectors bend with

---

[3] As described in the glossary, a stratified set consists of interconnected *manifolds* of different codimensions.

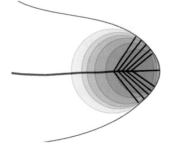

Fig. 1.8 Medial geometry of end atoms. The continuous relationship between a point on a medial curve and the points of contact between a disk inscribed at the point and the boundary of the object asymptotes at the end point of the curve: equal steps along the medial curve result in increasing steps along the boundary

respect to the skeletal set. These operators not only describe the local properties of the radial vector field but can also be used to express the local differential geometry of the boundary.

Another way to view the skeletal set for an object is that the skeletal locus is a fully cyclic piece of plastic wrap fit on both sides of the sheets forming the medial counterpart and that each point on the plastic wrap has a single spoke emanating from it, with none of the spokes crossing. Thus, at points that are neither branch points nor end points, there are two sheets of the plastic wrap touching each other and thus two spokes emanating, one from each sheet. These spokes need not be of equal length. At branch points there are three pieces of plastic wrap touching, and at end points the plastic wrap doubles back on itself and there is a single spoke, as illustrated in Fig. 1.8. Damon has shown that a natural measure on the skeletal sheet is the product of ordinary Riemannian distance on the sheet and the sine of the angle between the spoke vector and the tangent plane of the sheet. In Chapter 3 it is shown that this measure and a skeletal generalization of the shape operator can be used to pull back integrals over the object interior or its boundary onto the skeletal sheet. The Blum medial locus can be constructed as a special case of the general skeletal set by requiring that the radial vectors at each point of the skeletal set be symmetric with respect to its tangent space.

1.2.4 Medial Atoms and M-Reps

In this section we show how medial atoms can be used as building blocks for a particular type of object representation called m-reps. Via this scheme a collection of medial atoms can be used to come up with an approximate representation of a graph of figures with fixed topology. Such a view can provide advantages when the task is to draw comparisons across a population of similar structures such as that obtained by drawing several instances from a particular class of objects.

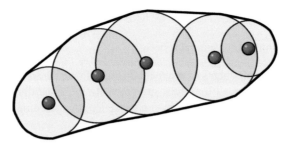

Fig. 1.9 Boundary of an object reconstructed by shrink-wrapping a collection of order 0 medial atoms

1.2.4.1 Medial Atoms

A medial atom is a modeling primitive that represents a place on the medial locus of an object. A medial atom describes such a place up to a specified differential order and with a specified level of tolerance. The continuous medial locus of an object can be discretely sampled into a set of medial atoms. However, medial atoms should be thought of as entities that exist independently of any medial locus, as demonstrated by the following definition.

Definition 1.2.10. An *n-dimensional medial atom of order 0* $(n = 2, 3)$ is a tuple $\mathbf{m} = \{\mathbf{p}, r\}$ that satisfies

$$\mathbf{p} \in \mathbb{R}^n, \ r \in \mathbb{R}^+. \tag{1.3}$$

Geometrically, a medial atom of order 0 is simply interpreted as a ball. Such medial atoms essentially correspond to maximal inscribed balls whose centers and radii form the medial loci of objects. Given a structured collection of medial atoms of order 0 sampled from the medial locus of an object, it is possible to approximately reconstruct the object's boundary by "shrink-wrapping" a sheet around the balls defined by the atoms, as shown in Fig. 1.9.

The use of the word "order" in the above definition refers to the fact that medial atoms can be used to approximate medial loci up to a given order. An order 0 atom describes zeroth order medial properties, which are position and radius. An order 1 atom, which is defined below, describes the derivatives of position and radius.

The shortcoming of medial atoms of order 0 and the shrink-wrap boundaries derived from them lies in the fact that while both the medial locus and the boundary can be approximately recovered from a set of order 0 medial atoms, the local symmetry relationships between pairs of medial involutes can not be directly expressed. However, medial atoms of higher order can be used to capture these symmetry relationships and are hence more useful for object representation.

Definition 1.2.11. An *n-dimensional medial atom of order* 1 is a tuple $m = \{\mathbf{p}, r, \mathbf{U}^{+1}, \mathbf{U}^{-1}\}$ that satisfies

$$\mathbf{p} \in \mathbb{R}^n, \; r \in \mathbb{R}^+, \; \mathbf{U}^{+1} \in S^{n-1}, \; \mathbf{U}^{-1} \in S^{n-1}, \tag{1.4}$$

where S^n is the unit n-dimensional sphere.

The additional components of the medial atom of order 1 are the two spoke orientations, $\mathbf{U}^{+1}, \mathbf{U}^{-1}$. Recall that these are precisely the same first order quantities that were used, after multiplication by the spoke length r to compute $\mathbf{S}^{\pm 1}$, to relate medial points to their boundary pre-images (Fig. 1.7).

The rowboat analogy used to describe the first order geometry at a point on the medial locus can be used equally well to describe medial atoms of order 1. The atom's position and radius correspond to the position of the rower and the length of the oars (spokes); the orientation corresponds to the direction of the boat's bow, and the object angle to the angle between the bow and the oars (spokes). Unlike their order 0 cousins, which one could visualize as an inflated rubber tube, medial atoms of order 1 explicitly define a pairs of points on the boundary of the object that they describe. These points are, of course, the tips of the spokes and are given by the following definition.

Definition 1.2.12. The tuples $\{\mathbf{b}^{-1}, \mathbf{U}^{-1}\}$ and $\{\mathbf{b}^{+1}, \mathbf{U}^{+1}\}$ given by $\mathbf{b}^{\pm 1} = \mathbf{p} + r\mathbf{U}^{\pm 1}$ are called the *implied boundary nodes* of the medial atom $m = \{\mathbf{p}, r, \mathbf{U}^{+1} \mathbf{U}^{-1}\}$. Here $\mathbf{U}^{\pm 1} = (cos(\theta), \pm \sin(\theta))^T$ in 2D and $(cos(\theta), \pm \sin(\theta), 0)^T$ in 3D in the coordinate system whose first basis vector is a unit bisector $\mathbf{U}^0$ of $\mathbf{U}^{+1}$ and $\mathbf{U}^{-1}$ and whose second basis vector is a unit vector in the direction of $\mathbf{U}^{+1} - \mathbf{U}^{-1}$, where θ is the object angle, i.e., half the angle between $\mathbf{U}^{+1}$ and $\mathbf{U}^{-1}$.

Medial atoms of order 1 correspond to medial points of type A_1^2, whose boundary pre-images contain two points.[4] Medial points of other types can be represented either using special medial atoms, as discussed later in Section 1.2.4.2, or as limit cases of A_1^2 medial points. In this section we deal only with order 1 medial atoms because these atoms contain all the information necessary to reconstruct the boundary pre-images of medial points and the symmetry relationships between pairs of medial involutes.

The agreement between a medial atom and an image is measured by a *medialness function*, the domain of which is the set of parameters defining a medial atom. Given a tuple of parameter values, the medialness function measures how well the position and orientation of the boundary nodes of a medial atom defined by these parameters match edge-like structures in the image.

As illustrated in Fig. 1.10, the medialness function most widely used in the core tracking literature is computed using the image intensity gradient at boundary node locations. Given an image I and an order 1 medial atom $\underline{\mathbf{m}} = \{\mathbf{p}, r, \mathbf{U}^{+1}, \mathbf{U}^{-1}\}$, this function is defined as

$$M(\underline{\mathbf{m}}) = \nabla_\sigma I(\mathbf{b}^{-1}) \cdot \mathbf{U}^{-1} + \nabla_\sigma I(\mathbf{b}^{+1}) \cdot \mathbf{U}^{+1}, \tag{1.5}$$

---

[4] The A_i^j taxonomy of medial points, as used by Giblin and Kimia (2000), was discussed in Section 1.2.1.

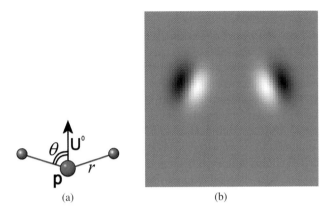

(a)   (b)

Fig. 1.10 Medial atoms used for core tracking. (**a**) A medial atom defined by position **p**, radius r, orientation determined by $\mathbf{U}^0$ and object angle θ. (**b**) A Gaussian derivative filter associated with the atom, as used for core tracking

where $\nabla_\sigma I(\mathbf{b})$ is the image gradient computed at the point $\mathbf{b}$ by convolution with the gradient of the isotropic Gaussian kernel with aperture σ:

$$\nabla_\sigma I(\mathbf{b}) = \int \nabla G_\sigma(\mathbf{b} - \mathbf{z}) I(\mathbf{z}) \, d\mathbf{z}. \tag{1.6}$$

The aperture σ is proportional to the radius of the medial atom. This proportionality makes the medialness function invariant to the magnification of the structures in the image.

1.2.4.2 M-Reps: A Medial Object Representation

First-order medial atoms serve as the building blocks for m-reps, which have the following distinguishing properties:

- The medial locus of an object is represented explicitly.
- A smooth approximate representation of the object's boundary, with tolerance, is implied by the medial locus representation.
- An accurate description of the boundary is given by a smooth fine-scale deformation of the implied boundary.

The explicit specification of an object's medial locus by m-reps makes it possible to compare similar objects in terms of symmetries. In contrast, medial loci yielded by applying a skeletonization method to similar objects can pose challenges for comparison because their branching topology may differ, as in the case of nearly circular objects.

M-reps come in different flavors. Discrete m-reps, due to Pizer et al. (1999), represent the medial locus using a structured sparse set of medial atoms (Fig. 1.11

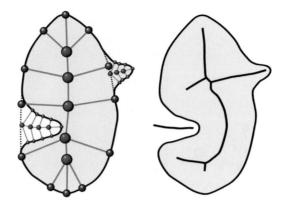

Fig. 1.11 (*Left*). Representation of an object using a discrete m-rep. The m-rep is organized into a hierarchy of figures based on structural properties of the object. At the root of the hierarchy lies the main figure whose implied boundary is indicated by the dotted curve. The children of the main figure are the protrusion and indentation figures. (*Right*). Continuous medial locus of the same object. Branches in the medial locus are determined by the geometry of the boundary and include branches that do not contribute to the structural description of the object

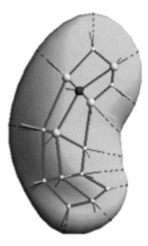

Fig. 1.12 A three-dimensional discrete m-rep figure of a kidney organized as a 3 × 5 quadrilateral mesh of medial atoms. The atoms with a larger light hub are the neighbors of the atom with a darker hub. The atom with a darker hub and the two atoms below it are interior medial atoms; the rest are end atoms.

shows a 2D example, and Fig. 1.12 shows a 3D example). Continuous m-reps, defined in Yushkevich et al. (2003), represent the medial locus as approximation curves or surfaces defined by a set of $\{\mathbf{p}, r\}$ control points. This section describes discrete m-reps.

M-reps specify the figural composition of an object explicitly. An m-rep representing a complex object contains multiple components which we shall refer to as *figures*. Each figure is a sheet of medial atoms. These figures are organized into a hierarchy of parent-child relationships, with parents representing the substantial parts of the object, such as the palm of the hand, and children representing protrusions and indentations, such as the fingers. The figural graph of an m-rep resembles the composition of a geometric object into figures with joints but does so only at a conceptual level. The manner in which an m-rep is organized into figures is guided by structural, conceptual, and populational properties, rather than by a desire to precisely mimic the medial branching topology of the objects being represented. The difference between the figural composition of m-reps and the branching topology of corresponding objects is illustrated in Fig. 1.11.

Various means for representing m-reps in a computer are discussed in Chapter 8.

In direct contrast to skeletonization methods, m-reps derive the boundary description of an object from its medial description. Since the composition of an m-rep into figures and atoms is explicitly imposed, one can describe different objects using m-reps with the same figural composition. This makes it possible to then compare such objects based on their medial properties. In particular, one only needs to compare the values of the medial atom parameters, assuming that medial atoms represent corresponding locations in the objects. This type of analysis makes an implicit assumption that the instances being compared are well described by the imposed m-rep fit. The alternative is to attempt to use the more standard skeletonization approaches to obtain a representation, but then one has to compare medial structures with different branching topologies. There is a growing literature on the use of graph theoretic methods for matching skeletal structures; some of this material is covered in Chapter 10. However, these techniques are more complex than the comparison of m-reps. For example, the use of m-reps with a common figural composition makes it possible to estimate probability distributions on the parameters of medial atoms and hence makes medial based shape characterization across a population of similar objects possible.

1.3 Psychophysical and Neurophysiological Evidence for Medial Loci

The medial representations described in Section 1.2 have not only found many applications by engineers but have also been found to be relevant components of human vision models. For example, there is literature which suggests that figural decomposition of objects using medial loci might correspond to the cognitive processing performed by the human brain. Rock and Linnett (1993) argue that figures are captured preattentively by the human visual system. Other studies have shown that figures which arise from locations of extrema of negative curvature on the boundary are often associated with the visual decomposition of objects (Hoffman and Richards, 1984; Biederman, 1987; Braunstein et al., 1989; Siddiqi et al., 1996; Singh and Hoffman, 1997). The junctions of figures can also have special importance, matching Biederman's work on the junctions of visual parts (Biederman, 1987). Whereas we shall not cover this literature in great detail in this book, in this section we review some of the findings that point to a role for medial loci in shape perception. We shall also cover the literature that provides neurophysiological evidence for medial axes.

Among the first reported psychophysical data is that of Frome (1972), who examined the role of medial axes in predicting human performance in shape alignment tasks. In these experime nts subjects were required to position an ellipse so as to be parallel to a reference line. It was found that for this task, the acuity with which the stimulus was placed could be explained by the length of the medial axis within the ellipse, i.e., the straight line connecting the centers of curvatures corresponding to the end points of the major axis of the ellipse.

Later, Psotka (1978) examined the role of medial loci in representing the perceived form of more complex outlines. Subjects were given outline forms (a square, a circle, a humanoid form and various rectangles) and were asked to draw a dot within each outline in the first place that came to mind. The superimposed dots for each outline were found to coincide well, for the most part, with the locations predicted by Blum's grassfire flow as opposed to locations suggested by other field theories of form perception that had been proposed, such as McCulloch's size constancy proposal (McCulloch, 1965).

Adopting a different experimental procedure, the effect of closure on figure-ground segmentation along with a possible role for medial loci was examined by Kovács and Julesz (1993, 1994) and Kovács et al. (1998). In the experiments reported in Kovács and Julesz (1993) subjects viewed a display of Gabor patches (GPs) aligned along a sampled curve presented in a background of randomly oriented Gabor patches playing the role of distractors. Using a two alternative forced choice paradigm, subjects were required to decide whether the display included an open curve or a closed one. The percentage of correct responses were recorded as a function of the separation distance between successive GPs. A significant advantage was found for the correct detection of configurations of closed (roughly circular) targets. In a second series of experiments subjects were required to detect a target GP of varying contrast placed either inside or outside a circular arrangement of GPs in a field of distractors. It was found that the contrast threshold at which the target could be detected was decreased by a factor of 2 when it was located at the center of the circle, as opposed to the periphery, suggesting a special role for a medial location.

This second finding of increased contrast sensitivity at the center was later examined more carefully in Kovács and Julesz (1994). For an elliptical configuration of GPs it was found that the peak locations of increased contrast sensitivity were in fact predicted by a type of medial model. Specifically, these locations coincided with the local maxima of a D_ε distance function, representing at each location the percentage of boundary locations over the entire outline that were equidistant within a tolerance of ε. Examples of the D_ε function for more complex forms were presented in Kovács et al. (1998) (see Fig. 1.13) along with the proposal that its maxima could play an important role in form perception tasks such as the processing of motion (see also Chapter 11, Section 11).

Whereas the above notion of an ε parameter is fixed as a global quantity, a different notion of scale provides the key motivation for medially subserved perceptual

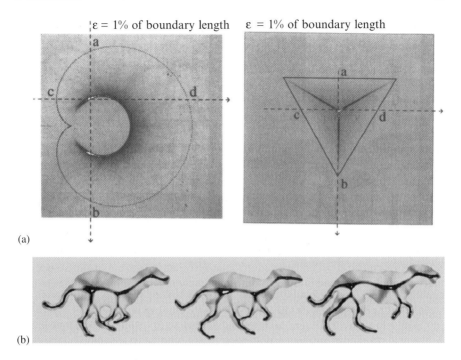

Fig. 1.13 (Adapted from (Kovács et al., 1998, Fig. 2, 8, 9 & 10), with permission from Ilona Kovács.) The function D_ε for various objects. Dark shading corresponds to increasing values of D_ε, and the "white spot" denotes its maximum. (**a**) D_ε for a cardioid on the left and a triangle on the right; cross-sections through maximum loci are indicated as dotted lines a–b and c–d. (**b**) D_ε for a few frames in a sequence depicting the movement of an animal

models based on *cores* (Burbeck and Pizer, 1995), which are discussed in Section 1.4.4. Underlying this model is the hypothesis that the scale at which the visual system integrates local boundary information is proportional to object width at medial loci. This view is corroborated by the findings of Burbeck et al. (1996), where the core model was shown to explain human performance in bisection tasks on shapes. In these experiments elongated stimuli were created by placing two sinusoidal waves in phase, side by side, and then filling in the region in between. An example of such a "wiggle" stimulus is shown in Fig. 1.14. The amount that it is perceived to bend depends on the frequency and amplitude of the sinusoids. For any given stimulus a subject was asked to judge whether a probe dot, placed between two sinusoidal peaks, appeared to the left or right of the object's center. By varying the position of the probe dot the perceived center was chosen to be the point about which a subject was equally likely to choose left or right in the task. The perceived central modulation was then defined as the horizontal distance between the centers for a successive left and right peak. The experiments revealed that for a fixed width the central modulation increased with increasing amplitude but decreased with increasing frequency. Furthermore, the modulation effects were greater for a narrow object,

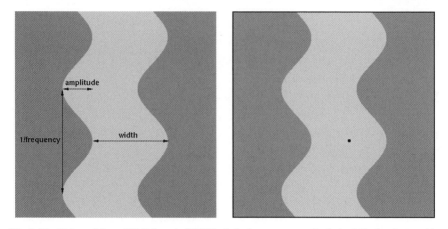

Fig. 1.14 (Adapted from Siddiqi et al. (2001)). *Left*: the geometry of a "wiggle" stimulus used in Burbeck et al. (1996). *Right*: the task is for the subject to judge whether the probe dot is to the left or to the right of the object's center

in a manner that was adequately explained by the linking of object boundaries at a scale determined by object width, as predicted by the core model.

These wiggle stimuli were revisited by Siddiqi et al. (2001), who showed that in fact the perceived centers in the study of Burbeck et al. (1996) were located precisely on the Blum medial axis, at locations that coincided with local maxima of the radius function. In several experiments using similar stimuli, but with varying degrees of translation between the sinusoidal boundaries, properties of medial loci were shown to account for human performance in shape discrimination tasks using a visual search paradigm.

Taken together, the above body of work provides a wealth of support for the role of medial loci in shape perception. Unfortunately, far less research has been carried out to provide neurophysiological support for medial axes. The one exception is the work of Lee (1995) and Lee et al. (1998) where neurons were isolated in the primary visual cortices of awake rhesus monkeys and their response to a set of texture images was examined. In the first study (Lee, 1995) the input images consisted of either a linear boundary with two regions of contrasting texture, or a rectangular strip or a square on a background of contrasting texture. In each case the texture was comprised of scattered bars in a vertical or horizontal orientation. The findings revealed a subset of neurons that had a peak response when their receptive fields were centered at the texture boundaries. Some of these neurons also had a sharp response when centered at the center of the rectangular strip or square, i.e., at locations predicted by the Blum medial axis. The subsequent more comprehensive results reported in Lee et al. (1998) revealed that the neurons with interior response peaks appeared to focus in the vicinity of the center of mass of compact shapes (squares and diamonds) but along the entire medial axis for elongated shapes (rectangles). This finding is consistent with the special status attributed to a local maximum of the radius function, e.g., in the context of the wiggle bisection experiments and the contrast sensitivity enhancement experiments discussed earlier. However, the neurophysiological data in support of medial axes has not yet been corroborated in the literature by other researchers.

1.4 Extracting Medial Loci of Objects

The human vision models described in Section 1.3 suggest that medial information is somehow extracted from the contrast available at boundaries or from the boundaries themselves. How do we get a computer to extract medial loci?

The computer vision literature describes a large number of *skeletonization* methods, which extract medial loci of objects starting from some boundary representation. In most practical applications the object and its boundary are represented discretely, for example as a set of pixels of the same intensity in a characteristic image or as a mesh of points. Skeletonization is made difficult by the inherent sensitivity of the medial locus to the fine details of the boundary representation. Given two different discrete representations of the same object, the true Blum medial loci of the two representations can have a different *medial branching topology*, i.e., a different number of figures and a different connectivity graph between the figures.

Hence, the challenge of skeletonization is not to find the precise medial locus of an imprecisely specified boundary, but rather it is to compute an approximation of the medial locus that is consistent with respect to different discrete representations of the same object. Moreover, a good skeletonization method should yield similarly structured medial loci for objects that are similar objects and for versions of the same object that have undergone similarity transformations or magnification.

This section introduces the notions of distance transforms, the Hessian, thinning and pruning, which are common ingredients to many skeletonization algorithms. It then overviews the various approaches to extracting medial loci in the literature. The approach of shocks of the grassfire flow, interpreted as singularities of the Euclidean distance function, is developed in more detail in Chapter 4. Methods based on digital distance transforms are described in Chapter 5, and those based on Voronoi techniques are detailed in Chapters 6 and 7. The boundary evolution and Voronoi methods are also compared to approaches based on height ridges of medialness (cores) in the overview paper by Pizer et al. (2003b). A special property of these three approaches is that they provide a scale parameter that makes it possible to tune the accuracy with which the result matches the precise medial locus of the input boundary. The loci computed at larger values of the scale parameter generalize better to different discrete representations of the input object as well as to other similar objects.

1.4.1 Distance Transforms, the Hessian, Thinning and Pruning

A *distance transform* of a boundary is obtained by assigning to each location in space its distance to the closest point on the boundary. Thus, points which lie on the boundary are assigned a value of zero. The distance transforms in use in the skele-tonization literature typically adopt a notion of Euclidean or approximate Euclidean distance. It is also common to distinguish locations which lie in the interior of a boundary from those that lie in the exterior by a sign change, leading to the notion of a signed distance transform. Distance transforms turn out to be very useful represen-tations since their level curves or surfaces represent the locus of positions reached by successive iterations of a grassfire flow. Furthermore, their singularities coincide with the Blum medial axis. Thus, numerous approaches to skeletonization use a dis-tance transform to simulate the grassfire flow along with techniques to locate its singularities. These approaches benefit from the fact the distance transform values at each point on the medial locus give the radius function.

The *Hessian* matrix of a smooth approximation to a distance function provides eigenvalues that can be used to characterize the type of local symmetry of its level sets: in 3D the slab, the tube, and the sphere. While the tube is non-generic and the sphere doubly non-generic, the Hessian eigenvalues give graded results such as "a slab but almost tubular." With $\lambda_1, \lambda_2, \lambda_3$ the eigenvalues of the Hessian and with $|\lambda_1| \leq |\lambda_2| \leq |\lambda_3|$, we have

- 0-D ridges (nearly spherical) when $\lambda_1 \approx \lambda_2 \approx \lambda_3 \gg 0$
- 1-D ridges (nearly tubular) when $\lambda_1 \approx 0, \lambda_2 \approx \lambda_3 \gg 0$ and
- 2-D ridges (very much a slab) when $\lambda_1 \approx \lambda_2 \approx 0, \lambda_3 \gg 0$

The same analysis can also be applied to pseudo-distance functions, such as the smooth edge strength v in Section 1.4.3.

There exist a number of techniques in the literature to compute the medial axis by successively peeling layers of the distance transform, e.g., by using morphological *thinning*. Binary mathematical morphology (Serra, 1982; Matheron, 1988; Jonker and Vossepoel, 1995) is based on the operations of erosion and dilation of an object via a structuring element. The general idea is to carry out a process of erosion by a ball to the binary image iteratively, thinning the object until it is one pixel thick. These methods take advantage of the fact that successive erosion by this small struc-turing element is equivalent to erosion by a larger disk or ball and that the discrete approximation to the larger disk improves when a small element is successively applied. As the successive erosions by the small structuring element are applied, the methods check for pixel or voxel removals that change the topology of the skeleton from that of the original object. These voxels are marked as skeleton voxels. This class of algorithms actually has a long history in the pattern analysis literature. The advantage of such methods is that they can be computationally very efficient. How-ever, the challenge faced by such methods derives from the fact that the resolution of both the original object and the skeleton is given by an underlying rectangular lattice: a voxel is either on the skeleton, or it is not. The resulting medial loci can be sensitive to the rasterization of the object and may have difficulty in discerning the local geometry of the medial locus near branch points, particularly in 3D. More-over, the resulting medial loci can be sensitive to the rotation and magnification of the object prior to the imaging process.

As discussed earlier, an inevitable difficulty faced by skeletonization algorithms is their inherent sensitivity to small perturbations of the boundary of an object. Thus many methods result in an initial coarse computation of the skeleton, followed by a second stage of *pruning* in which components which are thought to represent insignificant boundary details are removed. The types of measures used for pruning are driven in part by the manner in which the original boundary is represented and the computational techniques that are used. For example, when the boundary is defined by a polygon or a mesh and Voronoi techniques are used to compute the medial locus, pruning functions are designed to measure the area or length contribution of a Voronoi edge or face to determine whether or not it should be kept (Chapters 6 and 7). On the other hand, when computations are carried out on a discrete lattice using digital distance transforms, pruning measures are designed to take into account invariance and/or stability properties of the reconstruction with or without a branch or sheet, as in Chapter 5. In both settings it is generally accepted that a pruning method should have the following properties:

1. It should preserve topology (homotopy type).
2. It should be continuous, i.e., small differences in the significance measure should result in small changes to the computed skeleton.
3. The significance measure should be local on the medial locus.

1.4.2 Skeletons via Shocks of Boundary Evolution

One class of approaches aims to simulate Blum's grassfire flow, as described by (1.1), and to then detect the locus of quench points. These approaches are distinct to the approaches of morphological erosion, in that the evolving curve is modeled as a partial differential equation. In practice this raises questions about how the flow should be numerically discretized and the singularities of the evolution detected, both of which are nontrivial issues.

Representative methods in this class are the techniques of Leymarie and Levine (1992); Tari et al. (1997); Siddiqi et al. (2002). In the first method the boundary of a 2D object is represented as an active contour, which is partitioned at locations of positive curvature maxima corresponding to the medial axis end points. The various segments of the active contour then propagate inwards driven by a potential function modeled by the negative gradient of the Euclidean distance function to the boundary. The active contours slow down in the vicinity of the medial axis, where the magnitude of the numerical gradient is small.

The skeletonization method in Siddiqi et al. (1999a, 2002); Dimitrov et al. (2003) is based on a novel characterization of singularities of the grassfire flow, which in turn lends itself to robust and efficient numerical implementations. The key insight is that in the limit as the area (2D) or volume (3D) within which average outward flux is computed shrinks to zero, the average outward flux of the gradient of the Euclidean distance function has different limiting behaviors at non-medial and medial points. This allows for a uniform treatment in 2D and 3D, along with associated skeletonization algorithms. These techniques, along with related methods based on the gradient of the Euclidean distance function, are discussed in Chapter 4. It turns out that for the case of shrinking circular neighborhoods the limiting values of the average outward flux reveal the object angle and hence allow for the explicit recovery of their boundary pre-image (Dimitrov et al., 2003; Dimitrov, 2003). An advantage of this approach is that the analysis extends to higher dimensions.

1.4.3 Greyscale Skeletons

It is well known that the grassfire flow is equivalent to a formulation where the time of arrival surface associated with the level curves of the flow satisfy an eikonal equation. A generalization of these results to the case of greyscale images is described in (Tari et al., 1997; Shah, 1996). The method is based on a linear differential equation that is developed from a model introduced by Modica and Mortola (1977) for approximating the characteristic function $\chi_{\partial\Omega}$ of the boundary $\partial\Omega$ of an object Ω. Given a boundary approximation $\chi_{\partial\Omega}$, possibly computed by thresholding a simple edge strength operator and thus not necessarily closed, it generates a scalar function v on space (image) giving modified edge strength, whose troughs yield a medial locus.

The modification is designed to produce an edge strength image that is smooth, i.e., has small gradient magnitude, and falls from a value 1 at $\partial\Omega$ towards 0 with distance from $\partial\Omega$. As such it minimizes the functional

$$E_\sigma(v) = \int_\Omega \left[\sigma \|\nabla v\|^2 + \frac{(v - \chi_{\partial\Omega})^2}{\sigma} \right] = \int_\Omega \left[\sigma \|\nabla v\|^2 + \frac{v^2}{\sigma} \right] \qquad (1.7)$$

with the constraint $v = 1$ along $\partial\Omega$. The minimizer of E_σ satisfies the elliptic differential equation

$$\nabla^2 v = \frac{v}{\sigma^2} \qquad (1.8)$$

with the constraint serving as a boundary condition. The parameter σ plays the role of a nominal smoothing radius. When σ is small compared to the local width of the shape and the local radius of curvature of $\partial\Omega$, the level curves of v capture the smoothing of $\partial\Omega$ by a curve evolution model with a combination of a constant motion (grassfire) and curvature motion term, as used in Kimia et al. (1995). In particular, as shown by Mumford and Shah (1989), when σ is small,

$$v(x,y) = \sigma \left(1 + \frac{\sigma\kappa(x,y)}{2} \right) \frac{\partial v}{\partial\eta}(x,y) + O(\sigma^3) \qquad (1.9)$$

where η is the direction of the gradient of v and $\kappa(x,y) = v_{\xi\xi} / \|\nabla v\|$ is the curvature of the level curve of v passing through the point (x,y) with $v_{\xi\xi}$ the second derivative of v in the direction ξ tangent to the level curve.

The skeleton is taken to be the troughs of v. These troughs are computed by finding the extrema of $|\nabla v|$ along the level curves of v. As indicated in (1.9), near the shape boundary the level sets of v mimic a curve (or surface) evolution process with a speed consisting of a constant component and a component proportional to curvature.

In mathematical terms the above extrema are the set of positions where

$$\frac{d\,\|\nabla v\|}{ds} = 0,$$

with s the arc-length along level curves of v. These are the points where the level curves are parallel and the gradient lines of v have inflection points. Geometrically this means that the gradient vector ∇v is an eigenvector of the Hessian of v at points on the axes of local symmetry. This method is developed in a series of interesting papers (Tari et al., 1997; Tari and Shah, 1998).

The above ideas have been applied to both the computation of medial loci of boundaries implicitly defined in greyscale images, and their segmentation into protrusions and indentations in (Shah, 2001, 2005a). The latter is accomplished by grouping points which share the same sign of the second derivative $\frac{d^2\,\|\nabla v\|}{ds^2}$. Furthermore, extensions of these ideas to higher dimensions have been explored (Tari and Shah, 2000; Shah, 2005b). The development of robust algorithms to compute these medial loci remains an area of investigation, and these computational methods are similar in spirit to the class of methods based on core tracking to be discussed next. A practical difficulty faced here is that the medial loci so obtained are disconnected. Nonetheless, it is possible to use the direction of ∇v to determine a hierarchical interpretation as a graph of components. The 2D case is detailed in Shah (2001).

1.4.4 Core Tracking

The greyscale skeletons of Shah and Tari can be seen as a special case of cores, with the function v, or more precisely its reciprocal, serving as a measure of medialness. Cores generalize the medialness to functions not just of position but also medial radius and spoke directions. *Cores* are medialness height ridges, i.e., maxima of dimension $p - k$ in a space of dimension p equal to that of medial atoms, i.e., 5 in 2D and 8 in 3D.

These medial loci are computed by tracking (Pizer et al., 1998; Furst and Pizer, 1998; Morse et al., 1998; Eberly et al., 1994; Fritsch et al., 1995a). These particular medial loci do not have the same branching properties as Blum medial loci, so special branch-searching and seeding strategies are needed (Fridman, 2004). Their strength is that they are derived directly from greyscale image data without the need of first extracting boundaries and that, as shown in the references just cited, they are quite insensitive to image noise.

Core tracking literature has made two different choices for defining the subspaces in which maxima are found, resulting in two classes of cores: *maximal convexity cores* and *optimal parameter cores*. In maximal convexity cores the sub-spaces are defined by the $p - 1$ directions of greatest second derivative of M, computed at each point as the unit eigenvectors of the Hessian matrix (Eberly, 1996). In optimal parameter cores, the atoms are required to attain a local maximum in radius, orientation, and object angle, as well as in the direction orthogonal to the Euclidean tangent space of the core, which is defined by the optimal orientation (Furst and Pizer, 1998; Fridman et al., 2004).

Core tracking methods work by following a core from a starting point in the parameter space. A user specifies a location, size, orientation and object angle of an initial medial atom and the algorithm searches for a ridge point in the vicinity of this initialization. The algorithm then tracks the core, taking small steps in the parameter space until some termination condition is met. Core tracking has been implemented using marching cubes methodology (Furst and Pizer, 1998; Lorensen and Cline, 1987) and using predictor-corrector methods (Fritsch et al., 1995b).

1.4.5 Skeletons from Digital Distance Transforms

The pattern analysis literature is replete with a class of methods which use digital distance transforms but are distinct from the methods of morphological erosion. These methods attempt to locate height ridges on a discrete lattice, using techniques from discrete geometry and topology. This is done either by iterative removal of pixels or voxels, or by marking centrally located elements, or a combination of these two ideas. The state of the art algorithms in this class are the subject of Chapter 5. These methods offer the advantage that they are computationally efficient and that formal guarantees on the quality of the results can be provided, including reversibility and homotopy preservation.

1.4.6 Voronoi Skeletons

Voronoi skeletons (Ogniewicz, 1993; Székely, 1996; Amenta et al., 2001b) are computed by calculating the *Voronoi diagram* of a set of points sampled from the boundary of an object. Figure 1.15a shows a Voronoi diagram of a set of six points on a plane. The diagram consists of *Voronoi regions*, which are sets of points located closer to a particular generating point than to any other generating point. The line segments in the diagram are called *Voronoi edges*; they separate Voronoi regions and are loci of points that are equidistant from a a pair of generating points. The points where Voronoi edges meet are equidistant from three or more generating points.

When the generating points of a Voronoi diagram are sampled from the boundary of an object, as shown in Fig. 1.15b, the similarity between Voronoi edges and the curves composing the medial locus becomes apparent. A circle of appropriate radius centered at a point on a Voronoi edge contains two generating points, i.e., has two

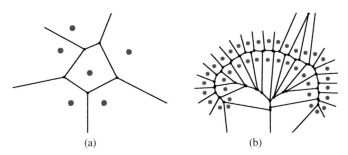

(a)   (b)

Fig. 1.15 Examples of Voronoi diagrams. (**a**) Voronoi Diagram of six points. (**b**) Voronoi Diagram of points sampled from the boundary of the corpus callosum. The skeleton of the object is just the internal portion of the Voronoi Diagram

points of contact with the boundary. A circle centered at an intersection of two Voronoi edges contains three generating points, i.e., has three points of contact with the boundary, as do disks centered at intersections of curves in the medial locus. The Voronoi diagram is also related to the grassfire analogy: if some points on the boundary are set on fire (as opposed to the whole boundary), the places where fire fronts meet and quench themselves are the edges in the Voronoi diagram of these points.

The Voronoi diagram of a set of boundary points contains edges that extend outside of the object, possibly to infinity. The discrete approximation of the boundary obtained by connecting the generating boundary points with line segments cuts the Voronoi diagram into internal and external parts. The internal part is called the *Voronoi skeleton*. The Voronoi skeleton generated by a discrete representation of an object's boundary is an approximation of that object's internal medial locus. Schmitt provides a proof that as the number of generating boundary points increases, the Voronoi skeleton converges in the limit to the continuous medial locus, with the exception of the edges generated by neighboring pairs of boundary points (Schmitt, 1989).

The Voronoi skeletons, such as the one shown in Fig. 1.15b, contain many branches, some of which are spurious and sensitive to the slightest changes to the generating boundary points. For instance, a Voronoi skeleton computed from the set of pixels forming the boundary of an object in an image can change significantly if the object in the image is rotated. In order to make Voronoi skeletons more robust to small boundary changes, researchers have proposed to isolate parts of the Voronoi skeletons that are most stable and significant. A number of measures of significance for edges and groups of edges in the Voronoi skeleton have been introduced in the literature (Ogniewicz and Kübler, 1995; Székely, 1996). The significance measures make it possible to establish trunk-branch relationships between connected edges in the Voronoi skeleton, and thus to establish a hierarchy of figures and sub-figures. The edges that fall far from the root of this hierarchy and have small significance values do not contribute to the descriptive ability of the Voronoi skeleton and are trimmed. Pruning on the basis of significance introduces a component of scale into

Segmentation via following a medial locus from image data, without needing an explicit prior probability, has also met with success. While many of the methods apply to both tubular and slab-shaped objects, they have been particularly success-ful in the case of trees of tubes whose branching structure is variable and thus not specifiable as a prior on a model with fixed branching topology. Aylward and Bullitt (2002) and Fridman et al. (2004) have produced core-based methods for this purpose. Lorigo et al. (2001) and Descoteaux et al. (2004) have produced tubular tree extraction methods leveraging the fact that skeletal curves are entities of co-dimension greater than 1 in three dimensions. The method of Descoteaux et al. is flux-based and follows the more general form described in (Vasilevskiy and Siddiqi, 2002). These applications are not detailed further in this book.

Because they represent objects as a graph of figures, with geometric descrip-tions of both the figures and their connections, medial representations are very well suited to object recognition and labeling of object parts. Several graph theoretic methods have been developed for such purposes and Chapter 10 covers this appli-cation with a focus on the problem of 3D model retrieval. These methods can also be applied to the analysis and comparison of anatomical structures as viewed in medical images, including vessel trees, bronchial trees (Tschirren et al., 2005), etc. Centerlines, which may be viewed as the limiting case of a medial manifold shrink-ing to a 3D curve, have also proven useful for the visualization of structures in virtual endoscopy (Deschamps and Cohen, 2001; Bouix et al., 2005b). These latter applications are not covered in this book.

Yet another application that has been developed for medial axes leverages the object-relative coordinates that they provide for applying physical simulations to objects and collections of objects. For example, one can do mechanical simulations based on medial models (Crouch et al., 2007), i.e., with spatial meshings based on medial coordinates. These partial differential equations can be applied to medial models directly in the object-based coordinates that the models provide, or even on the curvilinear high-dimensional spaces that describe medially represented objects (see Chapter 8). However, since there are only a few such applications thus far in the literature, they are not covered in this book.

We conclude the book with Chapter 11, which provides an overview of appli-cations of medial representations in different fields at scales ranging from the very large to the very small.

Part I
Mathematics

Chapter 2
Local Forms and Transitions of the Medial Axis

Peter J. Giblin and Benjamin B. Kimia

Abstract We define the concept of contact between a plane curve and a circle, and between a surface in 3D and a sphere, and describe all the contacts which generically occur for curves and surfaces, or 1-parameter families of such. We then list the local forms of the symmetry set and medial axis in 2D and 3D and briefly describe the local reconstruction of a boundary surface from its symmetry set or medial axis. Next, we describe the transitions which occur on the medial axis and symmetry set in a 1-parameter family of plane curves and which occur on the medial axis in a 1-parameter family of surfaces in 3D. Finally we make some remarks on the constraints which the medial axis must obey at special points such as Y-junctions. Some applications are given to illustrate these ideas.

2.1 Introduction

In this chapter given a plane curve or surface $\mathscr{B}$, we shall explain the local structure of the medial axis and also the symmetry set of the region Ω bounded by $\mathscr{B}$. This will be done for the case where $\mathscr{B}$ is a generic curve in the plane or surface in 3-space. We shall also describe the transitions occurring in a generic 1-parameter family of curves or surfaces, though we shall not list all the transitions of the symmetry set in 3-space. The organization of the chapter is as follows.

In Section 2.2 we recall the definitions and in Section 2.3 we study the contact of curves and surfaces with, respectively, circles and spheres. It is this contact which enables us to make precise the 'generic' cases which we consider from then on.

P.J. Giblin
Department of Mathematical Sciences, University of Liverpool, UK,
e-mail: pjgiblin@liv.ac.uk

B.B. Kimia
Division of Engineering, Brown University, USA,
e-mail: kimia@lems.brown.edu

K. Siddiqi and S. Pizer (eds.) *Medial Representations – Mathematics, Algorithms and Applications*.
© Springer Science + Business Media B.V. 2008

Roughly speaking, the 'higher' the contact between, say, a circle and a curve, the more special is the resulting point on the medial axis or symmetry set, and only certain orders of contact are stable for a single curve, or in a 1-parameter family of curves. A 'higher' contact than this can always be eliminated by an arbitrarily small deformation of the curve. In Sections 2.4 and 2.5 we give the local forms which the symmetry set and medial axis have in 2D and 3D respectively (see Pollitt, 2004).

In Sections 2.6 and 2.7 we study the local reconstruction of the boundary $\mathscr{B}$ from the medial axis M. The information needed to reconstruct the boundary $\mathscr{B}$ is of two kinds: geometrical information on the medial axis and 'dynamic' information on the radius function. In the next chapter James Damon will show how to combine these two sources of information into the 'radial vector field' and the 'radial shape operator', but for the time being we shall keep the two things separate.

In Section 2.8 we turn to a plane curve evolving in a family of nonsingular curves—that is, simple closed smooth curves. The local transitions (changes) on the medial axis and symmetry set will be explained there, and we shall also give some global examples which illustrate these transitions. The importance of these transitions is that when a region changes shape its medial axis and symmetry set can be assumed to undergo only a small number of local changes. These have to be responsible for all global changes when they occur in a suitable sequence. The considerably harder case of surfaces is covered in Section 2.9, and we restrict there to the medial axis, where the generic evolutions were found in (Bogaevsky, 1990, 2002). The list of evolutions of the symmetry set in 3D is a long one, and we do not attempt to cover it here.

Finally in Section 2.10 we shall give a brief introduction to consistency conditions (constraints) which a medial axis must satisfy at special points such as Y-junctions. These arise because each component of the boundary $\mathscr{B}$ is reconstructed not from one but now from *two* pieces of the medial axis. The fact that the same boundary has to be constructed in two ways gives a consistency relation. The simplest examples of these relations tell us that the detailed geometry of the medial axis at, for example, Y-junctions, is far from arbitrary.

2.2 Definitions

For the convenience of the reader we recap in a suitable form the definitions which will be used in this chapter. The concept of 'contact' will be expanded in the next section.

The *symmetry set* of a curve (resp. surface) $\mathscr{B}$ is the closure of the locus of centers of circles (resp. spheres) which are tangent to $\mathscr{B}$ in more than one place.

A circle (resp. sphere) is called *maximal* relative to $\mathscr{B}$ if its radius equals the minimum distance from its center to $\mathscr{B}$.

The (Blum) *medial axis* of a curve (resp. surface) $\mathscr{B}$ in the plane (resp. in 3-space) is the closure of the locus of centers of maximal circles (resp. spheres) which are tangent to $\mathscr{B}$ in more than one place.

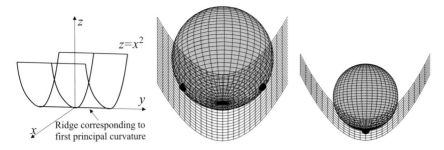

Fig. 2.3 *Left*: the parabolic cylinder $z = x^2$. *Center*: a view from 'beneath' the surface of a bitangent sphere whose two points of contact are marked by dark blobs. *Right*: the two contact points have coincided in a ridge (A_3) contact point and the sphere has shrunk to become a principal sphere of curvature at the ridge point. This is a typical way in which ridge points are formed; the limiting direction in which the two points of contact approach one another is in fact that of the principal direction corresponding to the ridge. That is, the corresponding principal curvature has an extremum along its line of curvature. In this very symmetric example, the two contact points actually lie on the line of curvature, which is the parabola $y = 0$ on the surface, but this is exceptional

curvature of $\mathscr{B}$ at the origin and the sphere is one of the two spheres of curvature there, having radius $1/\kappa_i$.

- **Ridge (or crest line) contact, A_3:** Let us take $i = 1$ above, so that $c = 1/\kappa_1$. Then we require that y, which is a factor of the quadratic terms of g, also divides the cubic terms, that is $b_0 = 0$. Another description of this (compare Section 6.4.2, Hallinan et al., 1999) is that, restricted to the principal curve on $\mathscr{B}$ tangent to the x-axis, the corresponding principal curvature has an extremum. Yet another description is that when two ordinary (A_1) points of contact come into coincidence they do so at an A_3 point, as in the curve case above. See Fig. 2.3.

- **Turning point contact, A_4:** (Symmetry set only.) Here the locus of points where there is A_3 contact—the ridge or crest line—is tangent to the corresponding principal curve on $\mathscr{B}$. The condition for this, in terms of the Monge coefficients in equation (2.1), is[2] $(\kappa_1 - \kappa_2)(8c_0 - \kappa_1^3) + 4b_1^2 = 0$, and hence depends on the fourth-order coefficient c_0. (Exactly A_4 contact requires that the ridge does not have inflectional (3-point) contact with the principal curve; the condition for this turns out to be $d_0(\kappa_1 - \kappa_2)^2 + b_1c_1(\kappa_1 - \kappa_2) + b_1^2b_2 \neq 0$.)

There are various combinations of these singularities which can occur on a generic surface, namely A_1^2, A_1^3, A_1^4, A_1A_2, $A_1^2A_2$, A_2^2, A_1A_3. This is the same as the list for curves *including* 1-parameter families: adding a dimension, passing from curves to surfaces, has the same effect as adding a parameter, passing from curves to families of curves. However for the symmetry set there is an entirely new contact singularity which does not arise from a family of curves. This is D_4, which

---

[2] The difference between this expression and the corresponding condition $P_1 = 0$ in (6.6) (Hallinan et al., 1999) arises because in the latter binomial coefficients are introduced into the Monge form (2.1).

corresponds to points of $\mathcal{B}$ which are *umbilics*, that is which are elliptic points at which the two principal curvatures κ_1 and κ_2 coincide.

Of course we can consider *families of surfaces*. This introduces many more cases, and in this book we shall only consider the additional cases which affect the medial axis, namely A_1^5, $A_1^2 A_3$, A_5, plus new varieties of A_1^4, $A_1 A_3$. See Section 2.9.

2.4 Local Forms of the Symmetry Set and Medial Axis in 2D

We now describe the local behavior of the symmetry set in the various situations of Section 2.3, restricting ourselves here to a single generic curve. We are interested only in circles which have contact at more than one point—referred to as *bitangent circles*—or limits of these. Thus we need to consider circles with contact of all types in Fig. 2.1 besides A_1 and A_2.

Some of these are easily seen on the so-called *pre-symmetry set*; we briefly describe this before proceeding. See also for example Kuijper and Olsen (2004). For ease of description let $\mathcal{B}$ be closed and parametrized as $\gamma(t)$, where the parameter t takes the values $0 \leq t \leq 1$, with $\gamma(0) = \gamma(1)$. We use a parameter square

$$\{(t_1,t_2) : 0 \leq t_1 \leq 1, 0 \leq t_2 \leq 1\}$$

to represent pairs of points of $\mathcal{B}$. Each pair (t_1,t_2) for which there is a circle tangent to $\mathcal{B}$ at $\gamma(t_1)$ and $\gamma(t_2)$ contributes the point (t_1,t_2) to the pre-symmetry set. We can equally well reverse t_1 and t_2 so that (t_2,t_1) is also in the pre-symmetry set. We also take limits as t_1 and t_2 coincide—as happens, for example, when the A_1^2 circle in Fig. 2.1 slides to the right and shrinks until it becomes the A_3 circle, tangent at P_3. These contribute 'diagonal points' (t_1,t_1) to the pre-symmetry set.

Remark 2.4.1. In order to obtain the pre-symmetry set we need to solve equations of the form

$$(\gamma(t_1) - \gamma(t_2)) \cdot (\mathbf{T}(t_1) - \mathbf{T}(t_2)) = 0 \tag{2.2}$$

where '·' is the scalar product of vectors and $\mathbf{T}$ is the unit tangent to the curve $\mathcal{B}$. There are three problems here: firstly, the solutions of this equation automatically include the trivial case $t_1 = t_2$, that is the diagonal of the parameter square. Secondly, the solutions include points where $\mathbf{T}(t_1) = \mathbf{T}(t_2)$, that is where the oriented tangents to $\mathcal{B}$ are parallel. This cannot occur for convex curves except on the diagonal $t_1 = t_2$, but for non-convex curves it occurs close to inflections, for example. Thirdly for the symmetry set (not the medial axis) we will want to include circles for which $(\gamma(t_1) - \gamma(t_2)) \cdot (\mathbf{T}(t_1) + \mathbf{T}(t_2)) = 0$. See Fig. 2.4.

See Fig. 2.5 for a simple example.

The following cases are easily visible on the pre-symmetry set:

- **A_3:** any circle with A_3 contact contributes a diagonal point to the pre-symmetry set, namely (t,t) where t is the parameter value of the unique point of contact

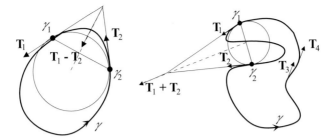

Fig. 2.4 *Left*: the more 'usual' situation, and the only one relevant to the medial axis, where the vector $\mathbf{T}_1 - \mathbf{T}_2$ is perpendicular to the chord $\gamma_1 - \gamma_2$. Here, $\mathbf{T}_1 = \mathbf{T}(t_1)$, etc. *Right*: for the symmetry set it can happen that $\mathbf{T}_1 + \mathbf{T}_2$ is perpendicular to the chord $\gamma_1 - \gamma_2$. This figure also illustrates the fact that, near an inflection on γ, we can have two parallel tangents to γ oriented the same way: $\mathbf{T}_3 = \mathbf{T}_4$, so that the (2.2) is automatically satisfied for such pairs

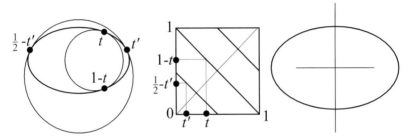

Fig. 2.5 An ellipse $\gamma(t) = (a\cos(2\pi t), b\sin(2\pi t))$, $0 \le t \le 1$ and two bitangent circles. The tangency points are at parameter values t and $1 - t$ for one circle, and t' and $\frac{1}{2} - t'$ for the other circle. On the right is the pre-symmetry set which contains all pairs (t_1, t_2) where there is a bitangent circle. Remembering that $t = 0$ is identified with $t = 1$, the two lines making up the pre-symmetry set are $t_1 + t_2 = 1$ and $t_1 + t_2 = \frac{1}{2}$. Any pair (t_1, t_2) also appears as (t_2, t_1), that is, the pre-symmetry set is symmetric about the diagonal (shown dashed). Points where the pre-symmetry set intersects the diagonal give A_3 circles where the contact points have coincided, and these give the four end points of the symmetry set. The four corners of the square all represent the same point $(0,0)$ or $(1,1)$, on the diagonal. Right: the symmetry set itself

of the circle with $\mathscr{B}$. The symmetry set has an end point at the center of the corresponding circle.

- $\mathbf{A_1A_2}$: a tangent to the pre-symmetry set parallel to the t_1 or t_2 axis (a 'horizontal or vertical tangent') indicates an A_1A_2 circle. More specifically, a tangent at (t_1, t_2) parallel to the t_2 axis indicates that there is a circle having A_1 contact at t_1 and A_2 contact at t_2. Thus the point of contact close to t_1 turns round at the A_1A_2 point. This causes the symmetry set itself to have a *cusp* at the center of the A_1A_2 circle. See Fig. 2.6.
- The A_1^3 case is not so easily seen on the pre-symmetry set, but we mention it here for future reference. There are now three points of contact and hence three pairs of points, (t_1, t_2), (t_1, t_3) and (t_2, t_3) say. This gives a *rectangle* of points as in Fig. 2.6, top right.

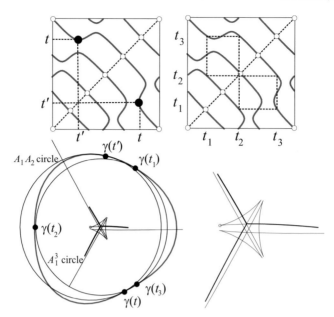

Fig. 2.6 *Top left*: the pre-symmetry set of the curve shown in Fig. 2.1 (and reproduced in the bottom left figure) exhibiting an A_1A_2 circle. The black dots mark the two symmetrical points (t,t') and (t',t) of the pre-symmetry set corresponding to the contacts of this circle and the curve. The diagonal of the square is shown dashed. There are six points (white dots) of the pre-symmetry set *on* the diagonal (one of these six points is at a corner of the square, and all the corners represent the same point $(0,0) = (1,1)$). These six diagonal points correspond to six vertices (A_3 points) of the curve and six end points of the symmetry set. *Top right*: consider the rectangle above the diagonal. It has three corners $(t_1,t_2), (t_1,t_3), (t_2,t_3)$ on pre-symmetry set and one (t_2,t_2) on the diagonal but not on the pre-symmetry set (this corner is *not* at the A_3 point on the diagonal!). This rectangle corresponds to an A_1^3 (tritangent) circle with tangency points t_1,t_2,t_3, and hence a triple crossing on the symmetry set or medial axis. *Bottom left*: the full symmetry set, with the A_1A_2 and A_1^3 circles drawn, and the center of the A_1A_2 circle marked by a small circle at one of the cusps. The medial axis is drawn heavily. On the symmetry set there is a cusp corresponding to each such horizontal or vertical tangent on the pre-symmetry set; allowing for symmetry about the diagonal this makes six cusps altogether. *Bottom right*: a closeup. Note that there are two triple crossings on the symmetry set and only one on the medial axis

Recapping from Chapter 1, if a circle with center c is bitangent to the boundary $\mathcal{B}$ at distinct points, the tangent or limiting tangent to the associated branch of the medial axis or symmetry set is the perpendicular bisector of the chord joining the two points. In the case where the two points coincide (A_3 or end point) the tangent to the symmetry set or medial axis is along the normal to $\mathcal{B}$. Thus for example at a triple intersection of the symmetry set (a Y-junction of the medial axis) there are three branches, and their tangents are the perpendicular bisectors of chords joining pairs of tangency points. An important feature of the geometry of this situation is that the three tangents pointing into the medial axis cannot lie within an angle of $180°$. The two possible cases are shown in Fig. 2.14e, f.

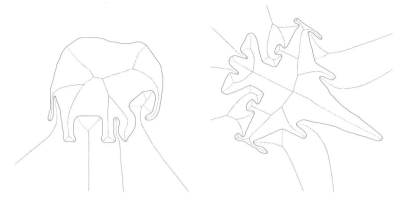

Fig. 2.7 Shock graph examples from Tek and Kimia (2003). Note that the direction has not been shown along each link

At a cusp of the symmetry set the 'cuspidal tangent' is likewise the perpendicular bisector of the chord joining the tangency points of the A_1A_2 circle.

Shock Graph: A dynamic view of the medial axis arises from an interpretation of the points of contact as the collision of wavefronts propagating from the boundaries and the formation of *shocks*. A classification of directions of flow at A_1^3 reveals that only two cases are possible: (i) all branches consist of shocks flowing inward, and, (ii) two branches flow outward while a third flows inward. The direction of flow at A_3 can only be outward into the branch. Finally, two types of A_1^2 points are identified, one with both sides either flowing outward ($A_1^2 - 2$), a source, and another where both branches flow inward ($A_1^2 - 4$), a sink. The two types of A_1^3, A_3, $A_1^2 - 2$ and $A_1^2 - 4$ gives five special points of the medial axis which are connected by monotonically flowing A_1^2 branches. The definition of the first five types as nodes and the connecting A_1^2 segments as link of graph gives a formal definition of a *shock graph* (Kimia et al., 1990, 1995; Siddiqi and Kimia, 1996; Sharvit et al., 1998). Examples of the shock graph obtained by an Eulerian-Lagrangian propagation method are shown in Fig. 2.7 (Tek and Kimia, 2001, 2003); see also Chapter 4. An alternative definition for a shock graph has been put forward by Siddiqi et al. (1999b).

2.5 Local Forms of the Medial Axis in 3D

We now turn to the medial axis of a surface in 3D; we shall also say something about the local form of the symmetry set but the emphasis will be on the medial axis. As in Section 2.3 there are the following cases:

(i) Analogous to the curve cases: A_1^2, A_1^3, A_1A_2, A_3
(ii) Analogous to the family of curves cases: A_4, A_1A_3, $A_1^2A_2$, A_1^4, A_2^2
(iii) New: D_4

For those in list (i) the local structure of the symmetry set or medial axis is essentially a *product*: take the corresponding 2D structure and move it to sweep out a surface in 3-space. Those in list (ii) have not yet appeared in the 2D case (they must wait for Section 2.8).

Only A_k with k odd can be a *minimum*, so for the medial axis the only relevant cases are A_1^2, A_1^3, A_3 from (i) and A_1A_3 and A_1^4 from (ii). The singularity D_4 can be reduced by changes of coordinates to $x^3 \pm xy^2$ and is never a minimum.

The full set of cases is detailed in the following.

- **A_1^2:** the symmetry set or medial axis is a smooth sheet.
- **A_1^3:** the symmetry set is three sheets intersecting along a smooth curve; the medial axis is three half-sheets meeting along a smooth curve (the A_1^3 curve or Y-junction curve). See Figs. 2.8, 2.9.
- **A_1A_2:** the symmetry set (only) is a cuspidal edge. See Fig. 2.8, right.
- **A_3:** the symmetry set or medial axis is a smooth surface with a boundary edge (the A_3 curve or rib line). See Fig. 2.8, left.
- **A_4:** the symmetry set (only) is a 'swallowtail'. We shall not illustrate this here; there is a diagram in (Giblin and Kimia, 2004, Fig. 11).
- **A_1A_3:** the symmetry set is a swallowtail and an extra sheet. See Fig. 2.8, right. On the medial axis this is called a *fin point*: only half the swallowtail and half the extra sheet are present. A fin point is an end point for both a Y-junction curve and an A_3 curve. See Figs. 2.9, 2.10. Figure 2.11 shows the possibly geometrical relationships between the A_1^3 (Y-junction) curve on the medial axis, the A_3 (edge, rib) curve and the normal to the surface, at an A_1A_3 (fin) point; compare (Giblin and Kimia, 2004).
- **A_1^4:** the symmetry set is six sheets meeting in four A_1^3 curves; the medial axis is six half-sheets meeting in four Y-junction curves. The A_1^4 point is also called a

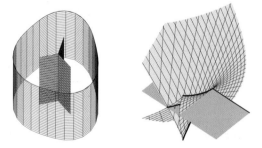

Fig. 2.8 *Left*: a Y-junction (A_1^3) curve (here a straight line) lying at the intersection of three medial axis sheets and representing centers of spheres which are tangent to the curved boundary surface in three points. The vertical edges of the three sheets are A_3 curves (also here straight) corresponding to three ridges on the boundary surface. On the symmetry set the three sheets would continue through the A_1^3 set. *Right*: the symmetry set at an A_1A_3 point: a 'swallowtail' surface with an additional surface, here a plane. The dark edge of the plane is the A_3 set and the dark 'cuspidal edges' on the swallowtail surface are the A_1A_2 sets. The dark line where the swallowtail surface and the plane intersect is the A_1^3 set. Compare with Fig. 2.10, where just the medial axis is shown

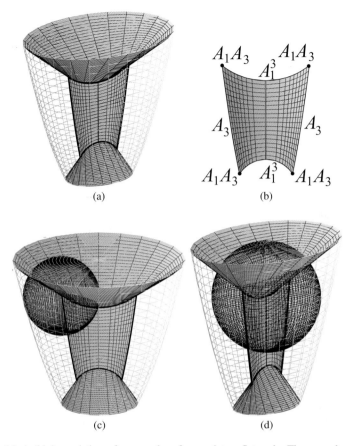

(a)   (b)

(c)   (d)

Fig. 2.9 (**a**) A 'bin' consisting of a curved surface and two flat ends. The curved surface has elliptical sections when sliced by horizontal planes and parabolic sections when sliced by vertical planes. The medial axis is shown inside the bin. (**b**) The flat sheet of the medial axis (centers of spheres tangent 'front and back') labeled with the types of the various contacts of spheres whose centers lie round its boundary edge. (**c**) An A_1A_3 sphere, centered at an intersection of the A_1^3 and A_3 curves, tangent to the top of the bin (A_1) and to the curved surface (A_3, ridge contact). (**d**) An A_1^3 sphere, centered on the boundary edge of the flat sheet and tangent to top, front and back of the bin

6-junction point; see Figs. 2.24, 2.10. Let **p** be the center of a sphere tangent to $\mathscr{B}$ in four points,[3] and consider three of the four tangency points. The tangent to the A_1^3 curve through **p** corresponding to these three points is perpendicular to the plane of the points. The direction into the medial axis is *away* from the fourth tangency point.

The three cases below are not illustrated here, nor will they be involved further. For illustrations, see Bruce et al. (1985).

---

[3] These points will generically be *non-coplanar*. The coplanar case is special and is dealt with in Section 2.9.

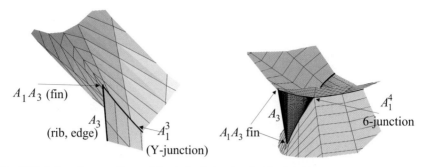

Fig. 2.10 *Left*: a closeup of one of the A_1A_3 (fin) points in Fig. 2.9. Note that three sheets of the medial axis meet along the Y-junction curve, and of these two merge into a single sheet and the other ends at the fin point. *Right*: a closeup of one of the A_1^4 (6-junction) points in Fig. 2.24. Note that there are four A_1^3 (Y-junction) curves, drawn heavily, and six sheets of the medial axis, meeting in the central A_1^4 point. This figure also contains two A_1A_3 (fin) points

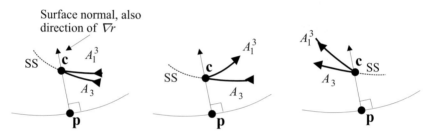

Fig. 2.11 An A_1A_3 (fin) point of the medial axis is the center **c** of a sphere which is a sphere of curvature at a ridge point **p** of the boundary surface $\mathscr{B}$, and which is also tangent to $\mathscr{B}$ elsewhere. In this diagram we show the normal to $\mathscr{B}$ at **p** and the A_1^3 (Y-junction) and A_3 (edge, rib) curves close to **c**. The *tangents* to these curves at **c**, and the normal line, are all coplanar. The A_3 curve continues on the symmetry set as shown by the dashed line marked 'SS'. The diagrams indicate the possible configurations, including the flow of the radius function r, indicated by arrows

- **$A_1^2A_2$:** (Symmetry set only)
- **A_2^2:** (Symmetry set only)
- **D_4:** (Symmetry set only)

Medial Scaffold: In analogy to the 2D case, the classification of the local form of the medial axis in 3D also leads to a graph-based hierarchical representation which is referred to as the *medial scaffold* (Leymarie and Kimia 2001, Leymarie 2003). The five types of medial axis points organize the locus into three separate groups: (i) the A_1^4 and A_1A_3 points are considered as nodes, (ii) the A_1^3 and A_3 curves are considered as links, and (iii) the A_1^2 sheets are considered as hyperlinks, which together form a hypergraph. Examples are shown in Fig. 2.12.

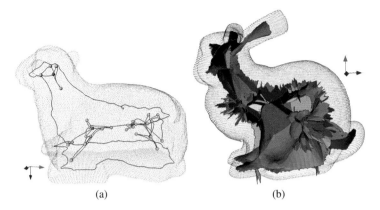

Fig. 2.12 (Adapted from (Leymarie, 2003; Chang et al., 2004)). (**a**) The medial scaffold is a hierarchical graph-based reorganization of the medial axis. The points and curves of the medial scaffold are shown. (**b**) An example showing the sheets as well

2.6 Local Reconstruction from the Symmetry Set or Medial Axis in 2D

In this section, preparing the way for Chapter 3, we describe briefly the way in which we can recover the boundary $\mathscr{B}$ from the symmetry set or medial axis and the radius function.

We begin with the 2D case, where we are given a smooth piece M of symmetry set or medial axis, as well as the radius function and we seek the local pieces of boundary curve $\mathscr{B}$ which gave rise to M. We suppose M is parametrized by $\gamma(s)$, where s is arclength. Further suppose that $r(s)$ is the radius of the circle centered at $\gamma(s)$. The idea is simply that $\mathscr{B}$ is the *envelope* of the circles centered at $\gamma(s)$ and of radius $r(s)$, that is, $\mathscr{B}$ is a curve tangent to all these circles. Figure 2.13 shows the basic setup: $\mathbf{T}$ is the unit tangent to γ, that is $\mathbf{T} = \gamma'(s)$ where the prime denotes $\frac{d}{ds}$, $\mathbf{N}$ is the unit normal (that is, $\mathbf{T}$ turned anti-clockwise through $90°$), and θ is the object angle shown, between $\mathbf{T}$ and the line from the center of the circle to either boundary point. A standard calculation of envelope points (Giblin and Kimia, 2003a, Section 3) then shows the following:

Proposition 2.6.1. *The boundary points corresponding to the point $\gamma(s)$ are*

$$\mathbf{b} = \gamma(s) - rr'\mathbf{T} \pm r\sqrt{1 - r'^2}\mathbf{N}$$
$$= \gamma(s) - r\cos\theta\mathbf{T} \pm r\sin\theta\mathbf{N}.$$

This proposition relates to material in Chapter 1 stating the following:

- $\cos\theta = -r'$ and $\sin\theta = +\sqrt{1 - r'^2}$ (we may assume $0 \le \theta \le \pi$). Equivalently $\cos\theta = -1/v$, where v is the radius-relative velocity ds/dr. If r is increasing

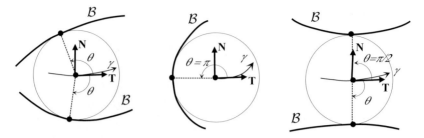

Fig. 2.13 Here γ, oriented as shown, is a smooth piece of the symmetry set or medial axis, with oriented unit tangent vector **T** and normal vector **N**. The boundary is $\mathcal{B}$ and the object angle θ is that between the tangent **T** and the vector from the center of the circle to one or other of the boundary points. *Left*: the general situation; *Center*: an end point of the medial axis, with $\theta = \pi$ (or 0); *Right*: an extremum of the radius function, here a minimum, when $\theta = \frac{1}{2}\pi$

along the oriented curve γ, then θ is obtuse, as in Fig. 2.13, while if r is decreasing then θ is acute.

- $r'^2 \leq 1$. If this condition[4] fails then the radii of the circles are changing too fast along γ and the circles fail to have an envelope. See Fig. 2.15. The special cases $r' = \pm 1$ and $r' = 0$ are of interest.

  - Let $r' = \pm 1$. Then $\theta = \pi$ or $\theta = 0$. These are the cases where the two boundary points **b** coincide; this occurs at an end point of the symmetry set or medial axis—an A_3 point in the terminology of Section 2.3.
  - Let $r' = 0$, i.e., the radius function has a maximum or minimum: the region Ω is locally fattest or thinnest at this place. It is instructive to see how the direction of increase of r behaves at the various special points of the symmetry set or medial axis, as listed in Section 2.3. The results are shown in Fig. 2.14. If we put an arrow on the symmetry set or medial axis to indicate the direction of increase of r then at these points the arrow will reverse. See Fig. 2.14g.

The reconstruction formulas together with the shock graph topology are useful for perceptual grouping. They partition the shape into a set of fragments (Fig. 2.16). As discussed in Chapter 11, the same idea can be applied to a non-closed contour such as the edge map of an image where the shape fragments together with their intrinsic distributions define *visual fragments*, which are used as canonical elements for perceptual grouping (Tamrakar and Kimia, 2004).

---

[4] The condition $r'^2 \leq 1$ is the only condition needed to obtain a smooth boundary $\mathcal{B}$; however if $\gamma(s)$ is to be a *medial axis* point then in addition $\sin\theta - r\theta' \geq r|\kappa|$, where κ is the curvature of the medial axis. Compare Giblin and Kimia (2003a, Lemma 2).

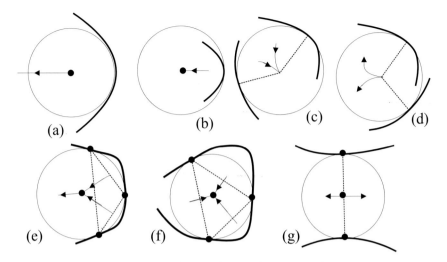

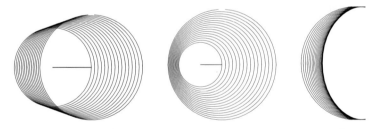

Fig. 2.14 The direction of increase of the radius function is shown in several cases. (**a**) A_3: a maximum of curvature on the boundary $\mathcal{B}$ (drawn heavily). The radius function r increases away from the end point, marked with a dot. (**b**) A_3: symmetry set only. A minimum of curvature on $\mathcal{B}$: r increases towards the end point. (**c**) and (**d**): A_1A_2: symmetry set only. The radius always has an extremum provided the ordinary tangency (A_1) and osculating tangency points (A_2) are not diametrically opposite. Whether maximum (**c**) or minimum (**d**) depends on the relationship between the A_1 contact point and the boundary $\mathcal{B}$ at the A_2 point. (**e**) and (**f**) show the two possible cases of A_1^3, depending on whether the three contact points lie in a semicircle or not. (**g**) shows a minimum of radius from which the arrows point outwards; for a maximum they would point inwards to the center of the circle

Fig. 2.15 *Left*: an envelope of circles centered on the x-axis, with radius $r = 1 + \frac{1}{3}x$, $0 < x < 1$. This interval of the x-axis is marked with a line. Here $r' = \frac{1}{3} < 1$ and the circles form an envelope. *Center*: here $r = 1 + \frac{3}{2}x$ and $r' > 1$ so that the circles do not form an envelope. *Right*: here the radius function $r = 1 + \frac{1}{8}x + \frac{7}{8}x^2$ has $r' = 1$ for $x = \frac{1}{2}$, so the envelope forms a curve with a vertex. The figure shows detail close to the vertex, and the circles are drawn only for $x \leq \frac{1}{2}$, where $r' < 1$

2.7 Local Reconstruction from the Symmetry Set or Medial Axis in 3D

The situation here is similar to that in Section 2.6. We suppose we are given a smooth piece M of the symmetry set or medial axis of a boundary surface $\mathcal{B}$, and the radius function r. Let ∇r be the gradient of the radius function at a point $\mathbf{p}$ of M, and let $\mathbf{T}$

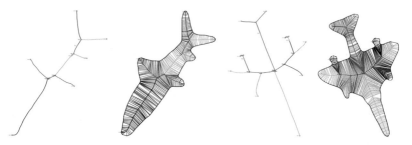

Fig. 2.16 The reconstruction of each shock branch gives an intuitive partition of the shape, as illustrated with two examples

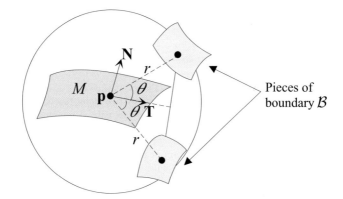

Fig. 2.17 A schematic drawing of local 3D reconstruction of the boundary $\mathcal{B}$ from a smooth sheet M of the medial axis or symmetry set. The vector $\mathbf{T}$ is a unit vector in the direction of ∇r

be the unit vector in this direction. (The formula below still works in the case when $\nabla r = 0$.) Let M be given a definite orientation and let $\mathbf{N}$ be the unit normal to M. Finally let θ be the object angle between $\mathbf{T}$ and the vector from $\mathbf{p}$ to either of the boundary points. Then we have the following (Giblin and Kimia, 2004, Section 3); see Fig. 2.17.

Proposition 2.7.1. *The boundary points corresponding to a point of M are*

$$\mathbf{b} = \mathbf{p} - r|\nabla r|\mathbf{T} \pm r\sqrt{1 - |\nabla r|^2}\mathbf{N}$$
$$= \mathbf{p} - r\cos\theta\mathbf{T} \pm r\sin\theta\mathbf{N}$$

In particular, the point $\mathbf{p}$, the two corresponding boundary points and the vectors $\mathbf{T}$, $\mathbf{N}$ all lie in one plane.

The gradient of r measures the rate at which r is changing with respect to arclength along the gradient lines of r, that is the orthogonal trajectories of the level sets $r = \text{constant}$. Note that if r has an extremum at $\mathbf{p}$, so $\nabla r = 0$, then the first equation still makes sense and the second one is to be interpreted as saying that

$\cos\theta = 0$, $\sin\theta = 1$. When $|\nabla r| = 1$, the two boundary points coincide; this is the A_3 case of Section 2.3, and $\mathbf{p}$ is a ridge point of the boundary surface $\mathcal{B}$.

As in the 2D case we can introduce velocity into the calculations: $\nabla r = (1/v)\mathbf{T}$ and as before $\cos\theta = -1/v$. The flow of the radius function on the symmetry set or medial axis in 3D is also of considerable interest. In Fig. 2.11 we show the possible flows near an A_1A_3 point. At an A_1^4 point either two, three or four arrows of increasing r point into the A_1^4 point; see Giblin and Kimia (2004, Section 4). There are other significant features; for example at a point of an A_3 curve where r has an extremum, the gradient of r is normal to the A_3 curve and points into the sheet of the medial axis whose boundary lies along this curve. It is also possible to divide the A_1^2 sheets of the medial axis into regions determined by the 'height ridges' of the radius function r, in the sense defined in Pizer et al. (1991); Eberly (1996); however we shall not pursue this topic here.

2.8 Symmetry Sets and Medial Axes of Families of Curves

In this section we give the changes ('transitions') which occur when the shape of the boundary curve $\mathcal{B}$ undergoes a deformation of a generic kind. That is, every deformation is arbitrarily close to one in which only the changes we list occur. (For proofs and further details see Bruce and Giblin, 1986 and Giblin and Kimia, 2003b.)

The generic cases to be covered here have already been mentioned in Section 2.3: A_1^4, A_1A_3, $A_1^2A_2$, A_2^2, A_4, of which only the first two can occur on the medial axis since only those involve exclusively A_k for k odd. However these transitions on the medial axis are always preceded by transitions on the symmetry set which are 'invisible' on the medial axis. We proceed to describe the transitions, indicating their effect on the pre-symmetry set (Section 2.4) as well as the symmetry set or medial axis. There are schematic pictures in Fig. 2.18 and actual examples of several transitions in[5] Figs. 2.20–2.22.

- A_1^4 This occurs when momentarily a circle is tangent to $\mathcal{B}$ in four points. The situation is shown in Fig. 2.19, where we also indicate the possible changes in the flow of the radius function. There are two geometrical cases here, depending on whether the four points lie in a semicircle or not.
- A_1A_3 This occurs when the circle of curvature at a vertex is tangent to the boundary $\mathcal{B}$ elsewhere, as in Fig. 2.2. The transition is sketched in Fig. 2.18. A branch of the symmetry set or medial axis penetrates another branch causing a Y-junction to appear on the medial axis and a triple crossing and two cusps on the symmetry set. Note that the cusps on the symmetry set can be seen as vertical (or horizontal) tangents to the pre-symmetry set, in Fig. 2.18. The transition occurs several times in the sequence of Figs. 2.20–2.22.

---

[5] This extended example was previously used in Giblin (2000).

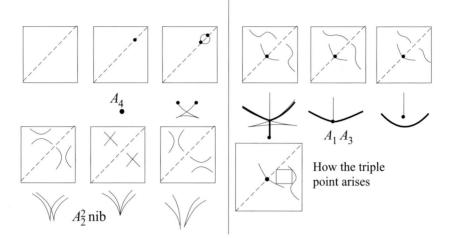

Fig. 2.18 Some schematic drawings of the transitions in the pre-symmetry set and symmetry set for the transitions A_4, A_1A_3 and A_2^2 nib. The diagonal of the pre-symmetry set diagram is drawn dashed; the pre-symmetry set is always symmetric about this diagonal. Dots represent end-points of the symmetry set and diagonal crossings of the pre-symmetry set. In the A_4 transition, two end points and two cusps are created on the symmetry set. In the A_1A_3 transition, two cusps and a triple crossing are created—on the medial axis, drawn heavily, one Y-junction point is created. Note the 'triple point rectangle', one of two such, as in Fig. 2.6. In the A_2^2 nib transition, cusps are exchanged between branches. All these transitions also occur in the sequence of Figs. 2.20–2.22

- $A_1^2A_2$ In this case a circle of curvature becomes momentarily tangent to the boundary curve $\mathcal{B}$ at two other points. This shows up on the pre-symmetry set by two vertical (horizontal) tangents at the same horizontal (vertical) level.
- A_2^2 A single circle has become the circle of curvature at two distinct points of $\mathcal{B}$. This very important transition occurs in two flavors, *nib* and *moth*, according to the way in which $\mathcal{B}$ crosses the circle at its two points of contact. Traveling along $\mathcal{B}$ from inside to outside the circle at an A_2 point assigns a local orientation to the circle. For the nib, these orientations are opposite (as in Fig. 2.2, left) and for the moth they are the same. The 'nib' transition, is illustrated schematically in Fig. 2.18 and is seen also in Fig. 2.21. The other flavor, the 'moth', creates a new isolated closed loop, away from the diagonal, on the pre-symmetry set and correspondingly creates a new isolated and closed component having four cusps on the symmetry set.
- A_4 This is the other transition whereby an additional branch can form on the symmetry set and eventually can affect the structure of the medial axis. The additional branch arrives with two end points and two cusps, and on the pre-symmetry set it appears as an isolated point on the diagonal, as in Fig. 2.18. The transition occurs twice at the beginning of the sequence in Fig. 2.20.

Remark 2.8.1. At the moment of A_4 transition there is an isolated point on the diagonal of the pre-symmetry set which grows into a small loop (Fig. 2.18). In the A_2^2

Tangents to $\mathcal{B}$

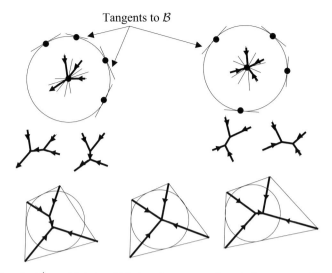

Fig. 2.19 *Top*: the A_1^4 transition, in which there is momentarily a circle tangent to $\mathcal{B}$ at four points, marked as dots on the circle. There are two cases, depending on whether the four points lie in a semicircle (left) or not (right). The thick lines show the local structure of the medial axis. The thin lines show the remainder of the symmetry set, namely (i) continuations of the four thick lines, (ii) two more lines whose direction is along the perpendicular bisectors of the non-adjacent pairs of boundary points. Thus *every* pair from the four boundary points contributes a branch to the symmetry set, but only adjacent pairs contribute a branch to the medial axis. Below each circle is shown the flow of r during the transition. In the right-hand diagram the flow along the short connecting lines can be in either direction. *Bottom*: an example of the A_1^4 transition where the four contact points are not within a semicircle. The circle shown only becomes maximal, and contributes to the medial axis, at the transition moment (center). An example of the other transition, where the four contact points lie within a semicircle, can be constructed similarly

'moth' transition, described above, there is an isolated point away from the diagonal. On the other hand in the A_2^2 'nib' transition there is a crossing on the pre-symmetry set away from the diagonal. Can such a crossing occur on the diagonal, creating a 'different' A_4 transition? The answer is no, and the geometrical reason is interesting. A circle having A_4 contact with the boundary $\mathcal{B}$ at $\mathbf{p}$ is tangent at a point where the first *two* derivatives of curvature are zero. This means that the curvature does not have an extremum at $\mathbf{p}$: the arc close to $\mathbf{p}$ has steadily increasing curvature[6] and therefore cannot support any bitangent circles. (This result was first proved by Leyton, 1987, p. 333.) So the pre-symmetry set corresponding to an A_4 point must be an isolated point, never a crossing.

As illustrated in Fig. 2.23, transitions can be understood as being between equivalence classes of shapes with the same shock graph topology. These ideas are developed in some detail and applied to object recognition in Sebastian et al. (2004).

---

[6] Such an arc with no extrema of curvature is called a *spiral*.

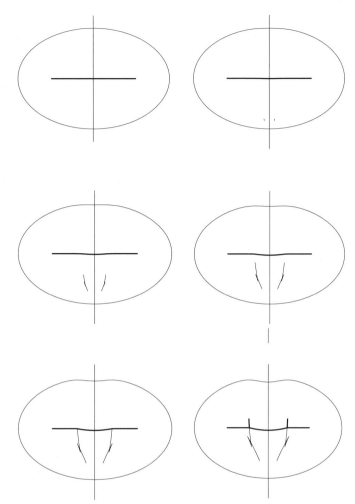

Fig. 2.20 A sequence illustrating several transitions on the medial axis (thick line) and the remainder of the symmetry set (thin line), in a family of curves starting from an ellipse which acquires a symmetrical Gaussian dent. By symmetry, two transitions happen simultaneously at each stage. *Top row*: A_4 transitions generate new branches on the symmetry set (as in Fig. 2.18, upper left). *Middle row*: the new branches grow, and the medial axis bends. *Bottom row*: $A_1 A_3$ transitions occur as one end point penetrates the medial axis (Fig. 2.18, upper right). This has created two A_1^3 points (Y-junctions) and, for each one, two cusps on the symmetry set. Throughout the sequence, the vertical line is the part of the symmetry set corresponding with circles outside the evolving curve. Continued in Fig. 2.21

2.9 Medial Axes of Families of Surfaces

The medial axis of a surface undergoes certain transitions as the surface changes shape in a generic way. These transitions were enumerated in (Bogaevsky, 1990, 2002); see also (Giblin and Kimia, 2002; Giblin et al., 2008) for a geometrical

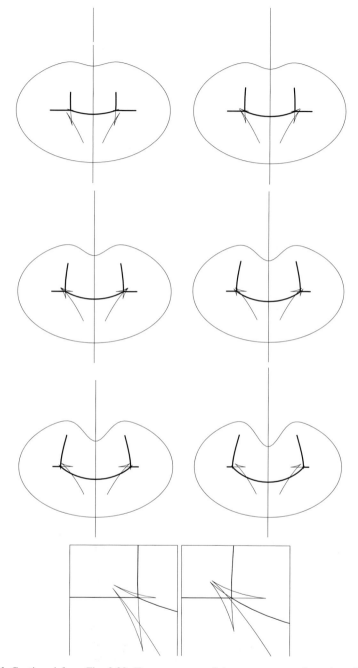

Fig. 2.21 Continued from Fig. 2.20. *Top row*: cusps of the symmetry set from the original A_4 transition and from the A_1A_3 transition are approaching each other. *Second row*: the cusps collide, and an enlargement is shown in the bottom row. These are A_2^2 nib transitions (Fig. 2.18, bottom left). The cusps change branches, as the enlargement shows. Third row: having exchanged cusps, branches are now preparing to withdraw towards the middle of the shape

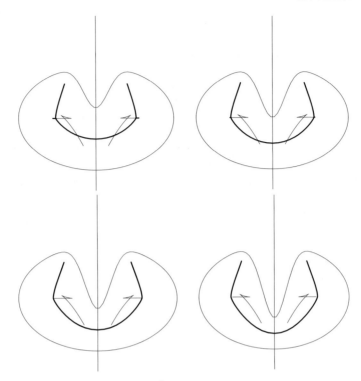

Fig. 2.22 Continued from Fig. 2.21. Two A_1^3 transitions which occur in the bottom left figure leave the medial axis (bottom right) as a U-shape

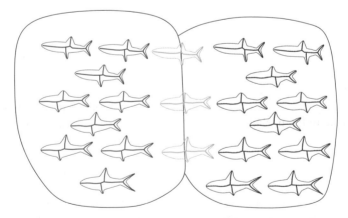

Fig. 2.23 Adapted from Sebastian et al. (2004). The set of shapes where continuous changes in the shape do not affect the medial axis topology define a *shape cell*. A transition shape defines the boundary between two shape cells. A traversal from any shape in the left shape cell to the right shape cell can be summarized by the transition itself, thereby bunching together numerous paths between two shape cells into a single event

approach and for many more details of the transitions. The additional transitions on the 3D symmetry set are very numerous, and we shall not attempt to describe them here. Some of them are studied in Pollitt (2004), and there is information on the pre-symmetry sets in 3D in Diatta and Giblin (2005).

The expected contacts which will yield transitions on the medial axis in 3D are combinations involving only A_k for k odd, and with the sum of the various k equal to 5 (compare to Section 2.8 where the corresponding list for transitions of the medial axis in a family of curves had '5' replaced by '4'). This would give $A_1^5, A_1^2 A_3, A_5$. However, there is an interesting phenomenon in 3D which did not occur in 2D. There are transitions which result from contacts of the boundary surface $\mathscr{B}$ with spheres which are no more 'degenerate' than those occurring for a generic surface but are in some way geometrically special. For example, one of the transitions occurs when $\mathscr{B}$ is tangent to a sphere at four points, that is the contact is A_1^4, but in the special case where these four points are *coplanar*. (The corresponding phenomenon in 2D, with three points of a circle collinear, is of course impossible.) The same can happen for $A_1 A_3$, which occurs generically on a single surface: in special circumstances this can cause a transition on the medial axis. See below.

- A_1^4 To illustrate this we can take the 'bin' example of Fig. 2.9, and shorten it; at a certain moment there will be a sphere *inside the bin* which is tangent to the top, the bottom and twice to the curved surface, once at the front and once at the back. Such spheres exist for a 'tall' bin, as in Fig. 2.9, but they are too large to be contained entirely inside, and therefore their centers do not contribute to the medial axis, only to the symmetry set. The four contact points of this sphere are *coplanar*, and a transition occurs. This transition can also be regarded as a collision between two A_1^3 (Y-junction) curves on the medial axis. The flat sheet of the medial axis depicted in Fig. 2.9b becomes shorter until the top and bottom curves, marked A_1^3, collide. After collision two ordinary A_1^4 (6-junction) points emerge; see Fig. 2.24. See also Figs. 2.26 and 2.27.
- A_1^5 At an isolated moment in a family of evolving surfaces, a sphere can become tangent at five distinct points. See Fig. 2.25 and also Fig. 2.27. At that moment, there are six A_1^3 (Y-junction) curves on the medial axis through the center of that sphere. On the symmetry set every subset of three of contact points out of five gives an A_1^3 curve, hence 10 such curves in all, but four of these subsets do not contribute to the medial axis. In fact the rule is that, for a given three contact points, the other two contact points must lie on the same side of the plane containing the given three points. For example, in Fig. 2.25, center, there is another branch of the symmetry set perpendicular to the top and bottom faces of the wedge, corresponding to spheres which remain tangent to the three rectangular faces of the wedge. But these spheres are not wholly inside the object.
- $A_1 A_3$ We have already met the $A_1 A_3$ case in Section 2.5 for a generic surface. The medial axis has a fin point and an A_1^3 (Y-junction) curve and an A_3 curve meet at a nonzero angle there. However there are two special cases of this contact where a transition is taking place. This occurs when the A_1^3 and A_3 curves are actually *tangent*. The geometrical condition is rather bizarre; we give it here. The bitangent sphere in this situation is (i) a sphere of curvature of $\mathscr{B}$ at a ridge point

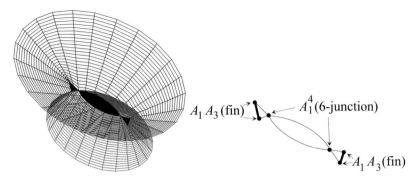

Fig. 2.24 *Left*: The medial axis of the 'bin' of Fig. 2.9 after it has shortened to the point where there are spheres inside it tangent to the top and the bottom faces of the bin. These spheres have their centers in the dark lips-shaped area in the middle of the figure; this is parallel to the top and bottom faces. The smaller dark areas at the ends of this are, on the contrary, in a vertical plane; they are centers of spheres tangent to the curved surface of the bin at front and back. *Right*: the A_1^3 (Y-junction) curves (thin lines) and the A_3 (rib or edge) curves (thick lines), together with the special points on them. Note that the 'triangular' pieces are in planes at right-angles to the central piece

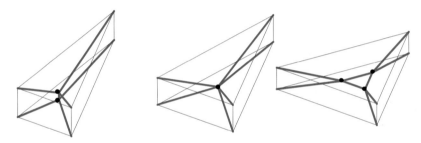

Fig. 2.25 An A_1^5 transition, illustrated through the change in the A_1^3 (Y-junction) curves, which are drawn heavily. At the transitional moment (center) there is a sphere tangent to all five faces of the boundary surface. The A_1^4 points (6-junction points) are indicated by dots; note that one is created in the process from left to right

$\mathbf{b}_1$, and (ii) tangent to $\mathscr{B}$ with ordinary A_1 contact at another point $\mathbf{b}_2$. At $\mathbf{b}_1$ there is a line of curvature γ (principal curve) on $\mathscr{B}$ associated with the ridge point: along γ the corresponding principal curvature has an extremum. The curve γ has an *osculating plane* Π at $\mathbf{b}_1$: the limiting plane through $\mathbf{b}_1$ and two nearby points of γ. Then we have (Pollitt, 2004; Diatta and Giblin, 2005, Chapter 4): *the $A_1 A_3$ transition occurs precisely when the point* $\mathbf{b}_2$ *lies in the osculating plane* Π. The two possible transitions[7] are illustrated in Fig. 2.27.

- In $A_1 A_3$-I, a new sheet of the medial axis penetrates an existing sheet, in much the same manner as the 2D equivalent $A_1 A_3$ transition (Figs. 2.18, 2.20).
- In $A_1 A_3$-II an A_3 curve and an A_1^3 curve collide and split apart generating two $A_1 A_3$ (fin) points.

---

[7] We do not know the geometrical condition which distinguishes these transitions.

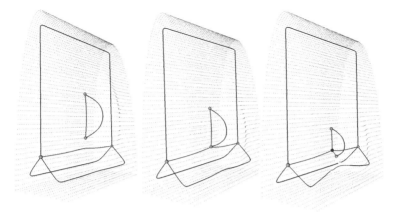

$$A_1^2A_3\text{-I}$$

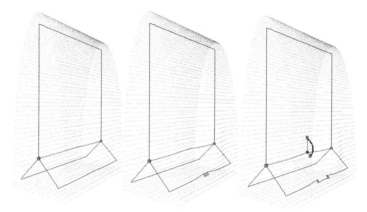

$$A_1^2A_3\text{-II}$$

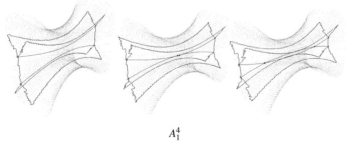

$$A_1^4$$

Fig. 2.26 Three of the generic transitions on medial axes. See Section 2.9

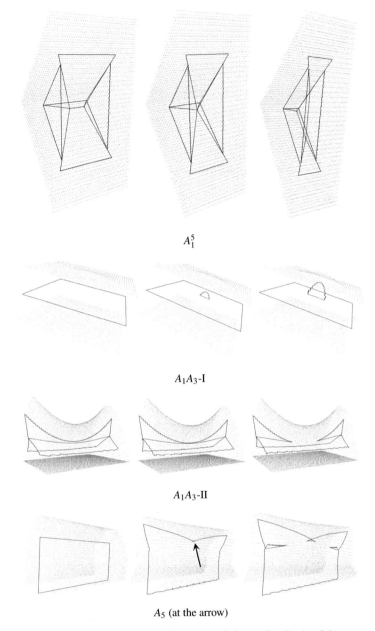

A_1^5

A_1A_3-I

A_1A_3-II

A_5 (at the arrow)

Fig. 2.27 The remaining four generic transitions on medial axes. See Section 2.9

- **$A_1^2A_3$** In a family of surfaces a sphere of curvature tangent at a ridge point $\mathbf{b}_1$ (A_3 contact) and at another point $\mathbf{b}_2$ (A_1 contact) can momentarily come into contact with the boundary $\mathscr{B}$ at a third point $\mathbf{b}_3$. This results in two possible transitions on the medial axis, which are distinguished by whether $\mathbf{b}_2$, $\mathbf{b}_3$ lie on the same or opposite sides of the osculating plane Π mentioned in the previous case. See Fig. 2.26.

  - In $A_1^2A_3$-I an A_1A_3 (fin) point collides with an A_1^3 (Y-junction) curve to generate an A_1^4 (6-junction) point.
  - In $A_1^2A_3$-II an A_1^3 (Y-junction) curve develops an A_1^4 (6-junction) point and an extra sheet of the medial axis results, bounded by two A_1^3 curves and an A_3 curve.

- **A_5** This is a particularly interesting transition, for a number of reasons. It is completely *local*: everything happens close to one point of $\mathscr{B}$; it results in the creation of a short piece of A_1^3 (Y-junction) curve, terminating in two A_1A_3 (fin) points. (Also two A_4 (turning) points are created, but these are visible only on the symmetry set and not on the medial axis.) See Fig. 2.27.

In analogy to the use of transitions of the medial axis in 2D in object recognition and perceptual grouping, the transitions in 3D (summarized in Figs. 2.26 and 2.27) have been used in regularizing the medial scaffold (Leymarie et al., 2004a), as illustrated by the examples in Fig. 2.28. Applications of these transitions to the problem of object registration (Chang et al., 2004) are discussed in Chapter 11.

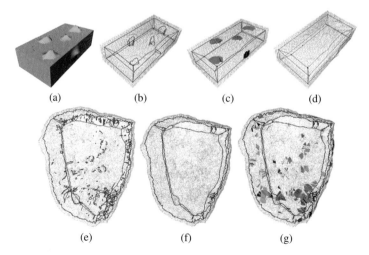

(a)   (b)   (c)   (d)

(e)   (f)   (g)

Fig. 2.28 (**a**) Example of a rectangular box uniformly sampled, but deformed by five protrusions (four on top, one on a side). (**b**) Medial scaffold left after surface meshing shown in (a) has been performed, with the three types of transitions due to surface perturbations. (**c**) The associated "cut off" patches. (**d**) Regularized shock scaffold after transition removal. (**e, f**) Pot sherd and its scaffold before and after transition removal. (**g**) Associated "cut off" patches

2.10 Consistency Conditions at Branches

Consider the medial axis in 2D, at an A_1^3 (Y-junction) point. There are three smooth branches which make certain angles with one another. Are these angles arbitrary? And is there any necessary relation between the angles and, say, the curvatures of the three branches? A similar question can be asked in 3D, where we have three smooth sheets meeting in an A_1^3 (Y-junction) curve. In this section we address these questions, which are relevant (in 2D) to asking whether we can start with an arbitrary 'graph' consisting of smooth arcs, end points and Y-junctions, and expect it to represent the medial axis of a 2D shape. We shall only give some basic results here; for many more results along the same lines see (Giblin and Kimia, 2003a; Pollitt et al., 2004; Pollitt, 2004). Note that the results here concern the smooth reconstruction of the boundary at special points, going beyond first order matching. There are more general first order results in Chapter 3.

In the 2D situation we consider a Y-junction on the medial axis, as in Fig. 2.29. Let the three branches of the medial axis (or symmetry set) be labeled 1, 2, 3, and let ψ_i be the angle between the two branches other than branch i. It follows from properties of angles in a circle that $\psi_i + \theta_i = \pi$ where θ_i corresponds to the angle in Fig. 2.13; one of the angles θ_i is also shown in Fig. 2.13, right. Let κ_i be the curvatures of the three branches at the Y-junction point. Then

$$\frac{\kappa_1}{\sin \psi_1} + \frac{\kappa_2}{\sin \psi_2} + \frac{\kappa_3}{\sin \psi_3} = 0. \tag{2.3}$$

Note that the angles ψ_i are always less than $180°$ (compare Fig. 2.14e, f). One consequence of this and the above equation is that the three curvatures κ_i cannot all have the same sign. Because $\theta_i = \pi - \psi_i$ the angles ψ_i in (2.3) can be replaced by θ_i.

Consistency relationships such as (2.3) are proved by reconstructing the three arcs of the boundary $\mathscr{B}$ close to the tangency points of the circle, as in Section 2.6. Each arc is reconstructed in two different ways, and identifying these two gives the consistency relation. In fact, recalling that $\psi_i + \theta_i = \pi$, we find the following (Giblin and Kimia, 2003a, (78)):

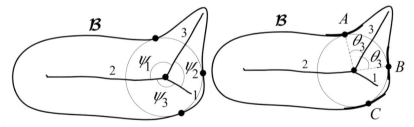

Fig. 2.29 *Left*: a Y-junction (A_1^3 point) and the associated branches and angles ψ_i. *Right*: one of the angles θ_i. Also, the arc A of the boundary $\mathscr{B}$ is reconstructed both from branch 2 and branch 3 of the medial axis; likewise B is reconstructed from branches 1 and 3, and C from branches 1 and 2. Identifying these different ways of reconstructing the same arcs yields results like (2.3)

$$\frac{\kappa_1}{\sin\psi_1} = -\frac{\psi_2'}{\sin\psi_2} + \frac{\psi_3'}{\sin\psi_3}$$

$$\frac{\kappa_2}{\sin\psi_2} = -\frac{\psi_3'}{\sin\psi_3} + \frac{\psi_1'}{\sin\psi_1}$$

$$\frac{\kappa_3}{\sin\psi_3} = -\frac{\psi_1'}{\sin\psi_1} + \frac{\psi_2'}{\sin\psi_2}.$$

Here $'$ denotes derivative with respect to arclength and adding these three equations gives (2.3). In addition, this shows that the tangents to the medial axis branches, and the ψ_i', determine the κ_i.

One can look for 'higher order' constraints, either on the geometry of the medial axis branches (that is, derivatives of the curvatures κ_i) or on the dynamics (that is, derivatives of the radius functions r_i, or of the object angles θ_i). This is investigated in Giblin and Kimia (2003a). The next constraint after (2.3) involves dynamics:

$$\sum \frac{\theta_i''}{\sin^2\theta_i} = \sum \frac{\cos\theta_i}{\sin^3\theta_i}(\kappa_i^2 + 2\theta_i'^2).$$

Thus the second derivatives of θ_i (or the third derivatives of r_i) are constrained by the lower order derivatives of curvature and θ_i. In general we find a constraint on the even order derivatives of κ_i ('geometry') and the odd order derivatives of r_i ('dynamics'), in each case expressed in terms of lower order derivatives of geometry and dynamics.

The situation when we are given two branches and one radius function is rather clearer. One of the results from Giblin and Kimia (2003a) is the following.

Proposition 2.10.1. *Suppose that two branch parametrizations γ_1, γ_2 and one radius function r_1, are given. Then the boundary $\mathcal{B}$ can be uniquely constructed close to the three contact points, and there are no constraints beyond the usual $|dr_1/ds_1| < 1$.*

When we consider consistency relations in 3D, the situation naturally becomes more complicated. We state here just one of the relationships which holds in the A_1^3 or Y-junction case. Here we have a curve of points in 3D along which three sheets of the medial axis or symmetry set intersect, as in Fig. 2.8. The following result is proved in (Pollitt et al., 2004; Pollitt, 2004).

Proposition 2.10.2. *Consider a point $\mathbf{p}$ on a Y-junction curve γ on the medial axis. There are three radius functions r_i, $i = 1, 2, 3$ corresponding to the three sheets M_i of the medial axis, and generically none of these will have an extremum on γ, so all of them will have a nonzero gradient vector ∇r_i. Let κ_i^r be the sectional curvature of the sheet M_i in the direction of this gradient vector, and let κ_i^t be the sectional curvature perpendicular to this, that is, in the direction of the level set $r_i = $ constant. For each sheet M_i there is an associated angle θ_i, defined as in Fig. 2.17 as the angle between the gradient direction ∇r_i and the line from $\mathbf{p}$ to either corresponding boundary point.*

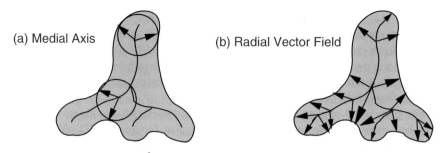

Fig. 3.3 Blum medial axis in $\mathbb{R}^2$ and associated radial vector field

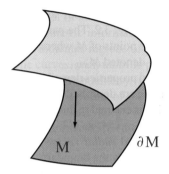

Fig. 3.4 Projection defining an edge parametrization for edge of Blum medial axis

In addition to these general properties, there are the specific *Blum properties* at smooth points $x \in M_{reg}$ (see Fig. 3.3): (i) the two values of $\mathbf{V}$ at x have the same length; (ii) the two values of $\mathbf{V}$ make equal angles (on opposite side) with the tangent space $T_x M$; and (iii) each value of $\mathbf{V}$ is orthogonal to the boundary $\mathscr{B}$ at the tangency point to which $\mathbf{V}$ points. Contrary to expectation, the first two of these three properties are not important for determining the geometry of the boundary; only the third property is necessary. We shall refer to property (iii) as the *partial Blum condition* for the pair $(M, \mathbf{V})$.

Remark 3.2.2. Because we shall be computing derivatives of functions or vector fields on M, we explain what exactly we mean. At smooth points of M, differentiability is in the usual sense for smooth curves or surfaces. For each smooth sheet of M meeting at a branch curve, differentiability is in the usual sense for a surface with a boundary edge. However, at edge points (including fin points), the usual notion of smoothness at an edge point of a surface no longer suffices. Instead "edge coordinates" are required for both $\mathbf{V}$ and r to be smooth at edge points (as explained in Damon, 2003). These coordinates correspond to the projection of a half–parabolic surface as shown in Fig. 3.4. For Blum medial axes these coordinates are not explicitly defined. Thus, although we use edge coordinates to state results at edges, we also explain how calculations at edge points can be alternatively carried out using standard coordinates for surfaces at points approaching edge points.

3.2.1.2 Skeletal Structures

If we deform a Blum medial axis M by stretching and/or bending, and also deform the direction and magnitude of the radial vector field $\mathbf{V}$, then we should not expect the resulting set and vector field to be a Blum medial axis and radial vector field of another region. Instead it will be a more general type of *skeletal structure* (which is formally defined in Damon, 2003). The enlarged class of pairs $(M, \mathbf{V})$ which we allow to be included should still retain certain properties of the Blum medial axis such as M being a Whitney stratified set and $\mathbf{V}$ having the same smoothness and multivalued properties. However, we do not require that any of the Blum conditions hold, nor do we require that only the standard local models for M are possible. This allows, for instance, nongeneric local models for M such as those given in Fig. 3.5.

However, care must be taken with an allowable radial vector field. It must satisfy a "Local Separation property" at a singular point x. This means that we can decompose M in a neighborhood of x into components M_i so that the radial lines from their boundary edges mapping into one local complementary component of M decomposes the complementary component into distinct regions corresponding to the M_i (see (b) of Fig. 3.6 versus its failing as in (c) of Fig. 3.6). Also, it must satisfy the "Local Edge Property" which requires that in some neighborhood of an edge point of M, not only must a smooth value $\mathbf{V}$ point to one side of the tangent space for that point, but it must also point to the same side of the tangent spaces for all points in that neighborhood. This can fail if M oscillates near the boundary so the radial lines keep intersecting M close to the edge as in (a) of Fig. 3.6.

For such a skeletal structure we can define the *associated boundary* as $\mathscr{B} = \{x + \mathbf{V}(x) : x \in M\}$ where we allow for each x all possible values of $\mathbf{V}(x)$. In the case that $(M, \mathbf{V})$ is a Blum medial axis of a region Ω, this associated boundary is the boundary of Ω. If $(M, \mathbf{V})$ is a more general skeletal structure, then $\mathscr{B}$ need not even be smooth, but may have singularities itself such as in Fig. 3.7. Then to understand the properties of $\mathscr{B}$, it is best to obtain it as the result of a flow from M.

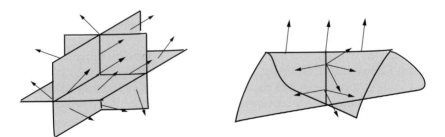

Fig. 3.5 Surfaces in $\mathbb{R}^3$ passing through each other in these ways, with their associated radial vector fields, are examples of nongeneric skeletal structures

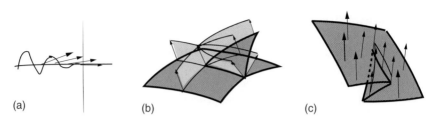

Fig. 3.6 (**a**) Failure of local edge property. (**b**) Illustration of the local separation property. (**c**) Failure of local separation property; only the spokes on one side of M are shown. Due to the creased form of the medial sheet, multiple spokes illegally account for the same points in the object

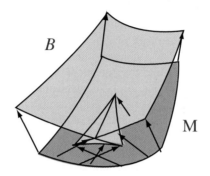

Fig. 3.7 Failure of smoothness for the boundary associated to a skeletal structure

3.2.2 Radial Flow Defined for a Skeletal Structure

In Chapter 2, Giblin and Kimia show how to recover the boundary of a region as the envelope of the set of tangent spheres on the medial axis. We take a different approach for constructing the boundary $\mathcal{B}$ from the medial axis M, using instead the *radial flow*, which is a backward version of the "grassfire flow" (Kimia et al., 1990, also see Siddiqi et al., 2002). In fact this method applies to any skeletal structure, and in the process, we also will obtain information about how the region is built up from the level sets of the flow.

Locally the radial flow is defined using a smooth value of $\mathbf{V}$ in some neighborhood of a point $x_0 \in M$. It is given by the *local radial flow* $\psi_t(x) = x + t\mathbf{V}(x)$.[1] The $t = 1$ value of this map defines a *radial map* ψ_1 (depending on the local choice of $\mathbf{V}$) from a region of M to a corresponding region of $\mathcal{B}$. Because $\mathbf{V}$ is multivalued, the radial flow cannot be globally defined on M to $\mathcal{B}$. However, there is a way to define a global flow from a mathematical object called the "double of M," denoted $\tilde{M}$. Briefly to construct $\tilde{M}$, we view M as having a slight thickness and separate the

---

[1] The flow variable t used here for radial flow is in some sense in reverse to the flow value t used in Chapter 2 for boundary flow, though the surfaces produced are not the same. Moreover, the variable t used here can be interpreted as a fraction of the distance along a spoke and and then used as a coordinate of the interior of the figure described by that medial sheet. When used in this way in Chapters 8 and 9, the variable is called τ.

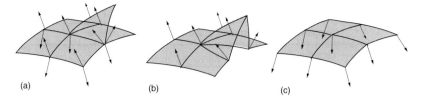

Fig. 3.8 (**a**) Neighborhood of a point in M and corresponding neighborhoods (**b**) and (**c**) in $\tilde{M}$ and the associated radial vector fields

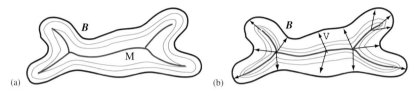

Fig. 3.9 Radial flow versus grassfire flow between the boundary B and the medial axis M (darkened branch curve). (**a**) Shows levels of the grassfire flow. (**b**) Shows levels $\mathscr{B}_t$ of the radial flow evolving along the multivalued radial vector field $\mathbf{V}$

two sides of M. For example, near a fin point M is separated into two sheets as shown in Fig. 3.8. Then $\tilde{M}$ provides a way to simultaneously consider both sides of M at once, and give a unique value for $\mathbf{V}$ at each point of $\tilde{M}$.

The local formula of the radial flow given by $\psi_t(x)$ above is satisfactory for doing local computations, even though the full radial flow is needed to establish its global properties, which we shall describe. The radial and grassfire flows are moving along the same radial lines, but in different directions. We may think of the grassfire flow as transmitting information from the boundary to the medial axis; and in the reverse direction, the radial flow allows us to transmit geometric information defined on the medial axis M back to the boundary $\mathscr{B}$. The flow also allows us to define the intermediate levels of the flow. For each **fixed** t, with $0 \leq t \leq 1$, we define the *level surface* $\mathscr{B}_t = \{x + t\mathbf{V}(x) : x \in M\}$ where we again allow for each x all possible values of $\mathbf{V}(x)$ (see Fig. 3.9). We shall see that the properties of the boundary are a consequence of controlling the level surfaces $\mathscr{B}_t$ during the radial flow.

We contrast the properties of the radial flow with those of the grassfire flow in Table 3.1.

3.2.2.1 Compatibility 1–Form and Compatibility Condition

Two key properties of the boundary are captured by a *compatibility condition* on the Blum medial axis. The *compatibility 1–form* $\eta_{\mathbf{V}}$ is defined by $\eta_{\mathbf{V}}(v) = \mathbf{U} \cdot v + dr(v)$; this is also multivalued because $\mathbf{V}$ is. M satisfies the *compatibility condition* at a point x_0 with smooth value $\mathbf{V}$ if $\eta_{\mathbf{V}} \equiv 0$ at x_0. The compatibility condition relates the (gradient of the) radius function with the unit radial vector field. As detailed in Damon (2003, Lemma 6.1) or Damon (2004, Lemma 3.1), the following criterion gives boundary properties resulting from this condition:

Moreover, because v_i and $\mathbf{n}$ are orthogonal, it follows that the matrix in (3.3) representing S is symmetric. Then the properties of symmetric matrices imply that the eigenvalues are real and the eigenvectors for different eigenvalues are orthogonal (these are the principal curvatures and principal directions of curvature). Then the Gauss and Mean curvatures are the determinant, respectively one half the trace, of the matrix in (3.3). From S we also define the second fundamental form of M by $II(v,w) = S(v) \cdot w$. The basic differential geometry of M is defined in terms of this information, all of which is contained in the shape operator.

Key geometric information is already contained in the sign of Gauss curvature. A negative sign indicates locally a saddle shaped surface, and positive Gauss curvature locally a concave or convex surface with the signs of the principal curvatures being positive in the concave case and negative in the convex case. These portions of $\mathscr{B}$ are separated by parabolic curves where at least one principal curvature is 0, and there are umbilic points where both principal curvatures are equal. Furthermore, there are principal curves whose tangent lines at points are principal lines, which further indicate the directions of extreme curvature.

3.2.3.2 Shape Operators and Radial Geometry

Although we are interested in the differential geometry of the boundary of the object, we shall introduce very different and seemingly unrelated "shape operators" for a skeletal structure which measures the "radial geometry" by describing the changes in $\mathbf{U}$ as one moves on the medial surface.

Radial Shape Operator

In fact we initially let $(M, \mathbf{V})$ denote a skeletal structure which relaxes the conditions for a Blum medial axis M with radial vector field $\mathbf{V}$. Suppose we choose in a neighborhood of a non–edge point $x_0 \in M$, a smoothly defined value for $\mathbf{V}$. Then the radial shape operator (for this value of $\mathbf{V}$) at x_0 is defined so that for a tangent vector $v \in T_{x_0} M$

$$S_{rad}(v) = -\mathrm{proj}_{\mathbf{V}} \left(\frac{\partial \mathbf{U}}{\partial v} \right)$$

with $\mathrm{proj}_{\mathbf{V}}$ denoting projection onto the tangent space $T_{x_0} M$ along $\mathbf{V}$ (in general, this is not orthogonal projection, see Fig. 3.10). Unlike the differential geometric shape operator, S_{rad} is in general not symmetric, so we are not even guaranteed that the eigenvalues are real. Nonetheless, for a basis $\mathbf{v}$ of $T_{x_0} M$, we let $S_{\mathbf{v}}$ denote the matrix representation of S_{rad}. The *principal radial curvatures* κ_i and *principal radial directions* are the eigenvalues and eigendirections of S_{rad}. These can be shown to be real-valued in the Blum case.

The definition we have given here is dimension independent and equally well applies for both two and three dimensional objects.

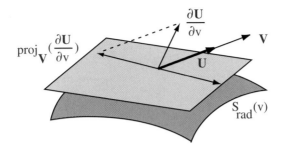

Fig. 3.10 Projection for defining the radial shape operator. The dashed line denotes projection onto $T_{x_0}M$ along $\mathbf{V}$

Example 3.2.2. We first consider M a 1–dimensional Blum medial axis of a 2–dimensional object in $\mathbb{R}^2$. If $\gamma(s)$ is a local parametrization of one of the smooth curve components of M, write

$$\frac{\partial \mathbf{U}}{\partial s} = a \cdot \mathbf{U} - \kappa_r \cdot \gamma'(s) \tag{3.4}$$

Then κ_r is the principal radial curvature and the radial shape operator is just multiplication by κ_r.

Example 3.2.3. Let M be a 2–dimensional Blum medial axis of an object in $\mathbb{R}^3$. Let $X(u_1, u_2)$ be a local parametrization of an open set W of one of the smooth sheets of M. Then $v_i = \dfrac{\partial X}{\partial u_i}, i = 1, 2$ gives a basis for $T_{x_0}M$ at each point $x_0 \in W$. We write

$$\frac{\partial \mathbf{U}}{\partial u_i} = a_i \cdot \mathbf{U} - b_{1i} \cdot v_1 - b_{2i} \cdot v_2 \quad i = 1, 2 \tag{3.5}$$

Then

$$S_{\mathbf{v}} = \begin{pmatrix} b_{11} & b_{12} \\ b_{21} & b_{22} \end{pmatrix} \tag{3.6}$$

The principal radial curvatures are the two eigenvalues κ_{r1} and κ_{r2} of $S_{\mathbf{v}}$.

If we had used a different basis $\mathbf{w} = \{w_1, w_2\}$, then if C denotes the transformation for the change of basis from $\mathbf{v}$ to $\mathbf{w}$, then $S_{\mathbf{w}} = C S_{\mathbf{v}} C^{-1}$.

Edge Shape Operator

Again we first give a dimension independent definition and then consider its meaning for 1D and 2D Blum medial axes. Let x_0 be an edge point of M, with a smooth value of $\mathbf{V}$ at x_0 corresponding to one side of M. Also, let $\mathbf{N}$ be a unit normal vector

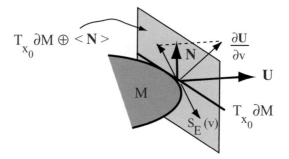

Fig. 3.11 Defining the edge shape operator. The dashed line denotes projection onto $T_{x_0} M \oplus \langle \mathbf{N} \rangle$ along $\mathbf{V}$

field to M pointing on the same side of M as a smooth value of $\mathbf{V}$. We define

$$S_E(v) = -\text{proj}' \left(\frac{\partial \mathbf{U}}{\partial v} \right)$$

Here proj' denotes projection onto $T_{x_0} \partial M \oplus \langle \mathbf{N} \rangle$ along $\mathbf{V}$ (again generally this is not orthogonal, see Fig. 3.11).

We emphasize that $\frac{\partial \mathbf{U}}{\partial v}$ has to be computed using edge coordinates. For $v \in T_{x_0} \partial M$, there is no problem; it is only for v pointing out from ∂M that we must be careful. For a matrix representation of S_E we use a special basis $\mathbf{v}$ for $T_{x_0} M$ consisting of a basis $\tilde{\mathbf{v}}$ for $T_{x_0} \partial M$ together with a vector in edge coordinates that maps to a multiple of $\mathbf{U}(x_0)$. For $T_{x_0} \partial M \oplus \langle \mathbf{N} \rangle$ we use for a basis $\tilde{\mathbf{v}}$ and $\mathbf{N}$. We denote the matrix by $S_{E\mathbf{v}}$.

To define the principal edge curvatures, we let $I_{n-1,1}$ denote the $n \times n$–matrix obtained from the identity matrix by changing the last entry to 0. For $n \times n$–matrices A and B, a generalized eigenvalue of (A, B) is a λ such that $A - \lambda B$ is singular. The *principal edge curvatures* are defined to be the generalized eigenvalues of $(S_{E\mathbf{v}}, I_{n-1,1})$.

Example 3.2.4. For a 1–dimensional Blum medial axis M of an object in $\mathbb{R}^2$, let $\gamma(s)$ be a local edge parametrization of M with say $\gamma(0) = x_0 \in \partial M$. Then we write

$$\frac{\partial \mathbf{U}}{\partial s} = a \cdot \mathbf{U} - c_{\mathbf{n}} \cdot \mathbf{N} \tag{3.7}$$

The edge shape operator is then multiplication by $c_{\mathbf{n}}$. As $I_{0,1}$ is the 1×1 0–matrix, provided $c_{\mathbf{n}} \neq 0$, there are no principal edge curvatures. Which is not surprising as the edge is 0–dimensional. The degenerate case would correspond to $c_{\mathbf{n}} = 0$, in which case all values are generalized eigenvalues. This cannot happen for by Proposition 4.7 of Damon (2003), the Blum medial axis satisfies the edge condition (see Section 3.3.1), which implies that all positive generalized eigenvalues are bounded from below by $\frac{1}{r}$.

Example 3.2.5. For M a 2–dimensional Blum medial axis of an object in $\mathbb{R}^3$, we let $X(u_1, u_2)$ be a local edge parametrization of an open set W with $X(0,0) = x_0 \in \partial M$.

We suppose X chosen so that if $v_i = \dfrac{\partial X}{\partial u_i}, i = 1, 2$, then $v_1 \in T_{x_0} \partial M$ and v_2 maps under the edge parametrization to $c \cdot \mathbf{U}_{tan}$, for $\mathbf{U}_{tan}$ the tangential component of $\mathbf{U}$ and $c \geq 0$. We write

$$\frac{\partial \mathbf{U}}{\partial u_i} = a_i \cdot \mathbf{U} - c_{\mathbf{n}i} \cdot \mathbf{N} - b_i \cdot v_1 \quad i = 1, 2 \tag{3.8}$$

Then the edge shape operator has the matrix representation

$$S_{E\mathbf{v}} = \begin{pmatrix} b_1 & b_2 \\ c_{\mathbf{n}1} & c_{\mathbf{n}2} \end{pmatrix} \tag{3.9}$$

When $c_{\mathbf{n}2} \neq 0$, the single principal edge curvature is the generalized eigenvalue κ_E of $(S_{E\mathbf{v}}, I_{1,1})$, which we can compute to be $\kappa_E = c_{\mathbf{n}2}^{-1} \cdot \det(S_{E\mathbf{v}})$.

We next explain how we may carry out this computation while avoiding the explicit use of edge coordinates. A way around this is to compute $S_{\mathscr{B}\mathbf{v}'}$ as a limiting value. Suppose v_1 is a smooth vector field on a neighborhood W of the edge point x_0 which is tangent to ∂M. Here smoothness is only in the sense of a surface with edge. We also let $v_2 = \mathbf{U}_{tan}$ the tangential component of $\mathbf{U}$. Then both $\mathbf{U}_{tan}$ and v_1 are smooth for edge coordinates. Then we can compute at smooth points $x \in W$ near x_0, a related operator S_E' extending S_E as follows:

$$S_E'(v) = -\text{proj}_{\mathbf{V}}'' \left(\frac{\partial \mathbf{U}}{\partial v} \right) \tag{3.10}$$

where now $\text{proj}_{\mathbf{V}}''$ denotes projection along $\mathbf{V}$ but onto L, the subspace spanned by $\{v_1, \mathbf{N}\}$. Then we can take the limit as $x \to x_0$ (using Proposition 3.9 of Damon, 2004) to conclude

$$S_E(v) = \lim_{x \to x_0} S_E'(v) \tag{3.11}$$

Although we cannot actually compute limits, by choosing x sufficiently close to x_0 we can obtain good approximations to S_E at x_0.

Computing Shape Operators Without Normalizing

For computational reasons, it would be preferable to compute directly from $\mathbf{V}$, rather than having to normalize $\mathbf{V}$ to first construct the unit vector field $\mathbf{U}$. From

$$\frac{\partial \mathbf{V}}{\partial u_i} = \frac{\partial r}{\partial u_i} \cdot \mathbf{U} + r \cdot \frac{\partial \mathbf{U}}{\partial u_i} \tag{3.12}$$

and (3.5), we obtain

$$\frac{\partial \mathbf{V}}{\partial u_i} = a_i' \cdot \mathbf{U} - b_{1i}' \cdot v_1 - b_{2i}' \cdot v_2 \quad i = 1, 2 \tag{3.13}$$

where $a_i' = \dfrac{\partial r}{\partial u_i} + r \cdot a_i$ and $b_{ij}' = r b_{ij}$. Hence, we may compute instead

$$-\text{proj}_{\mathbf{V}}\left(\frac{\partial \mathbf{V}}{\partial v}\right) = r \cdot S_{rad}(v)$$

Then the eigenvalues of $r \cdot S_{rad}$ are $r \kappa_{ri}$, with the same eigenvectors as for S_{rad}. We shall see that all of the conditions which we shall obtain can be equally well be expressed in terms of $r \cdot S_{rad}$ and its eigenvalues $r \kappa_{ri}$ and eigenvectors.

In a similar fashion, for the edge shape operator, we can compute instead

$$-\text{proj}'\left(\frac{\partial \mathbf{V}}{\partial v}\right) = r \cdot S_E(v)$$

Then we write

$$\frac{\partial \mathbf{V}}{\partial u_i} = a_i' \cdot \mathbf{U} - c_{\mathbf{n}i}' \cdot \mathbf{N} - b_i' \cdot v_1 \quad i = 1, 2. \tag{3.14}$$

Hence, the matrix representation $r \cdot S_{E\mathbf{v}}$ is given by

$$S_{E\mathbf{v}} = \begin{pmatrix} b_1' & b_2' \\ c_{\mathbf{n}1}' & c_{\mathbf{n}2}' \end{pmatrix} \tag{3.15}$$

Also, the generalized eigenvalue of $(r \cdot S_{E\mathbf{v}}, I_{1,1})$ is $r \kappa_E$.

3.2.4 Level Set Structure of a Region and Smoothness of the Boundary

We give the first consequence of computing the radial geometry of a skeletal structure $(M, \mathbf{V})$. We can determine when the associated boundary $\mathscr{B}$ is smooth, and in the course of doing it we also are able to give a level set decomposition of the associated region Ω using the level sets $\mathscr{B}_t$ of the radial flow.

We define

1. (*Radial Curvature Condition*) For all points of M off ∂M

$$r < \min\left\{\frac{1}{\kappa_{ri}}\right\} \quad \text{for all principal positive radial curvatures } \kappa_{ri}$$

2. (*Edge Condition*) For all points of $\overline{\partial M}$

$$r < \min\left\{\frac{1}{\kappa_{Ei}}\right\} \quad \text{for all positive principal edge curvatures } \kappa_{Ei}$$

3. (*Compatibility Condition*) For all singular points of M (which includes edge points), $\eta_V \equiv 0$.

Remark 3.2.4. In the case (M, V) is a Blum medial axis, it satisfies the radial curvature and edge conditions (by Proposition 4.7 and Lemma 6.1 of Damon, 2003) and the compatibility condition at all points of M.

We also note that the radial curvature and edge conditions relate the pure radial distance information given by r with the pure radial direction information computed from U (and in the Blum case the distance and direction properties are further related by the compatibility condition holding on all of M).

Then the smoothness of the boundary and level set decomposition is guaranteed by the following criterion, which is a consequence of (Damon, 2003, Theorem 2.3).

Criterion for Level Set Structure and Boundary Smoothness

Let (M, V) be a skeletal structure in $\mathbb{R}^2$ or $\mathbb{R}^3$ which satisfies the Principal Curvature Condition, Edge Condition, and Compatibility Condition. Then

(i) For M in $\mathbb{R}^2$, the associated boundary $\mathscr{B}$ is a C^1 curve which is smooth at all points corresponding to smooth points of M and which only has nonlocal intersections from distant points in M. If there are no nonlocal intersections then $\mathscr{B}$ is an embedded curve. Furthermore the region Ω is decomposed into the disjoint union of the level curves $\mathscr{B}_t$ of the radial flow.

(ii) For M in $\mathbb{R}^3$:

  1. The associated boundary $\mathscr{B}$ is an immersed surface which is smooth at all points except possibly at those corresponding to points of M_{sing}.
  2. At points corresponding to points of M_{sing}, $\mathscr{B}$ is weakly C^1, which means that it has a well defined limiting tangent space at such boundary points (this implies that it is C^1 except possibly at points coming from fin or 6–junction points).
  3. Also, if there are no nonlocal intersections, $\mathscr{B}$ will be an embedded surface, and the region Ω is decomposed into the disjoint union of the level surfaces $\mathscr{B}_t$ of the radial flow. These surfaces are only stratified rather than being smooth everywhere.

For both cases, at smooth points, the projection along the lines of V will locally map $\mathscr{B}$ diffeomorphically onto the smooth part of M.

Also, when (M, V) is a Blum medial axis, we obtain a level set decomposition of the region Ω by the radial flow.

Example 3.2.6. For a 1–dimensional Blum medial axis M (or more generally a skeletal structure), by Example 3.2.2 the radial curvature condition becomes

$$r < \frac{1}{\kappa_r} \quad \text{if } \kappa_r > 0, \text{ and no condition otherwise.}$$

The edge condition reduces to $c_{\mathbf{n}} \neq 0$ (otherwise all values are generalized eigenvalues). Finally, if $\gamma(s)$ is a unit speed parametrization of a smooth component of M with say $\gamma(0) = x_0 \in M$, then the compatibility condition becomes $\mathbf{U} \cdot \gamma'(0) + \dfrac{\partial r}{\partial s} = 0$. As x_0 approaches an edge point, the compatibility condition has limiting form $1 + \dfrac{\partial r}{\partial s} = 0$, which is a well-known property of the Blum medial axis. Here there is the proviso that the meaning of the derivative $\dfrac{\partial r}{\partial s}$ has to be reinterpreted using edge coordinates.

Example 3.2.7. For a 2–dimensional Blum medial axis M, by Example 3.2.3 the radial curvature condition becomes:

$$r < min\left\{\frac{1}{\kappa_{ri}}\right\} \qquad \text{for those } \kappa_{ri} > 0 \; i = 1, 2.$$

The edge condition becomes by Example 3.2.5, for $\kappa_E = c_{\mathbf{n}2}^{-1} \cdot \det(S_{E\mathbf{v}})$

$$r < \frac{1}{\kappa_E} \quad \text{if } \kappa_E > 0, \text{ otherwise no condition.}$$

For example, if both $\kappa_{ri} < 0$ then as we see in Section 3.3.1, the boundary is convex and there is no condition. If instead both $\kappa_{ri} > 0$ then (by Section 3.3.1) the boundary is concave and both κ_{ri} place restrictions on r.

Remark 3.2.5. The radial curvature and edge conditions can be summarized in terms of the alternate computations in (3.13) and (3.14) by the condition that all of the eigenvalues of $r \cdot S_{rad}$ and the generalized eigenvalue of $(r \cdot S_E, I_{1,1})$ are less than 1.

3.3 Local and Relative Geometry of the Boundary

3.3.1 Intrinsic Differential Geometry of the Boundary

One natural approach to determining the local differential geometry of the boundary of a region has been to seek the relation between the differential geometry of the boundary and that of the (smooth part of the) medial axis using derivatives of the radius function r. Results obtained by this approach include: curvature of boundary curves in the 2D case, originating with Blum and Nagel (1978); the Gaussian and Mean curvatures of boundary surfaces in 3D by Nackman and Pizer (Nackman, 1982; Nackman and Pizer, 1985); and in the opposite direction, deriving differential geometric properties of the medial axis from the differential geometry of the boundary by Siersma (1999); Sotomayor et al. (1999) and van Manen (2003a,b). In both of the surface cases, the relationship actually involves the differential geometry of a parallel surface of the boundary (rather than the boundary itself).

We shall use a different approach (taken from Damon, 2003 and Damon, 2004), which is somewhat counterintuitive in that it does not explicitly involve the differential geometry of the medial axis. Instead, we shall see that the radial geometry of the medial axis plays a much more direct role.

3.3.1.1 Differential Geometric Shape Operator in Terms of the Radial Shape Operator

The computation of the differential geometry of the boundary $\mathcal{B}$ naturally breaks into two cases: those points not on a crest curve of $\mathcal{B}$ which correspond to non–edge points of M, and those points on crest curves which correspond to edge points of M.

We begin by expressing the differential geometric shape operator for the boundary $\mathcal{B}$ at a point on the boundary $x'_0 \in \mathcal{B}$ associated to a non–edge point $x_0 \in M$. We let $\mathbf{v}'$ be the image under the radial map $d\psi_1$ of a basis $\mathbf{v} = \{v_1, \ldots, v_n\}$. $n = 1$ or 2, of $T_{x_0}M$ (in case $x_0 \in M_{sing}$, the value of $\mathbf{V}$ is determined by x'_0). We can apply Theorem 3.2 of Damon (2004) to the special case of Blum medial axes.

Local Differential Geometry from Radial Geometry

Suppose $(M, \mathbf{V})$ is a skeletal structure which satisfies the partial Blum condition as defined in 3.2.1.1 (this includes the special cases of Blum medial axes). Let $x'_0 \in \mathcal{B}$ correspond to the non–edge point $x_0 \in M$ as above.

1. The differential geometric shape operator $S_{\mathcal{B}}$ of $\mathcal{B}$ at x'_0 has a matrix representation with respect to $\mathbf{v}'$ given by

$$S_{\mathcal{B}\mathbf{v}'} = (I - r \cdot S_{\mathbf{v}})^{-1} S_{\mathbf{v}} \tag{3.16}$$

2. There is a one-to-one correspondence between the principal curvatures κ_i of $\mathcal{B}$ at x'_0 and the principal radial curvatures κ_{ri} of M at x_0 given by

$$\kappa_i = \frac{\kappa_{ri}}{(1 - r\kappa_{ri})} \quad \text{or equivalently} \quad \kappa_{ri} = \frac{\kappa_i}{(1 + r\kappa_i)} \tag{3.17}$$

3. The principal radial directions corresponding to κ_{ri} are mapped by the derivative $d\psi_1$ to the principal directions corresponding to κ_i

Remark 3.3.1. A consequence of the radial curvature condition is that in the correspondence in (3.17), that κ_i and κ_{ri} always have the same sign and one is 0 exactly when the other is.

Relation Between Radii of Curvature

A simpler way to express (3.17) is in terms of the signed radii of curvatures and radial curvatures. The *signed radii of curvature* $r_i \stackrel{def}{=} \frac{1}{\kappa_i}$ and the corresponding *signed radii of radial curvature* $r_{ri} \stackrel{def}{=} \frac{1}{\kappa_{ri}}$. In terms of the signed curvatures, (3.17) can be rewritten in the following simple form.

$$\text{radii of curvature equation:} \qquad r_{ri} = r + r_i \qquad \text{for all } i \qquad (3.18)$$

For instance, this gives us an immediate comparison of principal curvatures κ_i and κ'_i at distinct points x'_0 and x'_1 of $\mathscr{B}$ where they have the same sign, in terms of the corresponding medial data.

$$\text{If } \kappa_{ri} < \kappa'_{ri} \text{ and } r \le r', \text{ then } \kappa_i < \kappa'_i.$$

This follows because (since they have the same sign) $r_{ri} > r'_{ri}$, so (3.18) implies $r_i > r'_i$, and hence $\kappa_i < \kappa'_i$.

The radii of curvature equation is also relevant for determining various "ridges of principal curvature" of $\mathscr{B}$, as considered by Bruce et al. (1996). These ridge-type curves concern the geometry of $\mathscr{B}$ as a surface in $\mathbb{R}^3$. One such is the crest curve, which corresponds to the edge of the medial axis. However, there are other ridge-type curves which are obtained as curves of critical points of κ_i along principal curves in $\mathscr{B}$.

A critical point x'_0 of κ_i along a curve $\gamma_1(s)$ is also a critical point for r_i (provided $\kappa_i \ne 0$). Suppose γ_1 is the image of γ under ψ_1 with $x'_0 = \psi_1(x_0)$. Then by (3.18), $\frac{\partial r_i}{\partial s} = 0$ iff $\frac{\partial r_{ri}}{\partial s} = \frac{\partial r}{\partial s}$. These points are thus identified from medial data using the principal radial curvatures and radial function.

Remark 3.3.2. Suppose $(M, \mathbf{V})$ satisfies the radial, edge, and compatibility conditions and as well the partial Blum condition. The associated boundary will be smooth as explained in Section 3.2.4. As a consequence of the partial Blum condition, the spheres of radius r at points M will be tangent to the associated boundary; however, they may not lie entirely in the region. However, the radial curvature condition does ensure, at the very least, that the spheres will be locally within the boundary in a neighborhood of the points of tangency. This will hold provided $r < |r_i|$ for all principal radii $r_i < 0$. To see this must hold, we note that if $r_i < 0$, then by (3.17) $r < |r_i|$ iff $r_{ri} < 0$. However, by Remark 3.3.1, $r_i < 0$ implies $r_{ri} < 0$.

Example 3.3.1. For a 1D Blum medial axis, we obtain that the curvature for the boundary curve $\mathscr{B}$ at a point which corresponds to a non–edge point (and particular value of $\mathbf{V}$) is given by

$$\kappa = \frac{\kappa_r}{1 - r\kappa_r}$$

Moreover, the sign of κ_r determines the sign of κ, which we recall from 3.2.3.1, by our convention, is negative when $\mathscr{B}$ curves toward the region, and positive when it curves away.

Example 3.3.2. For a 2D Blum medial axis, a non–crest point x'_0 of $\mathscr{B}$ corresponds to a non–edge point x_0 of M. We compute $S_{\mathbf{v}}$ at x_0 with value of $\mathbf{V}$ pointing toward

x'_0, as in (3.6) by the 2×2–matrix (b_{ij}). We obtain the principal radial curvatures and directions from the eigenvalues and eigenvectors of $S_{\mathbf{v}}$. Then we obtain the principal curvatures of $\mathscr{B}$ by applying (3.17), and the principal directions by applying $d\psi_1$ to the eigenvectors. Finally, we obtain the matrix representation $S_{\mathscr{B}\mathbf{v}'}$ for the differential geometric shape operator by (3.16).

This calculation requires that we only compute two first directional derivatives, followed by simple linear algebra operations. In doing this, we determine the differential geometry using what is essentially the minimal amount of medial information possible.

We also note that from $S_{\mathscr{B}\mathbf{v}'}$ we deduce the *Second Fundamental Form* $II(v'_i, v'_j) = v'_i \cdot S_{\mathscr{B}}(v'_j)$. A matrix representation of II with respect to the basis $\mathbf{v}'$ is given by $G \cdot S_{\mathscr{B}\mathbf{v}'}$ where $G = (g_{ij})$ is given by the metric on $\mathscr{B}$ by $g_{ij} = v'_i \cdot v'_j$. G can be computed using $d\psi_1$ by (Damon, 2003, Section 4). However, it follows from the formula (3.16) that shape operators give the most direct relation between the differential geometry of $\mathscr{B}$ and medial data.

3.3.1.2 Differential Geometry of the Boundary at Crest Points

Next we compute the differential geometric shape operator at a crest point x'_0 on $\mathscr{B}$ (corresponding to an edge point x_0) using the edge shape operator.

For a special basis $\mathbf{v}$ at x_0 as explained in Section 3.2.2 with corresponding basis $\mathbf{v}'$ for $T_{x_0}\mathscr{B}$, we may apply (Damon, 2004, Corollary 3.6).

Local Differential Geometry at Crest Points

Suppose again $(M, \mathbf{V})$ is a skeletal structure satisfying the partial Blum condition (which includes a Blum medial axis of a generic region with smooth boundary). Then the differential geometric shape operator for $\mathscr{B}$ at x'_0 has a matrix representation with respect to $\mathbf{v}'$ given in terms of the edge shape operator for $n = 1, 2$ by

$$S_{\mathscr{B}\mathbf{v}'} = (I_{n-1,1} - r \cdot S_{E\mathbf{v}})^{-1} S_{E\mathbf{v}} \qquad (3.19)$$

Hence, the principal curvatures κ_i and principal directions of $\mathscr{B}$ at x'_0 are the eigenvalues and eigendirections (after identification by $d\psi_1$) of the RHS of (3.19).

Example 3.3.3. For a 1D Blum medial axis, at a point of the boundary curve $\mathscr{B}$ corresponding to an edge point we obtain from (3.19) using the calculation in (3.8) of Section 3.2 that

$$\kappa = (0 - rc_{\mathbf{n}})^{-1} c_{\mathbf{n}} = \frac{-1}{r}.$$

This is the curvature of the osculating circle of radius r (with minus sign resulting from our sign convention).

Example 3.3.4. For a 2D Blum medial axis, at a crest point of the boundary surface $\mathcal{B}$ we computed the edge shape operator in Example 3.2.5. Thus, by (3.19) we compute the differential geometric shape operator. Let $K_E = \det(S_{E\mathbf{v}})$. Then a computation shows

$$S_{\mathcal{B}\mathbf{v}'} = \begin{pmatrix} \frac{K_E}{c_{\mathbf{n}2} - rK_E} & 0 \\ \frac{c_{\mathbf{n}1}}{r(rK_E - c_{\mathbf{n}1})} & -\frac{1}{r} \end{pmatrix} \tag{3.20}$$

Hence, using the calculation of κ_E in Example 3.2.5, we see that

$$\text{the principal curvatures are } -\frac{1}{r} \text{ and } \frac{\kappa_E}{1 - r\kappa_E} \tag{3.21}$$

We note the special case where $\frac{\partial \mathbf{U}}{\partial v_1}$ is orthogonal to $\mathbf{N}$. This implies $c_{\mathbf{n}1} = 0$, so in (3.20) $K_E = b_1 c_{\mathbf{n}2}$. Then by (3.20), $S_{\mathcal{B}\mathbf{v}'}$ becomes diagonal with eigenvalues $\frac{1}{\frac{1}{b_1} - r}$ and $-\frac{1}{r}$.

To actually compute the differential geometric shape operator at a crest point, we must overcome, in a practical way, the use of edge coordinates. However, we can use the remark in Section 3.2.3.2, which explains how to compute $S_{\mathcal{B}\mathbf{v}'}$ as a limiting value via (3.11) of Section 3.2. Hence, the differential geometric shape operator $S_{\mathcal{B}\mathbf{v}'}$, at a crest point x_0', with respect to the basis $\mathbf{v}'$ associated to $\mathbf{v}$ is given by

$$S_{\mathcal{B}\mathbf{v}'} = \lim_{x \to x_0} (I_{1,1} - r \cdot S_{E\mathbf{v}}')^{-1} S_{E\mathbf{v}}' \tag{3.22}$$

Hence, the principal curvatures at x_0' are the limits as $x \to x_0$ of the eigenvalues of the RHS of (3.22). Moreover, if the principal curvatures at x_0 are distinct, then the principal directions are the limits of the eigendirections of the RHS of (3.22) as $x \to x_0$. Although computationally we cannot really take the limit, by choosing x sufficiently close to x_0, we can compute the RHS of (3.22) to determine good approximations to both the eigenvectors and eigenvalues for the crest point x_0'.

3.3.1.3 Differential Geometry Versus Radial Geometry of the Medial Axis

We conclude this section by returning to a point raised at the beginning, namely, why would it not be better to directly relate the differential geometry of the boundary with that for the smooth part of M using the derivatives of r? As already mentioned the results of Nackman and Pizer compute the Gauss and Mean curvature of the boundary by first computing them for the parallel surfaces of the boundary. To compare the two approaches we give a relation between the radial shape operator S_{rad} with the differential geometric shape operator S_{med} for the medial axis (at smooth points). As the derivatives of r must enter into any such relation, it is not surprising that the radial Hessian operator H_r is involved. There is yet one other operator which must be included in the relation. We let $\mathbf{U} = \rho \mathbf{N} + \mathbf{U}_{tan}$ denote the decomposition of $\mathbf{U}$

into normal and tangential components (this ρ will also play a fundamental role for the integration formulas we obtain in Section 3.4). Then define

$$Z(v) = \rho^{-1} \left(\frac{\partial \mathbf{U}}{\partial v} \cdot \mathbf{N} \right) \mathbf{U}_{tan} \tag{3.23}$$

Z does not have an obvious geometric meaning. Also, in contrast to S_{med} and H_r, Z need not be symmetric. However, both Z and H_r enter into the relation between the radial and the differential geometric shape operators for the medial axis. This relation is given in Damon (2004) or Damon (2005) by

$$S_{rad} = \rho \cdot S_{med} + H_r + Z. \tag{3.24}$$

If we combine this equation with our results from Section 3.2.4, we can express the differential geometric shape operator of the boundary in terms of S_{med} by substituting the expression for S_{rad} from (3.24) into (3.16). While the resulting expression will involve H_r as is expected, it will also involve Z. Thus, barring some remarkable unexpected identities, the representation in this form will be considerably more complicated than that just in terms of S_{rad}.

3.3.2 Geometric Medial Map

Next we use the results in the preceding section to construct the intrinsic portion of the *geometric medial map* on the Blum medial axis M for a region Ω with generic smooth boundary $\mathscr{B}$. This geometric medial map will give on M a collection of geometric data which will capture both the local and relative geometry of the boundary $\mathscr{B}$. The intrinsic portion will correspond under the radial map to the corresponding objects for the differential geometry of the associated boundary $\mathscr{B}$. The relative part will consist of a system of curves which include ridges of thickness and valley curves of thinness for the region Ω. This geometric medial map will be defined solely in terms of operations on the unit radial vector field $\mathbf{U}$ without any reference to the radial function r.

3.3.2.1 Intrinsic Part of the Geometric Medial Map

In the previous section we expressed the differential geometric shape operator in terms of the radial or edge shape operators by formulas which also involved r. However, by Remark 3.3.1, the sign of a principal curvature κ_i agrees with that of the corresponding principal radial curvature κ_{ri}; and $\kappa_i = 0$ iff $\kappa_{ri} = 0$. Hence, for a point $x_0' \in \mathscr{B}$ which corresponds to $x_0 \in M_{reg}$ via the radial map, any property at x_0' which can be given in terms of the signs of the principal curvatures can be expressed in terms of the same conditions for the signs of the corresponding principal radial curvatures.

Table 3.2 Table 3.2 Relation between radial geometry on the medial axis and differential geometry of the boundary

Radial geometry on medial axis	$\overset{\psi_1}{\Longleftrightarrow}$ Differential geometry of boundary
(i) Regions of positive (negative) radial curvature	(i) Regions of positive (negative) Gaussian curvature
(ii) Parabolic radial curves	(ii) Parabolic curves
(iii) Radial umbilic points	(iii) Umbilic points
(iv) Signs of principal radial curvatures	(iv) Signs of principal curvatures
(v) Principal radial directions	(v) Principal directions
(vi) Principal radial curves	(vi) Principal curves

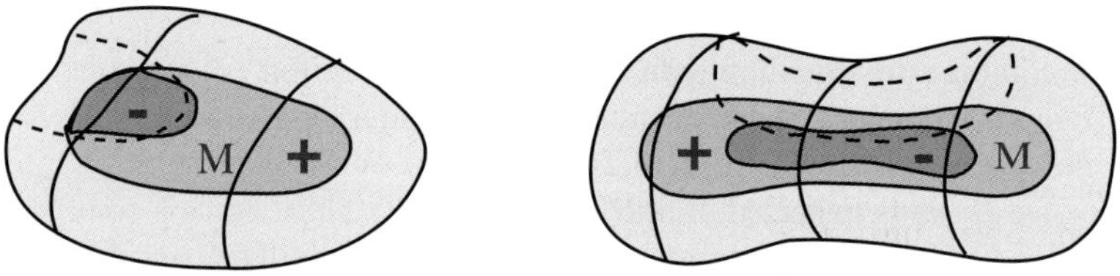

Fig. 3.12 Illustrating geometric medial maps capturing distinct properties of boundaries (in this case regions where Gaussian curvature changes sign)

We use S_{rad} to define on M the "radial analogues" of the corresponding objects for classical differential geometry of surfaces. For example, we define $K_{rad} = \det(S_{rad})$ as the *radial curvature*; the curve where $\det(S_{rad}) = 0$ as the *radial parabolic curve*; etc. We define the intrinsic part of the *geometric medial map* to consist of the radial versions of geometric objects given in the left hand column of Table 3.2. Then the relation between the two columns of Table 3.2 is provided by the radial map ψ_1.

Intrinsic Part of Geometric Medial Map

For the Blum medial axis $(M, \mathbf{V})$ of a region Ω with smooth boundary $\mathscr{B}$, the radial map sends the radial objects in the geometric medial map (in the first column of Table 3.2) to the corresponding differential geometric objects for $\mathscr{B}$ in the second column of Table 3.2.

In Fig. 3.12, we see that distinct properties of boundaries can be detected by the intrinsic part of the geometric medial map.

3.3.2.2 Relative Geometry of the Boundary via Relative Critical Sets

We have seen in the previous section that the radial function r plays an important role in the relative geometry of the boundary when we compare geometry at distinct points of the boundary in terms of medial data. We now turn more generally to the relative geometry as captured by properties of r on the medial axis. To capture such relative geometry, we use the "relative critical set of r". This will consist of a network of curves which detect ridges of r along which r is largest and valleys of r along which r is smallest, connected by certain intermediate curves where mixed behavior occurs.

We first explain how the relative critical set captures geometry of a function f on $\mathbb{R}^2$ (or more generally $\mathbb{R}^n$), and then explain how it extends to r on the medial axis.

Relative Critical Sets on $\mathbb{R}^2$

Suppose $f : W \to \mathbb{R}$ is a smooth function on an open subset W of $\mathbb{R}^2$. For a point $x_0 \in W$, let $\lambda_1 < \lambda_2$ denote the eigenvalues of the Hessian $H(f)(x_0)$, with eigenvectors e_1 and e_2. First, x_0 is called a *(height) ridge point* of f if $\nabla f(x_0)$ is orthogonal to e_1 and $\lambda_1 < 0$. The (height) ridge, which we henceforth call the ridge of f, is the set of (height) ridge points of f. A ridge curve represents the points along which the function is decreasing most rapidly in the directions orthogonal to the gradient direction. It was introduced by Pizer and Eberly to investigate properties of grey–scale boundary strength or medial strength functions (Pizer et al., 1998; Eberly, 1996). The ridge will generally consist of pieces of smooth curves. These curves carry information about the graph of f, viewed as a surface, but where the direction of the dependent variable remains distinguished.

However, the ridge curves consist of disjoint pieces without any structure to relate them. This is because they are only part of the complete structure needed to reveal the full geometry of f. This structure is called the *relative critical set of f*. We consider in addition to the ridge set the following sets of points consisting of those x_0 which are:

1. *Valley points* for which $\nabla f(x_0)$ is orthogonal to e_2 and $\lambda_2 > 0$
2. *r–connector points* for which $\nabla f(x_0)$ is orthogonal to e_1 and $\lambda_1 > 0$ and
3. *v–connector points* for which $\nabla f(x_0)$ is orthogonal to e_2 and $\lambda_2 < 0$

Then the relative critical set, denoted $\mathscr{RC}(f)$ is the closure of the four ridge, valley, r–connector, and v–connector sets. In addition, it contains critical points, singular Hessian points (where one of the $\lambda_i = 0$), and (partial) umbilic points (where $\lambda_1 = \lambda_2$). For higher dimensions one can analogously define relative critical sets, except they become increasingly more varied as the dimension increases (see Damon, 1998; Damon, 1999; Miller, 1998; and Keller, 1999). As an example of a relative critical set for a function, see Fig. 3.13, which exhibits ridge and valley curves and v-connector curves.

Remark 3.3.3 (Generic Properties of the Relative Critical Set). For generic f on $\mathbb{R}^2$, the relative critical set has the following generic properties: each of the four types form smooth curves; these curves only cross at critical points at which the Hessian is nonsingular; the types of curves which can cross are determined by the type of the critical point (see Fig. 3.14); the curves can change from one type to another as they pass through singular Hessian or (partial) umbilic points; and the specific changes

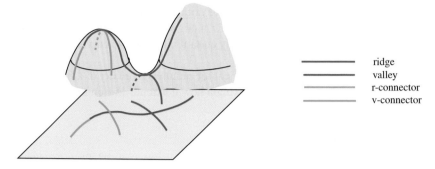

Fig. 3.13 Graph of a function, with the relative critical set exhibiting ridge, valley, and v-connector curves, with the corresponding curves on the graph

Local Maximum   Local Minimum   Saddle Point

Fig. 3.14 Crossings of relative critical set curves at critical points

Singular Hessian

(partial) Umbilic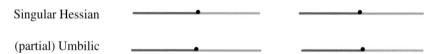

Fig. 3.15 Changing type for relative critical set at singular Hessian and (partial) umbilic points

are uniquely determined (see Fig. 3.15). This network of curves does not end (as, for example, Blum medial axes do) but continues to the end of the open set.

These properties are stable under small perturbation of f (see e.g. Damon, 1998) (and the changes which can occur in a generic one parameter deformation are also determined (Damon, 1998; Keller, 1999)).

Relative Critical Sets for Functions on Surfaces and the Medial Axis

We extend the preceding to r on the medial axis. As an intermediate step, we consider a function $f : N \to \mathbb{R}$ where N is a smooth surface (in $\mathbb{R}^3$). In Eberly (1996), Eberly used generalized eigenvalues and tensor index notation to define the (height) ridge of f on any Riemannian manifold. We are going to give a formulation of the relative critical set for f which for the ridge part will be equivalent to that given by Eberly.

We let ∇f be the *Riemannian gradient* so that $\nabla f(x_0) \cdot v = df_{x_0}(v)$ for all tangent vectors $v \in T_{x_0}N$. Then ∇f is a vector field on N which is orthogonal to the level curves of f on N. Then the *Riemannian Hessian* is defined by $H(f)(v,w) \overset{def}{=} \nabla_v(\nabla f) \cdot w$. Here "$\nabla_v$" denotes the covariant derivative of the vector field ∇f (we note that $\nabla_v X(x_0) = \text{proj}_{\mathbf{N}}(\frac{\partial X}{\partial v})$ for a vector field X, where $\text{proj}_{\mathbf{N}}$ denotes orthogonal projection onto $T_{x_0}N$).

By properties of the covariant derivative, $H(f)$ is symmetric in v and w. We define the *Hessian operator* $H_f : T_{x_0}N \to T_{x_0}N$ by $H_f(v) = \nabla_v(\nabla f)$. As $H_f(v) \cdot w = H(f)(v,w)$, it follows that H_f is symmetric, so it has real eigenvalues, and the eigenvectors for distinct values are orthogonal.

We again let $\lambda_1 < \lambda_2$ denote the eigenvalues of the Hessian operator H_f at x_0, with eigenvectors e_1 and e_2. Then we repeat the definition for ridge, valley, r–connector, and v–connector sets. It is shown in Damon (2007a) that the same generic properties for functions on $\mathbb{R}^2$ which we listed in Remark 3.3.3 continue to hold for generic smooth functions on a given smooth surface N. This allows us to define the relative critical set of r on (the smooth part of) the medial axis M.

The compatibility condition places restrictions on r, so it is not an arbitrary function even on M_{reg}. However, it can be shown for generic medial axes that the relative critical set of a generic r possesses the same generic properties (on M_{reg}) as do functions on $\mathbb{R}^2$. This allows us to define the relative part of the geometric medial map.

3.3.2.3 Relative Part of the Geometric Medial Map

By the compatibility condition, $\nabla r = -\mathbf{U}_{tan}$, the tangential component of $\mathbf{U}$. Thus, the Hessian operator for r takes the form $H_r(v) = -\nabla_v \mathbf{U}_{tan}$. Thus, $\nabla r = -\mathbf{U}_{tan}$ being orthogonal to an eigenvalue e_i is equivalent to $\mathbf{U}_{tan}$ being an eigenvector (a multiple of e_j), with eigenvalue $\lambda (= \lambda_j)$ for H_r. Hence we have the following description of the relative critical set of r on M_{reg} (which uses only properties of $\mathbf{U}_{tan}$).

Relative Part of Geometric Medial Map

For $(M, \mathbf{V})$ a Blum medial axis and radial vector field for a generic region with smooth boundary $\mathscr{B}$, the relative part of the geometric medial map is the network of curves given by the relative critical set of r on M_{reg}. It consists of those $x_0 \in M_{reg}$ such that

$$-\nabla_{\mathbf{U}_{tan}}(\mathbf{U}_{tan}) = \lambda \cdot \mathbf{U}_{tan} \tag{3.25}$$

If μ is the other eigenvalue of $H_r(v) = -\nabla_v \mathbf{U}_{tan}$, then the different types of points are characterized as follows:

1. For ridge points $\mu < 0$ and $\mu < \lambda$
2. For valley points $\mu > 0$ and $\mu > \lambda$

Then Q_φ is multiplication by q.

For a 2D medial axis M, let $X(u_1, u_2)$ denote a parametrization of a smooth component of M with say $x_0 = X(0,0)$. Then $X_1(u_1, u_2) = \varphi \circ X(u_1, u_2)$ is a parametrization of M' near $x_0' = \varphi(x_0)$. As in Example 3.2.3, we let $v_i = \dfrac{\partial X}{\partial u_i}, i = 1, 2$ denote a basis for $T_{x_0} M$ at each point x_0 in the parametrized region. We write

$$d^2 \varphi_{x_0}(v_i, \mathbf{U}) = a_i \cdot \mathbf{U}' - q_{1i} \cdot v_1' - q_{2i} \cdot v_2', \quad i = 1, 2 \tag{3.28}$$

where $v_i' = d\varphi(v_i) = \dfrac{\partial X_1}{\partial u_i}$. Then $Q_{\varphi \mathbf{v}}$, the matrix representation of Q_φ with respect to the basis $\mathbf{v} = \{v_1, v_2\}$ is given by

$$Q_{\varphi \mathbf{v}} = \begin{pmatrix} q_{11} & q_{12} \\ q_{21} & q_{22} \end{pmatrix} \tag{3.29}$$

Edge Distortion Operators

To determine the effect of φ on edge shape operators, there is an analogous edge distortion operator $Q_{E,\varphi}$. In addition, we must also take into account the failure of $d\varphi$ to send $\mathbf{N}$ to $\mathbf{N}'$ at edge points of M via an operator $E_{\varphi \mathbf{v}}$.

3.3.3.2 Radial and Edge Shape Operators for Deformed Structures

Although $(M', \mathbf{V}')$ need not be a medial axis, the radial and edge shape operators are still defined. We can compute them by the following formulas involving the original operators for $(M, \mathbf{V})$ and the distortion operators described in Damon (2004, Theorem 4.5).

Shape Operators for Deformed Structures

Suppose $(M', \mathbf{V}')$ is the image of a 1D or 2D medial axis $(M, \mathbf{V})$ under the local diffeomorphism φ. Then with the preceding notation

1. For a non-edge point $x_0 \in M$, the radial shape operator $S_{\mathbf{v}'}$ at $x_0' = \varphi(x_0)$ (for the basis $\mathbf{v}'$ from the parametrization $X_1 = \varphi \circ X$) is given by

$$S_{\mathbf{v}'} = \sigma(S_\mathbf{v} + Q_{\varphi \mathbf{v}}) \tag{3.30}$$

2. For a 2D medial axis and a point $x_0 \in \partial M$, we may compute the edge shape operator $S_{E \mathbf{v}'}$ at $x_0' = \varphi(x_0)$ by

$$S_{E \mathbf{v}'} = \sigma(S_{E \mathbf{v}} + Q_{E \varphi, \mathbf{v}} + E_{\varphi \mathbf{v}}) \tag{3.31}$$

As an illustration of the preceding, we give a corollary ensuring that the image $(M', \mathbf{V}')$ satisfies the radial curvature condition. First, if M is 1–dimensional then $Q_{\varphi \mathbf{v}} = (q)$. If M is 2–dimensional, we let $b_i, i = 1, 2$ denote the eigenvalues of $S_{\mathbf{v}} + Q_{\varphi \mathbf{v}}$.

Radial Curvature Condition for Deformed Structures

In the preceding situation, if M is a 1D medial axis, then $(M', \mathbf{V}')$ satisfies the Radial Curvature Condition iff at all non-edge points of M

$$ r < \frac{1}{\kappa_r + q} \quad \text{if } \kappa_r + q > 0 \text{ and no condition otherwise} \tag{3.32} $$

If M is 2–dimensional, then $(M', \mathbf{V}')$ satisfies the Radial Curvature Condition iff at all non-edge points of M

$$ r < \min \left\{ \frac{1}{b_i} \right\} \quad \text{for all positive eigenvalues } b_i \text{ of } S_{\mathbf{v}} + Q_{\varphi \mathbf{v}} \tag{3.33} $$

There is an analogous version for the edge condition (see Damon, 2004, 2005).

3.4 Global Geometry of a Region and Its Boundary

3.4.1 Skeletal and Medial Integrals

Given a skeletal structure $(M, \mathbf{V})$ which defines a region with smooth boundary, and a multivalued function h on M, We shall introduce the *skeletal integral of h over M*. By a multivalued function h on M, we mean a function which at a point $x \in M$ has several values of h at x, with exactly one for each value of $\mathbf{V}$ at x. Hence, for example, at a smooth point x of M, We shall have a value of h at x for each side of M. We say h is a continuous multivalued function if for a continuously varying value of $\mathbf{V}$, the corresponding values of h also vary continuously. Then h being multivalued and continuous can be understood in terms of h extending to a function h' on the double $\tilde{M}$ and h' being continuous. For example, if $g : \mathscr{B} \to \mathbb{R}$ is a continuous function on $\mathscr{B}$, then the composition with the radial map $g \circ \psi_1$ defines a continuous multivalued function $\tilde{g}$ on M.

3.4.1.1 Definition of Skeletal and Medial Integrals

First, we explain how to define a skeletal integral for continuous multivalued continuous functions on M. This will be equivalent to defining the integral of a continuous function on $\tilde{M}$. This integral satisfies the usual properties and can be expressed as an integral with respect to a measure (in the mathematical sense). Then integrals are

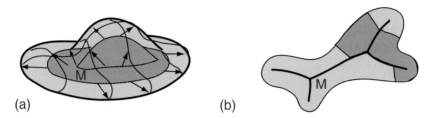

Fig. 3.18 Shaded regions indicating "regions in $\tilde{M}$" and their images under the radial flow

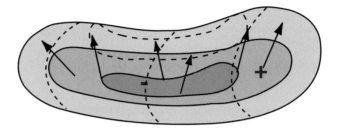

Fig. 3.19 Region in M where $\mathscr{B}$ has positive Gauss curvature

actually defined for the large class of "Borel measurable functions", over regions of $\tilde{M}$ which are "Borel sets". These include all possible sets and functions of interest. We use the nontechnical terms "reasonable functions" and "reasonable sets".

These will include, for example, piecewise continuous functions and regions of $\tilde{M}$ with piecewise differentiable boundaries. By a piecewise differentiable region R of $\tilde{M}$, we mean that we can represent R using a finite union of subsets $R_i \subset M$ with a specific choice of smooth values of $\mathbf{V}$ on each R_i so that any common point in distinct R_i and R_j with the same smooth value belongs to the boundary of both. Fig. 3.18 illustrates such a region where the shaded areas indicate the image under the radial flow of the corresponding region of $\tilde{M}$.

A multivalued function g on M is piecewise continuous if we can decompose $\tilde{M}$ into a union $\cup S_j$, of subsets S_i with a choice of a smooth value of $\mathbf{V}$ on them, so each S_j is a region with piecewise smooth boundary, and for S_i and S_j on the same side of M, they only intersect at boundary points; and the restriction of g to the interior of S_j has a continuous extension to S_j.

The reader is free to think of the "reasonable functions and regions" in terms of the special cases. The only reason that this is not sufficient for all considerations is that if $g : \mathscr{B} \to \mathbb{R}$ is a piecewise continuous function on $\mathscr{B}$, then the composition $g \circ \psi_1$ need not define a piecewise continuous function on $\tilde{M}$; but it does define a Borel measurable function.

Heuristically we view a reasonable region of $\tilde{M}$ as associating a region of a smooth stratum of M to one side of M, and then taking a finite disjoint union of such regions. For example, consider in Fig. 3.19 the region of $\tilde{M}$ consisting of points where at the corresponding points on $\mathscr{B}$, the Gaussian curvature is positive. It consists of the bottom side of M and part of the top side as indicated in Fig. 3.19.

To actually define the integral of a reasonable multivalued function, we may decompose M into a finite union of sets $\{W_i\}$ which satisfy the following properties: (i) for each W_i, each value of $\mathbf{V}$ defined at a point of W_i extends smoothly to values of $\mathbf{V}$ on all of W_i; (ii) the closure of each W_i can be smoothly parametrized by an interval if M is 1D or by a rectangle if M is 2D, and distinct W_i only meet at a point, respectively an edge. We refer to such a decomposition as a *paving of M* (see, for example, the earlier Fig. 3.8). The W_i are "curvilinear intervals or rectangles".

We label copies of W_i for the j-th value of $\mathbf{V}$ by W_{ij}. For a multivalued function h, we denote the function values of h on W_i corresponding to the value $\mathbf{V}_j$ by h_{ij}. We let $\rho_j = \mathbf{U}_j \cdot \mathbf{N}_j$, where $\mathbf{U}_j$ is a value of the unit vector field for each of the smooth values of $\mathbf{V}$ and $\mathbf{N}_j$ is the normal unit vector pointing on the same side as $\mathbf{U}_j$. Lastly we let dV_j denote the Riemannian length form (in the 1D case) or area form (in the 2D case) with W_i oriented by $\mathbf{N}_j$. Then we define the integral of a multivalued function h on M by

$$\int_{\tilde{M}} h \, dM = \sum_i \sum_j \int_{W_{ij}} h_{ij} \, \rho_j \, dV_j \qquad (3.34)$$

This is independent of how we decompose M into the curvilinear intervals or rectangles. We refer to the integral $\int_{\tilde{M}} h \, dM$ as a *skeletal integral*. In the special case that $(M, \mathbf{V})$ satisfies the partial Blum condition, We shall refer to it instead as a *medial integral*. When this integral is finite, we say f is *integrable*. The integrable functions include the large class of piecewise continuous functions.

Remark 3.4.1. When we decompose into curvilinear intervals or rectangles, at edge points we must use "edge coordinates". Under these the parametrization of a neighborhood of an edge is one–one differentiable and a local diffeomorphism off the edge. Because a continuous function for edge coordinates is still continuous for regular surface coordinates, this does not cause any problem with integration.

Then the integral has the usual properties: it is linear and if $h \geq 0$, then $\int h \, dM \geq 0$.

3.4.1.2 Skeletal and Medial Integrals on Regions

Let $R \subset \tilde{M}$ be a reasonable subset. We let χ_R denote the characteristic function of R. This is a multivalued function on M. Then for a multivalued reasonable function f on M, we define

$$\int_R f \, dM = \int_{\tilde{M}} \chi_R \cdot f \, dM$$

Remark 3.4.2. By invoking the Riesz Representation Theorem from measure theory, it follows there is a positive Borel measure dM on $\tilde{M}$ so that integration is with respect to this measure (the details are explained in Damon, 2007b).

3.4.1.3 Medial Measure and Blum Medial Axis as a Measure Space

We refer to the measure $dM = \rho \, dV$ on $\tilde{M}$ as the *medial measure*.[2] It corrects for the failure of $\mathbf{V}$ to be orthogonal to M. In the case of the Blum medial axis, dM is actually defined on M.

This changes our perspective on the Blum medial axis from just being a stratified set to being as well a measure space, where the significance of parts of the space are determined by their medial measure. For example, the introduction of a small bump on the boundary $\mathcal{B}$ leads to the creation of another sheet of M. However, the bump only introduces a small change in the volume. We shall see from the volume formulas in Section 3.4.3, that the additional integral on the added sheet represents this small change. The smallness of the bump forces $\mathbf{V}$ to be close to being tangent to M on the additional sheet. This implies that the medial measure is small on the added sheet. Thus, although set theoretically the added sheet is a significant alteration of the medial axis, from the point of view of measure theory the added sheet is very small.

3.4.2 Global Integrals as Skeletal and Medial Integrals

We now suppose that $(M, \mathbf{V})$ is a skeletal structure in $\mathbb{R}^2$ or $\mathbb{R}^3$ which defines a region Ω with smooth boundary $\mathcal{B}$. In this section, we explain how to compute integrals of functions over the boundary $\mathcal{B}$ in terms of medial integrals in the case that $(M, \mathbf{V})$ satisfies the partial Blum condition, and integrals of functions over Ω in terms of skeletal integrals over M in general without any Blum condition.

3.4.2.1 Boundary Integrals as Medial Integrals

We know that $\mathcal{B}$ is smooth off the image $\psi_1(M_{sing})$ of the singular set of M, where we only know it is weakly C^1. The images of the strata of M_{sing} are still smooth submanifolds of $\mathcal{B}$, and using the radial map we see that points in $\psi_1(M_{sing})$ have paved neighborhoods. Then $\mathcal{B}$ is piecewise smooth and thus has a Riemannian length or area form, denoted by dV. Hence, the same argument used for M allows us to define the integral $\int_{\mathcal{B}} g \, dV$ for a continuous function g. Then even if $\mathcal{B}$ is not smooth, integrals are still defined on $\mathcal{B}$ (the Riesz Representation Theorem applies so integrals are defined for general Borel measurable functions and regions on $\mathcal{B}$).

Surface Integrals via Medial Integrals

Suppose $(M, \mathbf{V})$ is a skeletal structure defining a region with smooth boundary $\mathcal{B}$ and satisfying the partial Blum condition. Let $g : \mathcal{B} \to \mathbb{R}$ be an integrable function.

---

[2] Here dV refers to a Riemannian measure of medial sheet area, i.e., of dimension $n-1$ when the objects reside in $\mathbb{R}^n$.

Then[3]

$$\int_{\mathscr{B}} g \, dV = \int_{\tilde{M}} \tilde{g} \cdot \det(I - r S_{rad}) \, dM, \tag{3.35}$$

where $\tilde{g} = g \circ \psi_1$.

By replacing g by $\chi_R \cdot g$, we obtain a version for an integral over a region R of $\mathscr{B}$.

Surface Integrals over Regions

Suppose $(M, \mathbf{V})$ is a skeletal structure defining a region with smooth boundary $\mathscr{B}$ and satisfying the partial Blum condition. Let R denote a reasonable subset of $\mathscr{B}$ and $g : R \to \mathbb{R}$ an integrable function. If $\tilde{R} = \psi_1^{-1}(R)$, then

$$\int_R g \, dV = \int_{\tilde{R}} \tilde{g} \cdot \det(I - r S_{rad}) \, dM, \tag{3.36}$$

where $\tilde{g} = g \circ \psi_1$.

3.4.2.2 Integrals over Regions as Skeletal Integrals

Second, we again consider a skeletal structure $(M, \mathbf{V})$ which defines a region Ω but we *do not assume* the partial Blum condition is satisfied. Even in the "non-Blum case", we still can represent integrals over a region Ω as skeletal integrals. We suppose that $g : \Omega \to \mathbb{R}$ is an integrable function. We let $g_1(x,t) = g(x + t\mathbf{V}(x))$. If $\tilde{\psi}(x,t) = \psi_t(x)$ denotes the radial flow from $\tilde{M}$, then $g_1 = g \circ \tilde{\psi}$. Provided the integral is defined, we let

$$\tilde{g}(x) = \int_0^1 g_1(x,t) \cdot \det(I - t \, r S_{rad}) \, dt \tag{3.37}$$

This is the integral of $g \cdot \det(I - t \, r S_{rad})$ restricted to the radial line determined by the value $\mathbf{V}$.

Remark 3.4.3. By a change of variables, we can alternately write (3.37) as

$$r(x) \cdot \tilde{g}(x) = \int_0^{r(x)} g_2(x,t) \cdot \det(I - t S_{rad}) \, dt$$

where $g_2(x,t) = g(x + t\mathbf{U}(x))$.

Then we can express the integral of g over Ω as a skeletal integral, as follows:

---

[3] Here dV refers to a Riemannian measure of object boundary area, i.e., of dimension $n - 1$ when the objects reside in $\mathbb{R}^n$.

If $n = 2$, then $\delta(x) = 1 - \frac{1}{2}r\kappa_r$. Alternately, for $n = 3$ we may write $\delta(x) = 1 - \frac{1}{2}Trace(B') + \frac{1}{3}\det(B')$, where B' is given by (3.13) in Section 3.2. Then we can compute the area or volume of Ω in terms of an integral of δ over $\tilde{M}$.

Medial Integral Formula for Areas or Volumes of Regions

Suppose $(M, \mathbf{V})$ is a skeletal structure which defines a region Ω with boundary $\mathscr{B}$ (without even being partially Blum). Then the area of Ω if $n = 2$, or the volume of Ω, if $n = 3$, is given by

$$\int_{\tilde{M}} \delta \cdot r \, dM. \tag{3.43}$$

The preceding formulas are the simplest to use for discrete approximations. We show in Section 3.4.4 an alternate way to compute these formulas for the length, area, and volume of the boundary or region by expanding them in terms of moment integrals and weight integrals of functions over M.

3.4.3.3 Medial Version of Gauss–Bonnet Theorem

In computing medial or skeletal integrals using discrete approximations, we would like to know how accurate our computations are. One way to answer this is to compute an integral for which we know the precise answer. As a third consequence of the medial integral formula, we give one such formula, namely, a medial version of the classical Gauss–Bonnet Formula. For closed surfaces N, the classical formula relates the integral of the Gauss curvature of the surface N with an intrinsic topological invariant of the surface, the Euler characteristic of the surface $\chi(N)$ (which equals $2 - 2g$ where g, the genus, is the number of handles out of which the surface is built). All spaces X, including the medial axis, have an Euler characteristic $\chi(X)$. For generic regions Ω with smooth boundary $\mathscr{B}$ and Blum medial axis M, the Gauss–Bonnet formula takes the following medial form.

Medial Version of Gauss-Bonnet Theorem

Suppose $\Omega \subset \mathbb{R}^n$, $n = 2$ or 3, is a region with smooth generic boundary $\mathscr{B}$ and Blum medial axis M. Then

$$\frac{1}{s_{n-1}} \cdot \int_{\tilde{M}} K_{rad} \, dM = \chi(\Omega) = \chi(M)$$

and if n is even

$$= \frac{1}{2}\chi(\mathscr{B}) \tag{3.44}$$

with $s_1 = 2\pi = \text{perimeter}(S^1)$ when $n = 2$ or $s_2 = 4\pi = \text{surface area}(S^2)$ if $n = 3$.

Since this is an absolute result for an integral on M, it allows us to check the accuracy of discrete methods for computing medial or skeletal integrals.

3.4.4 Expansion of Integrals in Terms of Moment Integrals

Both of the global formulas for integrals over the boundary of the region $\mathscr{B}$ or the entire region Ω as integrals over M are in terms of integrals involving one of the terms $\det(I - rS_{rad})$ or $\det(I - t\, rS_{rad})$. We can carry out the expansion of these terms and formulate the integrals as sums of moment integrals of the original function g. If g is defined on Ω, then we can define

$$m_i(g)(x) = \int_0^1 g(x + t\mathbf{V}(x)) \cdot t^i \, dt$$

which is the i–th radial moment of g along the radial line $\{x + t\mathbf{V}(x) : 0 \le t \le 1\}$. This then becomes a multivalued function on M.

Second, given a multivalued function h on M, we can define the ℓ–th weighted integral of the multivalued function h by

$$I_\ell(h) = \int_M h \cdot r^\ell \, dM.$$

Using the weighted integrals and radial moments, we can expand both of the integral formulas over the medial axis.

We expand the integrand in the RHS of (3.35). If $n = 2$, then S_{rad} is multiplication by κ_r. For $n = 3$, we let $H_{rad} = \frac{1}{2}Trace(S_{rad})$ and $K_{rad} = \det(S_{rad})$ denote the mean radial curvature and Radial curvature. Then we may expand $\det(I - rS_{rad})$.

$$\det(I - rS_{rad}) = \begin{cases} 1 - r\kappa_r & n = 2 \\ 1 - 2rH_{rad} + r^2K_{rad} & n = 3 \end{cases} \qquad (3.45)$$

If we carry out a similar expansion for $\det(I - trS_{rad})$ and substitute into the formula for $\tilde{g}$ in (3.38), we obtain

$$r \cdot \tilde{g} = \begin{cases} r \cdot m_0(g) - r^2 m_1(g) \cdot \kappa_r & n = 2 \\ r \cdot m_0(g) - 2r^2 m_1(g) \cdot H_{rad} + r^3 m_2(g) \cdot K_{rad} & n = 3 \end{cases} \qquad (3.46)$$

If we substitute (3.45) into the integral formula (3.35), we obtain the following:

Expansion of boundary integrals

$$\int_{\mathscr{B}} g \, dV = \begin{cases} I_0(g) - I_1(g\kappa_r) & n = 2 \\ I_0(g) - 2I_1(gH_{rad}) + I_2(gK_{rad}) & n = 3 \end{cases} \qquad (3.47)$$

Likewise, we can substitute (3.46) into (3.38) and obtain instead the following:

Expansion of integrals over regions

$$\int_{\Omega} h \, dV = \begin{cases} I_1(m_0(h)) - I_2(m_1(h) \cdot \kappa_r) & n = 2 \\ I_1(m_0(h)) - 2I_2(m_1(h) \cdot H_{rad}) + \\ \qquad I_3(m_2(h) \cdot K_{rad}) & n = 3 \end{cases} \tag{3.48}$$

If we consider special cases of these integrals, we obtain expansions for the length, areas, or volumes of boundaries or regions:

3.4.4.1 Expansion of the Boundary Length or Area

In the special case that $g \equiv 1$ we obtain the following formulas for the length or area of the boundary.

Example 3.4.1.

$$\text{length}(\mathscr{B}) = \int_{\tilde{M}} dM - \int_{\tilde{M}} r \kappa_r \, dM \tag{3.49}$$

The first integral on the RHS of (3.49) is $2\tilde{\ell}(M)$, where $\tilde{\ell}(M)$ is the length of M, but with respect to the "Medial length" using $dM = d\tilde{s} = \rho \cdot ds$.

For the second case of $\Omega \subset \mathbb{R}^3$, we have

$$\text{area}(\mathscr{B}) = \int_{\tilde{M}} dM - 2 \int_{\tilde{M}} r H_{rad} \, dM + \int_{\tilde{M}} r^2 K_{rad} \, dM \tag{3.50}$$

Again the first integral on the RHS represents twice the area of M but measured using the "Medial Area form" $dM = \rho \cdot dA$.

Weyl Expansion of Areas or Volumes for General Regions

Lastly, let $\Omega \subset \mathbb{R}^2$ or $\mathbb{R}^3$ be a region with smooth boundary defined by a skeletal structure $(M, \mathbf{V})$. We expand the integral for area or volume of the region by letting $h \equiv 1$ in (3.48).

Example 3.4.2. We may compute the area of a 2-dimensional region Ω by

$$\text{area}(\Omega) = \int_{\tilde{M}} r \, dM - \frac{1}{2} \int_{\tilde{M}} r^2 \kappa_r \, dM. \tag{3.51}$$

For a 3–dimensional region Ω, the volume is computed by

$$\text{volume}(\Omega) = \int_{\tilde{M}} r \, dM - \int_{\tilde{M}} r^2 H_{rad} \, dM + \frac{1}{3} \int_{\tilde{M}} r^3 K_{rad} \, dM. \tag{3.52}$$

Remark 3.4.6. In general r is not constant and cannot be taken outside the integrals above. If r is constant and if the partial Blum condition holds, then $\mathbf{V}$ is normal at all points. Then M must be a closed submanifold without boundary. Thus, this is the case of a "generalized annulus of constant radius r with respect to the manifold M" (mathematicians have called these entities *tubes*). In fact, as M is smooth with $\mathbf{V}$ normal of constant length, we have two consequences. First, as r is constant it can be taken outside the integrals, so the expressions become polynomials in r. Second, the radial shape operator is then the differential geometric shape operator of M. However, there is one for each side of M at a point x_0, and the $\mathbf{U}$ on one side is the negative of that on the other. Hence, the principal curvatures computed for each side differ by signs. Thus, the H_{rad} for each side differ by -1, while K_{rad} agree. Thus the integrals of H_{rad} on each side will cancel, while for K_{rad} they double, giving a global curvature invariant of M. This formula becomes a case of Hermann Weyl's "Volume of Tubes Formula". Thus, the formula for volume of regions can be viewed as an extension of the Weyl formula from "annular tubes" to general regions.

There are further extensions of Weyl's theorem such as for partial tubes and offset regions (in the form of Steiner's formula). A number of applications of the skeletal integral formula and medial Crofton's formula to these are further explained in Damon (2007b).

3.4.5 Divergence Theorem for Fluxes with Discontinuities Across the Medial Axis

A further application of medial and skeletal integrals involves an extension of the classical divergence theorem to the case of a vector field F defined on a region Ω in $\mathbb{R}^n$ with smooth boundary $\mathscr{B}$ which is defined by a skeletal structure $(M, \mathbf{V})$. For example, M could be the Blum medial axis M of Ω in the case of a smooth generic boundary $\mathscr{B}$. If the vector field F has discontinuities across M, then the standard divergence theorem does not apply. A correction term is needed which is supplied by an appropriate skeletal integral. Then we indicate how this result applies to the case of the vector field corresponding to the grassfire flow. Using this result we finally explain how a computation of the "limiting average flux" of this vector field allows the identification of the medial axis points in the algorithm due to Siddiqi et al. in Chapter 4.

By a vector field F being smooth on Ω with discontinuities across M, we mean that the vector field is smooth when approaching M from each side of any point, but the limiting values from each side may differ.

For example, we may translate the radial vector field $\mathbf{V}$ along each radial line to obtain a vector field (again denoted by $\mathbf{V}$) on Ω, which is multivalued on M but smooth on $\Omega \backslash M$. Thus, this $\mathbf{V}$ is smooth with discontinuities across M. Likewise the corresponding unit vector field $\mathbf{U}$ analogously obtained by translation is also smooth with discontinuities across M. In the Blum case, $-\mathbf{U}$ is the vector field

algorithm given in Chapter 4 uses a discrete version of the average flux to determine the Blum medial axis. To see why this algorithm works, we use the modified divergence theorem. We allow regions in $\mathbb{R}^n$ for $n = 2, 3$.

We limit our discussion to Γ which is either a disk or square if $n = 2$ or a 3D ball or cube if $n = 3$.

We let $\text{vol}_{n-1}(\tilde{\Gamma}_\varepsilon(x))$, resp. $\text{vol}_{n-1}(\partial \Gamma_\varepsilon(x))$, denote the $n-1$–dimensional volume of $\tilde{\Gamma}_\varepsilon(x)$, resp. $\partial \Gamma_\varepsilon(x)$, for $n = 2, 3$. Now we consider the average flux across $\Gamma_\varepsilon(x)$ to be

$$\text{the average flux across } \Gamma_\varepsilon(x) = \frac{1}{\text{vol}_{n-1}(\partial \Gamma_\varepsilon(x))} \cdot \int_{\partial \Gamma_\varepsilon(x)} G \cdot \mathbf{n}_{\partial \Gamma_\varepsilon(x)} \, dS.$$

We are interested in the limit of the average flux as $\varepsilon \to 0$. In particular, by the divergence theorem it will again be zero off the medial axis. However, now for points on the medial axis it will not vanish (except at edge points). The nonvanishing on the medial axis is due to the medial integral term in the modified divergence theorem. We explain the contribution of this term to the limiting average flux.

The value of the limiting average flux can vary for points on the medial axis; however, it can be bounded in terms of two invariants associated to each point x on the Blum medial axis M. First, we recall from Section 3.4.1 that $\rho(x) = \mathbf{U} \cdot \mathbf{N}$ is a piecewise smooth multivalued function on M which has values at x corresponding to each local component of M of x. We let $\min(\rho)(x)$ denote the minimum nonzero value of $\rho(x)$ for the multiple values at x.

Second, we define for each possible generic type T for points of M a *medial density* $m_T \Gamma$. We recall that the generic types for the 1-dimensional medial axis are: smooth points, branch points, and end points; for the 2-dimensional medial axis they are: smooth points, Y–branch points, edge points, fin points and 6–junction points. The medial density $m_T \Gamma$ has the property that for any point $x \in M$ of generic type T,

$$m_T \Gamma \leq \lim_{\varepsilon \to 0} \frac{\text{vol}_{n-1}(\tilde{\Gamma}_\varepsilon(x))}{\text{vol}_{n-1}(\partial \Gamma_\varepsilon(x))}$$

and this is the largest constant with this property for all generic regions (see Fig. 3.22). Here $\text{vol}_{n-1}(\tilde{\Gamma}_\varepsilon(x))$ denotes the integral $\int_{\tilde{\Gamma}_\varepsilon(x)} dV$ for the usual length form ($n = 2$) or area form ($n = 3$) dV.

Remark 3.4.9. We note that this constant can differ for different Γ such as a disk versus a cube. In the case of a disk, this gives, up to a constant factor, the local density defined in a purely theoretical context for studying singular spaces by Kurdyka and Raby (1989).

A computation shows that the values for the cases of disks, squares, 3D balls and cubes are given in Table 3.3.

Then we define

$$M_\Gamma = \min_{\text{non-edge } T} \{m_T \Gamma\}$$

Finally, the values for the limiting flux at nonedge points of M are bounded in terms of $\min(\rho)$ and M_Γ by the following for $x \in \Omega \setminus \mathcal{B}$:

Table 3.3 Medial densities for the 1 and 2–dimensional cases, for Γ an n–disk or n–cube

1–dimensional case			2–dimensional case		
Type	2-disk	square	Type	3-disk	cube
Smooth pt.	$\frac{2}{\pi}$	$\frac{1}{2}$	Smooth pt.	$\frac{1}{2}$	$\frac{1}{3}$
Branch pt.	$\frac{3}{\pi}$	$\frac{3}{4}$	Y–branch pt.	$\frac{3}{4}$	$\frac{1}{2}$
End pt.	$\frac{1}{\pi}$	$\frac{1}{4}$	Fin pt.	$\frac{1}{2}$	$\frac{1}{3}$
M_Γ	$\frac{2}{\pi}$	$\frac{1}{2}$	6–junction pt.	$\geq \frac{1}{2}$	$\geq \frac{1}{3}$
			End pt.	$\frac{1}{4}$	$\frac{1}{6}$
			M_Γ	$\frac{1}{2}$	$\frac{1}{3}$

$$\lim_{\varepsilon \to 0}(\text{avg. flux across } \Gamma_\varepsilon(x)) \begin{cases} = 0 & x \notin M \text{ or } x \in \partial M \\ < -M_\Gamma \cdot \min(\rho)(x) & x \in M \backslash \partial M \end{cases} \quad (3.55)$$

where M_Γ is positive and depends on Γ and $\min(\rho)(x) > 0$. Hence, as asserted in Chapter 4, the points on the Blum medial axis (except ∂M) are detected by the nonvanishing of the limiting average flux as $\varepsilon \to 0$.

Example 3.4.3. For the special case of the disk for $n = 2$, a more precise calculation for the limiting average outward flux can be given (Dimitrov et al., 2003) and is presented in Chapter 4. For example, for a medial curve near the point x_0, $\rho(x_0)$ will be approximately $\cos(\theta_i)$ where θ_i is the object angle, the angle between $\mathbf{V}$ and $\mathbf{N}$ at x_0 (or $\sin(\alpha_i)$, where α_i is the angle between $\mathbf{V}$ and the tangent line for the appropriate medial curve at x_0). Using this, a calculation yields for the limiting average flux for a Y-junction point, $\frac{-1}{\pi} \sum_{i=1}^{3} \sin(\alpha_i)$, summed over the three angles α_i for the Y-junction point. This can be compared with the weaker bound $\frac{-3}{\pi} \min\{\sin(\alpha_i)\}$ given by (3.55). Similar results can be obtained for the case of a sphere for $n = 3$.

3.5 Global Structure of the Medial Axis

3.5.1 Graph Structure for Decomposition into Irreducible Medial Components

In addition to determining the local, relative, and global geometry of the boundary $\mathscr{B}$ of a region Ω from medial structure $(M, \mathbf{V})$, we also ask how the global structure of the medial axis is related to the properties of the region Ω. We describe how the answer for 2–dimensional regions has a counterpart for 3–dimensional (objects or) regions. However, this counterpart turns out to be considerably more complicated.

In the 2–dimensional case, the medial axis is a collection of branched curves. We can assign a graph to describe the structure of the medial axis. We assign vertices

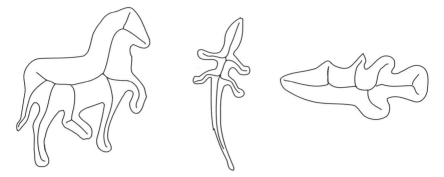

Fig. 3.23 Contractible regions in $\mathbb{R}^2$ with their Blum medial axes

to each branch point and each end point of the curves. For each curve segment we assign an edge between the two vertices corresponding to the ends of the curve segment, whether they be end points or branch points. This graph contains all of the topological information about the medial axis.

For example, in $\mathbb{R}^2$, if the boundary $\mathscr{B}$ is a single closed curve, then the "Jordan Curve Theorem" and "Schoenflies Theorem" from topology together assert that Ω is topologically equivalent (homeomorphic) to the standard 2–disk. Such regions are called "contractible" because they can be shrunk down (contracted) within themselves to a point. Most outlines of objects have this property. Several examples given in Fig. 3.23 (taken from Pizer et al., 2003b) were computed by Siddiqi and coworkers using their algorithm described in Chapter 4.

For such contractible regions the graph is always an (unrooted) tree. Hence, various search algorithms can be successfully applied to such graphs in polynomial time. An example of a noncontractible region would be the image of a doughnut, whose medial axis would encircle the hole. Hence, it would be described by a graph which is not a tree.

For 3–dimensional objects or regions a similar description would be highly desirable. Let us concentrate on 3–dimensional contractible regions Ω with generic smooth boundary $\mathscr{B}$. Such regions do not have holes through them and are without cavities in the interior. Many objects can be understood to have this form. We shall characterize the Blum medial axis M of such Ω (based on results in Damon, 2007c, with a general region investigated in Damon, 2006).

By analogy with the 2D-case, we might initially expect a simple description for the Blum medial axis M of Ω, where we replace curve segments of the 2D–case with pieces of surfaces which are topologically 2–dimensional disks, with some of these two disks attached along Y-branch curves ending in fin points as shown in Fig. 3.24. While such a Blum medial axis does correspond to a contractible region, it represents only a very small portion of the possibilities and fails to illustrate the intricate structure that is possible. We shall explain just how complicated the structure can be.

The structure of the medial axis can be represented by a collection of "irreducible components" which are attached to each other along "fin curves". The irreducible

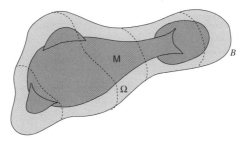

Fig. 3.24 Example of a Blum medial axis for a contractible region in $\mathbb{R}^3$

components are themselves without fin curves. There are two crucial characteristics of the general contractible case which are absent in Fig. 3.24. First, unlike the simple model in Fig. 3.24, the components themselves can have a complicated structure, having more in common with the component M_1 of the medial axis in Fig. 3.32. Second, a component may be attached to more than one other component along a single fin curve.

Our goal is twofold: to give a concise description of the structure of each irreducible medial component which captures its topological structure; and second, to provide an algorithm for reconstructing M from the M_i by attaching along fin curves. We shall see that not all of the irreducible components are uniquely defined because the presence of certain types of fin curves requires us to make choices, leading to different possible representations. Under different geometric conditions for the equivalent medial axes, different choices may seem "more natural". Also, the attaching process is actually an inductive process and requires considerably more data than is needed for the 2D–case.

However, we can simplify the form of the attachings to give a simplified medial axis which is equivalent to M in a weaker topological sense (homotopy equivalence). These attachings can be described by a graph $\Gamma(M)$, which turns out to be a tree. Second, We shall also describe the structure of each irreducible medial component M_i by a secondary graph $\Lambda(M_i)$ which describes how the smooth surface sheets are attached to the network of Y–branch curves. In the contractible case, each $\Lambda(M_i)$ will also be a tree. Finally, if Ω and hence M are contractible, each M_i must also be contractible. It is then somewhat counterintuitive that the smooth medial surface sheets which make up the M_i need not be contractible, nor must the connected components of the network of Y–branch curves form trees. Exactly how complicated a particular medial surface sheet is and how it is attached to the Y–network curves is part of the data attached to the secondary graph.

3.5.1.1 Decomposing Along Fin Curves

We first explain how to decompose M into pieces by cutting along "fin curves". To understand fin curves, we note that certain Y–branch curves end at fin points. At a fin point, there is locally identified a distinguished sheet, "the fin sheet". We can follow along the Y–branch curve keeping track of the fin sheet. Even if the Y–branch

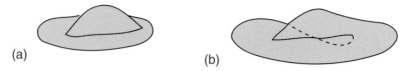

Fig. 3.25 Two possibilities for fin curves on a medial sheet: (**a**) essential fin curve (**b**) inessential fin curve

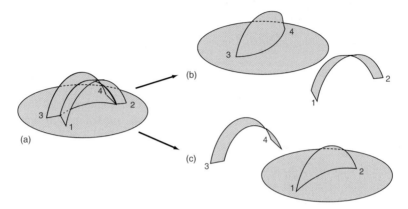

Fig. 3.26 Nonuniqueness of medial decomposition resulting from type-2 essential fin curves: (**a**) is a contractible medial axis with only inessential fin curves; and (**b**) and (**c**) illustrate the results from cutting along the fin curves 1-2 or 3-4

curve meets a 6–junction point, we can still follow the fin sheet to the other side of the 6–junction point, identifying the continuation of the Y–branch curve. We can continue following the fin sheet and Y–branch curve to which it attaches until we finally must reach another fin point where the Y–branch curve that we are following ends. We refer to the Y-branch curve so identified as a "fin curve".

At the end of the fin curve, what was identified as the fin sheet from the beginning may or may not be the fin sheet for the end point. If it is only a fin sheet at one end, then we refer to the fin curve as "inessential", while if the sheet if a fin sheet at both ends, then we refer to the fin curve as being "essential" (later discussion will explain the reason for these labels). Examples of these are shown in Fig. 3.25.

First, we can cut the fin sheet along a fin curve. Then we can take the two remaining sheets still attached along the fin curve and smooth them to form a smooth sheet along the curve, with former 6–junction points on the fin curve becoming Y–branch points (for another Y–branch curve). The result depends upon a further distinction for essential fin curves. A *type-1 essential fin curve* will be one which only intersects other essential fin curves at 6–junction points; otherwise, it shares a segment of Y–branch curve with another essential fin curve, and it will be *type-2 essential fin curve* (see, for example, Fig. 3.26). If we cut along a type-1 essential fin curve, then the fin sheet becomes disconnected from the other sheets (at least along the curve) and this does not alter any other essential fin curve. If we cut along a type-2 essential fin curve, then it will alter the structure of the other essential fin curves sharing a segment of Y–branch curve with it.

With these observations in mind, we prescribe the following algorithm:

Algorithm for Decomposing Medial Axis into Irreducible Components

1. Identify all type-1 essential fin curves and systematically cut along essential fin curves (in any order).
2. After cutting along all type-1 essential fin curves, we may change certain inessential fin curves to type-1 essential ones. If so, return to step 1.
3. There only remain type-2 essential fin curves and inessential fin curves. Choose a type-2 essential fin curve, and cut along it. If a type-1 essential fin curve is created, return to step 1. Otherwise, repeat this step until no essential fin curves remain.
4. At this point no other essential fin curves remain. Choose an inessential fin curve which crosses a 6–junction point, and cut it from one side until we cut across one 6–junction point.
5. Check whether we have created an essential fin curve. If so then we cut along it, and repeat the earlier steps 1 and 2.
6. If no essential fin curve is created, we repeat step 3 until there are only inessential fin curves which do not cross 6–junction points.
7. Finally we can contract each such inessential fin curve to a point, producing part of a smooth sheet (i.e., in effect the fin curve disappears).
8. The remaining connected pieces are the "irreducible medial components" M_i of M.

Remark 3.5.1. The distinct connected pieces created following steps 1 and 2 are intrinsic to M, while those created using steps 3 and 4 are not because choices are involved. Which choices are made typically depends on the given situation and the importance we subjectively assign to how sheets are attached.

Example 3.5.1. In panel a of Fig. 3.26, we have a contractible medial axis with a pair of type-2 essential fin curves, which we refer to via the two numbers labeling their end points. Depending on which essential fin curve we choose, 1-2 (which passes from point 1 along the base plate, then onto a seam with fin curve 3-4 and the base plate, and then along the base plate only) or 3-4, we choose to cut along in step 3, we obtain two possible situations: see panels b and c. The two situations lead to different attachings (and hence top level graph) for the irreducible medial components. An alternate possibility would be to cut each fin sheet along the fin curves and view them as attached partially along the edge of a fourth sheet.

Example 3.5.2. For example, in Fig. 3.27a we have a contractible medial axis with 10 fin points 1-10, and all fin curves are inessential. Depending on how we choose cuts in step 4 of the algorithm, we can end up with 1, 2, or 3 irreducible medial components.

If we cut from 4 through the first 6–junction point, then 3-6 becomes an essential fin curve, and we cut away the fin sheet M_1 as in Fig. 3.27b. Then further cutting from 8 through the first 6–junction point, we create another essential fin curve 2-9.

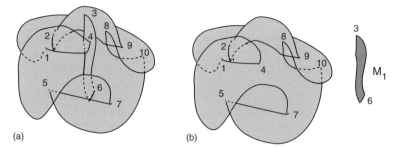

Fig. 3.27 Nonuniqueness of medial decomposition resulting from inessential fin curves: (**a**) is a contractible medial axis with only inessential fin curves; and (**b**) illustrates the cutting of irreducible medial component M_1 after cutting from fin point 4

Cutting along it creates a second fin sheet M_2. The remaining inessential fin curves 1-4, 5-7, and 8-10 can be contracted to points on edges of the third sheet M_3. Each of these three medial sheets are then irreducible components.

Alternatively, after the first cut, we could have instead cut from 9, and then from 4 again, and then only inessential fin curves remain without 6–junction points, so they contract to a second sheet, and we only obtain two irreducible components. Thirdly, we could have begun cutting from 7, then 8, and then 4 twice and we would obtain only a single medial sheet with inessential fin curves, leading to a single irreducible component.

To reverse the algorithm and reconstruct M from the M_i requires that we create the appropriate inessential fin curves from appropriate edges of the M_i and then reverse the steps by attaching the M_i along edges to fin curves which can cross multiple components. The attaching is an inductive process which for a given component requires the list of successive components and the embedded curves in each component along which the attaching will occur.

There is a simplified version of the attachings of the M_i which sacrifices some of the detailed structure of M but retains the topological structure (for homotopy equivalence). This simplified form is described by the graph structure $\Gamma(M)$. To obtain $\Gamma(M)$, instead of cutting along a fin curve, we alternately slide the sheet along the fin curve, passing all of the 6–junction points, as in Fig. 3.28. The sheet is then still attached along a fin curve, but only to a single smooth sheet (and without 6–junction points). Again after sliding the sheet, we can smooth the remaining two sheets attached along the curve. Also, we can still contract inessential fin curves to edge points of the M_i. Note that if we slide along the fin curves and then cut, we obtain the same components as we would have had by just cutting as originally described.

Then after sliding sheets along fin curves and contracting inessential fin curves, we define $\Gamma(M)$ to be the graph which has a vertex for each irreducible component M_i and directed edges from the vertex M_i to M_j for each edge of M_i attached to M_j along a fin curve.

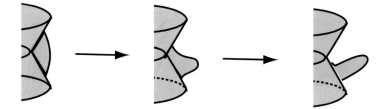

Fig. 3.28 Deforming by sliding a fin curve so it has no 6–junction points

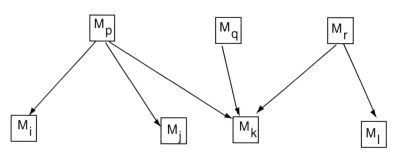

Fig. 3.29 Tree structure $\Gamma(M)$ given by decomposition into irreducible medial components M_i with edges indicating the attaching along "fin curves"

Importantly, as shown in Damon (2006, 2007c), *for a contractible region* Ω, $\Gamma(M)$ *is an (unrooted) directed tree*. This is illustrated in Fig. 3.29.

Because medial components can be attached to a number of components along a single fin curve, there are choices involved in deciding to which component we slide the fin curve. These choices mean that in the absence of external criteria for preferring a given component in terms of say importance, there is a nonuniqueness in the graph $\Gamma(M)$. The contractible medial axis given in panel a of Fig. 3.30, has medial sheets M_2, M_3 and M_4 attached to each other and to M_1. The graph of all attachings is shown in panel b. One way to slide the sheets only leaves the attaching to M_1, yielding the tree in panel c.

This has relevance to the perceptual grouping and recognition applications based on medial descriptions discussed in Chapters 2 and 10. For these applications one may choose to make certain classes of descriptions as equivalent, or one may wish to combine topological and geometric criteria. However, while means of doing this in 2D have been developed, the complications just described in 3D leave these possibilities as open questions for research.

3.5.2 Graph Structure of a Single Irreducible Medial Component

Next, to describe the structure of a single irreducible medial component M_i, we also use another graph (with data attached to each vertex and edge).

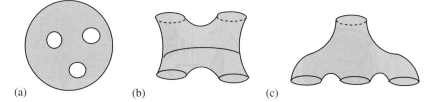

Fig. 3.33 Surface which is topologically a 2–disk with three holes (**a**) with several different geometric forms (**b**) and (**c**)

Characterization of Contractible Regions

Suppose a generic region Ω has a Blum medial axis M whose graph $\Gamma(M)$ is a tree and whose irreducible medial components each have graphs $\Lambda(M_i)$ which are also trees. If furthermore, the four characteristic properties above are satisfied, then the region Ω is contractible.

3.5.3.1 The General Classification

Suppose we now pass from the simpler contractible regions to more general regions Ω which we still suppose are bounded, connected, and have generic smooth boundaries $\mathcal{B}$. Now the local models for the singular points of the Blum medial axis M are still valid. Also, we can still decompose M into irreducible medial components as earlier. However, the resulting structures are no longer as simple. We briefly summarize the main extra complications as follows (see Damon, 2006 for more details).

Structure of the Medial Axis for General Regions

1. At the top level, $\Gamma(M)$ is a directed graph, but no longer need be a tree.
2. At the second level the directed graph $\Lambda(M_i)$ still has the same general form of S–vertices and Y–nodes, with edges only going from S–vertices to Y–nodes. However, again the graph need not be a tree.
3. Now each medial sheet S_{ij} can be any surface with boundary and can have varying genus and need not even be orientable. Also, it may have a multiple number of boundary circles which are edge curves of the medial axis.
4. The components of the Y-network will still have the same general form. However, there may be more than one boundary circle of S_{ij} being attached to the same component $\mathcal{Y}_{ik}$.
5. There are numerical relations between the the number of sheets, edge curves, etc.; except now these relations also involve topological invariants of the region.
6. Likewise, there is an analogue of the fundamental group relation, except now it involves the fundamental group of the region.

3.6 Summary

In this chapter we have explained a number of ideas and tools for analyzing the structure of the Blum medial axis M and determining its relation with the local, relative, and global geometry of the region Ω and its boundary $\mathscr{B}$. We have analyzed the geometric structure of $\mathscr{B}$ in terms of the radial geometry of the associated multi-valued vector field on M via radial and edge shape operators, a compatibility 1–form and the radial flow. These invariants defined on M allowed us to define a geometric map on M which directly corresponds to geometric properties of $\mathscr{B}$. These were further used to express global geometric invariants of Ω and $\mathscr{B}$ in terms of integrals on M which include expressions involving the radial shape operator.

As well we have analyzed the structure of the Blum medial axis in terms of irreducible components, their structure and how they are attached. This structure is related to the topological structure of Ω.

Acknowledgment This work was partially supported by National Science Foundation grant DMS-0405947, National Science Foundation/DARPA grant CCR-0310546 and DARPA grant HR0011-05-1-0057.

Part II
Algorithms

Chapter 4
Skeletons via Shocks of Boundary Evolution

Kaleem Siddiqi, Sylvain Bouix, and Jayant Shah

Abstract In this chapter we develop the Hamiltonian formulation of the eikonal equation and then relate it to the specific case of Blum's grassfire flow, which gives the level sets of the Euclidean distance function to the boundary. This view provides an explicit association between medial loci and the singularities of this flow. In order to detect these singularities we consider the average outward flux of the gradient of the Euclidean distance function. This measure has very different limiting behaviors depending upon whether the region over which it is computed shrinks to a singular point or a non-singular one. At medial loci the limiting values are related to the object angle. We combine the flux measurement with a homotopy preserving thinning process applied in a discrete lattice. This leads to a robust algorithm for computing skeletons in 2D as well as 3D, which has low computational complexity. We also discuss a related approach to associate medial loci with locations where the gradient of the Euclidean distance function is multi-valued, which uses the object angle as a measure of salience. Both approaches are illustrated with computational examples.

4.1 Overview

The grassfire flow is a special case of the eikonal equation which arises in geometric optics and has an associated Hamiltonian formulation. In this chapter we develop this formulation in Section 4.2. Whereas this presentation is somewhat detailed, the

K. Siddiqi
School of Computer Science & Centre for Intelligent Machines, McGill University, Canada,
e-mail: siddiqi@cim.mcgill.ca

S. Bouix
Psychiatry Neuroimaging Laboratory, Harvard University Medical School, USA,
e-mail: sylvain@bwh.harvard.edu

J. Shah
Department of Mathematics, Northeastern University, USA,
e-mail: shah@neu.edu

K. Siddiqi and S. Pizer (eds.) *Medial Representations – Mathematics, Algorithms and Applications*.
© Springer Science + Business Media B.V. 2008

development provides a comparison with the more standard Lagrangian formulation and suggests an alternative method for simulating the grassfire flow. It also shows that in this general setting medial loci may be viewed as the locus of positions that correspond to shocks of this boundary flow. In order to detect these shocks we consider the average outward flux ideas introduced in Chapter 3, Sections 5 and 6 and provide an implementation based on digital distance functions in Section 4.2.3. In Section 4.3 we combine this implementation with a homotopy preserving thinning process and illustrate the overall algorithm with 2D and 3D examples. We then discuss a related approach which uses the object angle as a measure of salience in Section 4.4, where we provide comparative 2D and 3D results between the two methods.

We begin with a brief overview of skeletonization methods based on properties of the Euclidean distance function, since these are most closely related to the approaches discussed in this chapter. As explained in Chapter 1, the Euclidean distance transform of a binary image is obtained by labeling each point in the domain by its Euclidean distance to the boundary of the object. The locus of skeletal points coincides with the singularities of this function. Thus, approaches in this class attempt to detect local maxima of this function, or the corresponding discontinuities in its derivatives (Arcelli and Sanniti di Baja, 1992; Leymarie and Levine, 1992; Kimmel et al., 1995; Gomes and Faugeras, 1999). A common approach is to find ridges in the digital distance map of the input image. Unfortunately, significant discretization problems are faced when working on a digital lattice, e.g., thresholding the Euclidean distance function is not sufficient to guarantee thin, homotopy preserving skeletons. Thus, more elaborate procedures have to be devised. For example, Malandain and Fernandez-Vidal (1998) obtain two sets based on thresholding a function of two heuristic measures to characterize the singularities of the Euclidean distance function. The two sets are combined using a topological reconstruction process. Pudney (1998) has also introduced a distance ordered homotopy preserving thinning procedure where points are removed in order of their distance from the boundary while anchoring end points and centers of maximal balls identified from a chamfer distance function. In Chapter 5 some of the state of the art algorithms for obtaining skeletons from digital distance transforms will be discussed. These algorithms are different from methods based on shocks of the grassfire flow, which are the focus of the present chapter, in that the latter are motivated by properties of the Euclidean distance function in the continuum along with the discretization of appropriate measures to locate its singularities.

As discussed in Chapter 1, Blum's (1973) definition of the medial set in terms of a *grass fire* has also been used as a model for skeletonization. Imagine the inside of the object to be a field of grass, and let the background be a non-flammable material. If the boundary of the object is lit, the wavefront originating from the boundary will meet exactly where the medial set of the object lies, assuming that the field of grass has constant refractive index. Instead of simulating the boundary erosion as a digital thinning, the motion of the boundary $\mathbf{W}$ under fire can be modeled as a continuous wavefront by the following partial differential equation:

$$\frac{\partial \mathbf{W}}{\partial t} = -\mathbf{n}, \tag{4.1}$$

where $-\mathbf{n}$ is the inward normal to the boundary. The medial axis of an object can be extracted by simulating the flow and detecting the points where the fronts meet (Xia, 1989; Leymarie and Levine, 1992; Kimmel et al., 1995). In these techniques, the contour has to be first segmented at curvature extrema, which is itself a challenging problem. Tek and Kimia (1999) have also proposed an interesting approach for calculating symmetry maps, which is based on the combination of a wavefront propagation technique with an exact (analytic) distance function. In their technique the representation must be pruned in order to distinguish salient branches from unwanted ones. These methods give good results and provide many of the properties the digital medial axis should have; however a generalization to the 3D case proves difficult.

4.2 Optics, Mechanics and Hamilton-Jacobi Skeletons

We now develop the connection between the grassfire flow and the formation of the skeleton. Let $\mathbf{W}(t)$ be the moving boundary (front) of the object (assumed to be a closed curve in 2D or a closed surface in 3D), $-\mathbf{n}$ the unit inward normal to $\mathbf{W}$ and t a parameter to denote the family of evolved fronts. Let $\mathbf{W}_0$ be the initial boundary of the object. The motion of the front is given by

$$\frac{\partial \mathbf{W}}{\partial t} = -f\mathbf{n}, \tag{4.2}$$

where the function $f = f(\mathbf{x})$ is the speed of the front and is related to the refractive index n of the medium. If n is a function solely of position $\mathbf{x}$, the medium is said to be *isotropic*. If it depends on both position and orientation, the medium is *anisotropic*. Furthermore, if n does not vary with position, the medium is said to be *homogeneous*. In the case of Blum's grass fire flow, f is the scalar constant $1/n$ and thus the medium is homogeneous and isotropic. Unfortunately, given an arbitrary boundary $\mathbf{W}_0$ no direct analytical method exists to detect the shocks of this equation.

4.2.1 Medial Loci and the Eikonal Equation

A numerical approach to simulating flows of the type in (4.2) while handling topological changes is to use level set methods developed by Osher and Sethian (1988); Sethian (1996). Below we focus the discussion on the case of a *monotonically advancing front* where $f = f(\mathbf{x})$ has fixed sign for all points $\mathbf{x}$ in the domain of $\mathbf{W}$. Let $\phi(\mathbf{x})$ be a graph of the solution, obtained by superimposing all the evolved fronts in time. In other words, $\phi(\mathbf{x})$ is the time at which the front crosses a point $\mathbf{x}$

in the medium. We thus have

$$\phi(\mathbf{W}(t)) = t. \tag{4.3}$$

Taking the total derivative of ϕ with respect to time we get

$$\frac{d}{dt}\phi(\mathbf{W}(t)) = 1 \tag{4.4}$$

$$\frac{\partial\phi}{\partial\mathbf{x}} \cdot \frac{\partial\mathbf{W}}{\partial t} = 1 \tag{4.5}$$

$$\nabla\phi \cdot \frac{\partial\mathbf{W}}{\partial t} = 1. \tag{4.6}$$

Now substitute for $\frac{\partial\mathbf{W}}{\partial t}$ using (4.2),

$$\nabla\phi \cdot -f\mathbf{n} = 1$$

$$\|\nabla\phi\|^2 = \frac{1}{f^2}. \tag{4.7}$$

Equation (4.7) is a form of the well known *eikonal equation*, a key concept of geometrical optics (Luneburg, 1964; Stavroudis, 1972). It is also at the core of a multitude of other scientific problems. A number of algorithms have been recently developed to numerically solve this equation, including Sethian's fast marching method (Sethian, 1996) which systematically constructs ϕ using only upwind values, Rouy and Tourin's (1992) viscosity solutions approach and Sussman et al.'s (1994) level set method for incompressible two-phase flows. However, none of these methods address the issue of shock detection explicitly, and more work has to be done to track shocks.

In the next section, we shall consider an alternate framework for solving the eikonal equation, which is based on the canonical equations of Hamilton. The technique is widely used in classical mechanics and rests on the use of a Legendre transformation (see Arnold, 1989; Shankar, 1994), which takes a system of d second-order differential equations to a mathematically equivalent system of $2d$ first-order differential equations.

4.2.2 Hamiltonian Derivation of the Eikonal Equation

4.2.2.1 Variational Principles

We begin by reviewing Lagrangian and Hamiltonian methods for solving minimization problems (Arnold, 1989; Shankar, 1994). Formally, let

$$\psi = \int_{t_0}^{t_1} \mathcal{L}(\mathbf{q},\dot{\mathbf{q}},t)\, dt, \tag{4.8}$$

be a functional over the space of curves $\{(\mathbf{q},t) : \mathbf{q}(t) = \mathbf{q}, t_0 \le t \le t_1\}$, where $\dot{\mathbf{q}} = \mathbf{q}'(t)$. $\mathscr{L}(\mathbf{q}, \dot{\mathbf{q}}, t)$ is called the *Lagrangian* of the problem. We are interested in finding the curve $\gamma = \{(\mathbf{q}, t) : \mathbf{q}(t) = \mathbf{q}, t_0 \le t \le t_1\}$, such that $\psi(\gamma)$ is an extremum. The procedure is well known and leads to the *Euler-Lagrange* theorem.

Theorem 4.2.1 (Euler-Lagrange Equation). *The curve γ is an extremal of the functional $\psi(\gamma) = \int_{t_0}^{t_1} \mathscr{L}(\mathbf{q}, \dot{\mathbf{q}}, t) \, dt$ on the space of curves joining $(\mathbf{q}_0, t_0)$ and $(\mathbf{q}_1, t_1)$, if and only if the following Euler-Lagrange equations*

$$\frac{d}{dt}\frac{\partial \mathscr{L}}{\partial \dot{\mathbf{q}}} - \frac{\partial \mathscr{L}}{\partial \mathbf{q}} = 0 \qquad (4.9)$$

are satisfied along the curve γ.

This is a system of d *second order* equations whose solutions depend on the $2d$ boundary conditions $\mathbf{q}(t_0) = \mathbf{q}_0$ and $\mathbf{q}(t_1) = \mathbf{q}_1$. This system of d second order equations can be transformed in a system of $2d$ *first order* equations which is usually easier to solve. The transformation makes use of the theory of *Hamiltonian Canonical Variables*.

The key to the method is to exchange the roles of $\dot{\mathbf{q}}$ by the canonical variable $\mathbf{p} = \frac{\partial \mathscr{L}}{\partial \dot{\mathbf{q}}}$, commonly referred to as the *momentum*, and replace the Lagrangian $\mathscr{L}(\mathbf{q}, \dot{\mathbf{q}}, t)$ with the function $\mathscr{H}(\mathbf{q}, \mathbf{p}, t)$, called the Hamiltonian, such that the velocities now become the derived quantities

$$\dot{\mathbf{q}} = \frac{\partial \mathscr{H}}{\partial \mathbf{p}}.$$

This can be done by applying the following *Legendre transformation*:

$$\mathscr{H}(\mathbf{q}, \mathbf{p}, t) = \mathbf{p} \cdot \dot{\mathbf{q}} - \mathscr{L}(\mathbf{q}, \dot{\mathbf{q}}, t) \qquad (4.10)$$

where the $\dot{\mathbf{q}}$'s are written as functions of $\mathbf{p}$'s. It is a simple exercise to verify that the above expression for the velocities $\dot{\mathbf{q}}$ then holds. This transformation is possible if the Lagrangian $\mathscr{L}$ is non-degenerate, i.e., the determinant of its Hessian is non zero. One can also take partial derivatives of the Hamiltonian with respect to the $\mathbf{q}$'s and verify that

$$\frac{\partial \mathscr{H}}{\partial \mathbf{q}} = -\frac{\partial \mathscr{L}}{\partial \mathbf{q}}.$$

Using (4.9), $\frac{\partial \mathscr{L}}{\partial \mathbf{q}}$ can be replaced with $\dot{\mathbf{p}}$ to give *Hamilton's canonical equations*:

$$\dot{\mathbf{p}} = -\frac{\partial \mathscr{H}}{\partial \mathbf{q}}, \qquad \dot{\mathbf{q}} = \frac{\partial \mathscr{H}}{\partial \mathbf{p}}. \qquad (4.11)$$

Thus, in the Hamiltonian formalism one starts with the initial positions and momenta $(\mathbf{q}(t_0), \mathbf{p}(t_0))$ and integrates (4.11) to obtain the phase space $(\mathbf{q}(t), \mathbf{p}(t))$ of the system.

Table 4.1 A comparison of the Lagrangian and Hamiltonian formalisms, taken from (Shankar, 1994)

The Lagrangian formalism	The Hamiltonian formalism
The state of the system is described by $(\mathbf{q}, \dot{\mathbf{q}})$	The state of the system is described by $(\mathbf{q}, \mathbf{p})$
The state may be represented by a point moving with a velocity in an d-dimensional configuration space	The state may be represented by a point in a 2d-dimensional phase space
The d coordinates evolve according to d second-order equations	The 2n coordinates and momenta obey 2d first-order equations
For a given $\mathcal{L}$ several trajectories may pass through a given point in the configuration space	For a given $\mathcal{H}$ only one trajectory passes through a given point in the phase space

Using (4.8) and (4.10) it is straightforward to see that

$$\psi(\gamma) = \int_{t_0}^{t_1} \mathbf{p} \cdot \dot{\mathbf{q}} - \mathcal{H} \, dt. \tag{4.12}$$

These results are summarized in the following theorem:

Theorem 4.2.2 (Hamilton's equations). *Let* $\mathbf{p} = \frac{\partial \mathcal{L}}{\partial \dot{\mathbf{q}}}$ *and substitute in (4.9). The system of d second order Euler-Lagrange equations* $\dot{\mathbf{p}} - \frac{\partial \mathcal{L}}{\partial \mathbf{q}} = 0$, *is equivalent to the system of 2d first order equations (Hamilton's equations)*

$$\dot{\mathbf{p}} = -\frac{\partial \mathcal{H}}{\partial \mathbf{q}}, \qquad \dot{\mathbf{q}} = \frac{\partial \mathcal{H}}{\partial \mathbf{p}},$$

where $\mathcal{H}(\mathbf{q}, \mathbf{p}, t) = \mathbf{p} \cdot \dot{\mathbf{q}} - \mathcal{L}(\mathbf{q}, \dot{\mathbf{q}}, t)$ *is the Legendre transform of the Lagrangian viewed as a function of* $\dot{\mathbf{q}}$.

A comparison of the Lagrangian and Hamiltonian formalisms is presented in Table 4.1.

In the case that the extremals emanating from the point $(\mathbf{q}_0, t_0)$ do not intersect elsewhere but instead form a so called *"central field of extremals"*, one can define the *action function* ϕ as the solution functional of our variational problem

$$\phi(\mathbf{q}, t) = \min_{\gamma}(\psi(\gamma))$$

$$= \int_{\gamma_*} \mathcal{L}(\mathbf{q}, \dot{\mathbf{q}}, t) \, dt$$

$$= \int_{\gamma_*} \mathbf{p} \cdot \dot{\mathbf{q}} - \mathcal{H} \, dt. \tag{4.13}$$

where γ_* is an extremal curve. It can be shown that $\mathbf{p} = \frac{\partial \phi}{\partial \mathbf{q}}$ and that the action function satisfies the *Hamilton-Jacobi equation* (Arnold, 1989)

$$\frac{\partial \phi}{\partial t} + \mathscr{H}\left(\mathbf{q}, \frac{\partial \phi}{\partial \mathbf{q}}, t\right) = 0. \tag{4.14}$$

We now have all the necessary tools to formalize the connection between geometric optics and the monotonically advancing front (4.2).

4.2.2.2 Fermat's Principle

Like most laws of classical physics, the equations of geometric optics can be derived from a *variational principle* (Stavroudis, 1972). In this context, the variational principle is called *Fermat's principle* which states that a ray always chooses a trajectory that minimizes the optical path length.[1] Consider a (possibly inhomogeneous) isotropic medium with $n(\mathbf{x})$ its refractive index. The following calculations can then be carried out in arbitrary dimensions, although for simplicity we focus on the 2D case. The (2D) trajectory $\gamma(t) = (x(t), y(t))$ connecting two points $\gamma(t_0) = \mathbf{q}_0$ and $\gamma(t_1) = \mathbf{q}_1$ in the medium minimizes the following integral

$$\psi(\gamma) = \int_{\mathbf{q}_0}^{\mathbf{q}_1} n \, ds$$

$$\psi(\gamma) = \int_{t_0}^{t_1} n(x,y) \sqrt{\frac{dx^2}{dt} + \frac{dy^2}{dt}} \, dt$$

$$\psi(\gamma) = \int_{t_0}^{t_1} L(x, y, \frac{dx}{dt}, \frac{dy}{dt}, t) \, dt, \tag{4.15}$$

where ds is the line element along the ray. In order to proceed with the derivation, we must ensure that the Lagrangian L is non degenerate. Unfortunately, the determinant of the Hessian of L is equal to zero due to the fact that the variational problem is independent of the parameter t. Fortunately, the analysis can be done by choosing the projected coordinate y as the new variable of integration.

$$ds = \sqrt{dx^2 + dy^2} = \sqrt{1 + \frac{dx^2}{dy}} \, dy \tag{4.16}$$

Equation (4.15) then becomes

$$\psi(\gamma) = \int_{y_0}^{y_1} n(x,y) \sqrt{1 + \frac{dx^2}{dy}} \, dy$$

$$= \int_{y_0}^{y_1} \mathscr{L}(x, x') \, dy \tag{4.17}$$

---

[1] More precisely the path must be a local extremum and in rare cases may in fact be a maximum (Luneburg, 1964).

where $x' = \frac{dx}{dy}$. $\mathscr{L}$ is sometimes referred to as the Fermat Lagrangian. We can apply the Euler-Lagrange equations (4.9) to the Fermat Lagrangian (4.17)

$$\frac{d}{dy} \frac{nx'}{\sqrt{1+x'^2}} - \frac{\partial n}{\partial x} \sqrt{1+x'^2} = 0 \tag{4.18}$$

This second order partial differential equation is called the *ray equation* and is related to the eikonal equation. In order to develop this connection we turn to the theory of *Hamiltonian Canonical Variables*.

First, we write the Hamiltonian using (4.10) and (4.17)

$$\mathscr{H}(x, p_x) = p_x x' - \mathscr{L}(x, x')$$
$$= p_x x' - n\sqrt{1+x'^2}, \tag{4.19}$$

with

$$p_x = \frac{\partial \mathscr{L}}{\partial x'} = \frac{nx'}{\sqrt{1+x'^2}}. \tag{4.20}$$

We may solve for x' in terms of p_x,

$$x' = \frac{\partial \mathscr{H}}{\partial p_x} = \frac{p_x}{\sqrt{n^2 - p_x^2}}, \tag{4.21}$$

which leads to the Fermat Hamiltonian

$$\mathscr{H}(x, p_x) = -\sqrt{n^2 - p_x^2}. \tag{4.22}$$

Now, let us assume a central field of extremals, and define $\phi(x,y) = \min_\gamma(\psi(\gamma))$. Substituting (4.22) into (4.14) yields

$$\frac{\partial \phi}{\partial y} = \sqrt{n^2 - \frac{\partial \phi}{\partial x}^2}. \tag{4.23}$$

Squaring both sides one obtains

$$\|\nabla \phi\|^2 = n^2, \tag{4.24}$$

which is the eikonal equation see (4.7) with speed $f(\mathbf{q}) = \frac{1}{n(\mathbf{q})}$. Observe that the speed of the front $f(\mathbf{q})$ is inversely proportional to the refractive index of the medium $n(\mathbf{q})$. We summarize the results in the following theorem.

Theorem 4.2.3 (Geometrical Optics in Isotropic Media). *According to Fermat's principle, the trajectory of a ray in an inhomogeneous isotropic medium minimizes the following integral*

$$\psi(\gamma) = \int_{t_0}^{t_1} n(\mathbf{q}) \|\dot{\mathbf{q}}\| \, dt = \int_{y_0}^{y_1} n(x,y) \sqrt{1 + \frac{dx^2}{dy}} \, dy.$$

Its corresponding d second order Euler-Lagrange equations lead to the ray equation

$$\frac{d}{dy} \frac{nx'}{\sqrt{1+x'^2}} - \frac{\partial n}{\partial x} \sqrt{1+x'^2} = 0.$$

Transforming the problem into Hamiltonian form we obtain a system of 2d first order equations

$$\frac{dp_x}{dy} = -\frac{\partial \mathcal{H}}{\partial x}$$

$$\frac{dx}{dy} = \frac{\partial \mathcal{H}}{\partial p_x}$$

where the Hamiltonian is given by $\mathcal{H}(x,p_x) = -\sqrt{n^2 - p_x^2}$. *Assuming a central field of extremals, a graph of the solution surface*

$$\phi(x,y) = \min_{\gamma}(\psi(\gamma)) = \int_{\gamma_*} n(x,y) \sqrt{1 + \frac{dx^2}{dy}} \, dy$$

exists and is called the action function. Furthermore

$$\frac{\partial \phi}{\partial y} = -\mathcal{H}(x, \frac{\partial \phi}{\partial x})$$

$$= \sqrt{n^2 - \frac{\partial \phi}{\partial x}^2}$$

which leads to the eikonal equation

$$\|\nabla \phi\|^2 = n^2.$$

Hence, there is an explicit connection between medial loci, a monotonically advancing front from an object's boundary, Fermat's principle and the eikonal equation. If one sets $n = 1/f = 1$, one gets the grass fire flow (4.1)

$$\frac{\partial \mathbf{W}}{\partial t} = -\mathbf{n},$$

whose shocks correspond to one of the Blum (1973) definitions of the medial set. Thus the medial set is the locus of positions where two or more front meets, i.e., where $-\mathbf{n}$ is not defined. If we turn back to the variational formulation we obtain

$$\psi(\gamma) = \int_{t_0}^{t_1} n(\mathbf{q})\|\dot{\mathbf{q}}\| \, dt \qquad (4.25)$$

$$= \int_{y_0}^{y_1} \sqrt{1 + \frac{dx}{dy}^2} \, dy, \qquad (4.26)$$

which is nothing other than the Euclidean distance $\|\mathbf{q}(t_1) - \mathbf{q}(t_0)\|$ between $\mathbf{q}(t_1)$ and $\mathbf{q}(t_0)$, with the extremal curve γ being a straight line. The associated action function is the *Euclidean distance function*

$$\phi(\mathbf{q}) = \min_{\mathbf{x} \in \mathbf{W}_0} \|\mathbf{q} - \mathbf{x}\|, \qquad (4.27)$$

where $\mathbf{W}_0 = \mathbf{W}(t_0)$ is the initial boundary of the object and the inward normals to the evolving front $\mathbf{W}$ are given by

$$-\mathbf{n} = \nabla \phi.$$

These constructs are illustrated for the outline of a panther shape in Fig. 4.1 (top and middle). Technically ϕ as an action function (4.13) is not defined at medial points, where the assumption of a central field of extrema is broken. Fortunately, ϕ in its form in (4.27) is defined and is continuous over $\mathbb{R}^n$. Thus we can conclude that the medial set corresponds to locations where the Euclidean distance function ϕ is singular and hence $\nabla \phi$ is multi-valued. This result is not novel; for example, Matheron (1988) noticed that ϕ is differentiable and that $\|\nabla \phi\| = 1$ for all points $\mathbf{q}$ not lying on the medial set. However, by interpreting the grass fire flow using Fermat's principle we provide a view that is motivated by considerations in classical mechanics.

The view developed thus far suggests that if the Euclidean distance function ϕ is available, Hamilton's equations can be used to simulate the grassfire flow while associating its singularities with the Blum skeleton. There are implementation choices for obtaining ϕ including the fast marching technique (Sethian, 1996) and the use of digital distance transforms (Borgefors, 1984). The method that we develop in Section 4.2.3 uses the latter approach and considers the vector field $\nabla \phi$ for all points in the interior of the object. It exploits the fact that the limiting behavior of the average outward flux of this vector field through a shrinking circular neighborhood can be used to distinguish medial points from non-medial ones. A more recent implementation is designed to handle an object whose boundary is given by a mesh (Stolpner and Siddiqi, 2006); it uses accurate techniques for computing point-to-mesh distances developed in computational geometry.

4.2.3 Divergence, Average Outward Flux and Object Angle

As explained in Chapter 3, Section 4.5, it is possible to extend the standard divergence theorem to the case of a vector field defined on a region Ω in $\mathbb{R}^n$ with smooth boundary $\mathcal{B}$ defined by a skeletal structure $(M, \mathbf{S})$. In particular, if $G = \nabla \phi$ denotes the unit vector field generating the grassfire flow for the region Ω with Blum medial axis M, then for a piecewise smooth region $\Gamma \subset \Omega$

$$\int_\Gamma \operatorname{div} G \, dV = \int_{\partial\Gamma} G \cdot \mathbf{n}_\Gamma \, dS + \int_{\tilde\Gamma} dM. \qquad (4.28)$$

Here $\mathbf{n}_\Gamma$ is the unit outward normal to Γ and $\tilde\Gamma$ consists of those $x \in \tilde{M}$ with smooth value $\mathbf{S}$ such that the radial line determined by $\mathbf{S}$ has non-empty intersection with Γ.

The first term on the right hand side of the above equation is the outward flux of the grassfire flow across $\partial\Gamma$. It differs from the divergence integral of G over Γ by the "medial volume of $\tilde\Gamma$". As explained in Chapter 3, Section 4.6, it is this medial volume term that allows for the discrimination of medial points from non-medial ones (see also Siddiqi et al., 2002; Dimitrov et al., 2003). The key idea is to consider the limiting behavior of the average outward flux (the outward flux normalized by the volume of $\partial\Gamma$) as Γ shrinks to a point $\mathbf{x}$. For points not on the medial locus the standard divergence theorem applies and it can be shown that the limiting value of the average outward flux is 0. However, for points on the medial axis the limiting average outward flux is non-vanishing, with the exception of edge points. Furthermore, the value that it attains can be bounded in terms of the quantities $\min(\rho)$ and M_Γ discussed in Chapter 3, Section 4.6. The quantity $\min(\rho)$ depends on both M and $\mathbf{S}$, and the quantity M_Γ is obtained from a medial density term that depends on the choice of Γ (see Chapter 3, Table 3.1 for the cases of Γ an n-disk or an n-cube, with $n = 2, 3$).

For the purposes of computation it is reasonable to choose Γ to be a n-disk, in which case a tighter bound can be obtained for the limiting values of the average outward flux. These calculations are presented in Dimitrov et al. (2003) and Dimitrov (2003) for the case $n = 2$; very similar results hold for $n = 3$. The remarkable fact is that for Γ a disk the limiting values actually provide a function of the object angle θ (the angle between $\mathbf{S}$ and M) at all generic points of M. These results are summarized in Table 4.2 for $n = 2$. For the case of the junction points the formula must be interpreted by considering the sum of the object angles for each of the three incoming branches at $\mathbf{x}$. Thus, an average outward flux computation not only allows medial points to be distinguished from non-medial ones, but also yields the object angle θ (and hence $\mathbf{S}$) as a by-product for each skeletal point $\mathbf{p}$. As explained in Chapter 1, Section 2.3, this in turn allows for boundary boundary reconstruction of the bi-tangent points $\mathbf{b}^{\pm 1}$ associated with each skeletal point, with the radius values obtained from the Euclidean distance function. In this sense the average outward flux of $\nabla\phi$ through a shrinking disk may be viewed as a type of flux invariant for detecting medial loci as well as characterizing the geometry of the implied boundary (Dimitrov et al., 2003; Dimitrov, 2003).

Algorithm 1 describes a discrete implementation of the average outward flux using a digital distance transform for ϕ. Figure 4.1 illustrates its computation on the silhouette of a panther shape, where values close to zero are shown in medium grey. All computations are carried out on a rectangular lattice, although the bounding surface is shown in interpolated form. Strictly speaking, the average outward flux is desired only in the limit as the region shrinks to a point. However, the average

Table 4.2 A summary of results from Dimitrov et al. (2003) relating the limiting values of the average outward flux of $\nabla\phi$ through a shrinking disk to the object angle θ for the case of $n = 2$. Here $\mathscr{F}_\varepsilon(\mathbf{x})$ denotes the outward flux at a point $\mathbf{x}$ of $\nabla\phi$ through $\partial\Gamma$, with Γ a disk of radius ε

Point type	$\lim\limits_{\varepsilon\to 0}\dfrac{\mathscr{F}_\varepsilon(\mathbf{x})}{2\pi\varepsilon}$
Regular points	$-\frac{2}{\pi}\sin\theta$
End-points	$-\frac{1}{\pi}(\sin\theta - \theta)$
Junction points	$-\frac{1}{\pi}\sum_{i=1}^{n}\sin\theta_i$
Non-skeletal points	0

Algorithm 1: Average Outward Flux

Data   : Object Ω.
Result   : Average Outward Flux Map.
Compute the digital Euclidean distance transform ϕ of the object (Borgefors, 1984);
Compute the gradient vector field $\nabla\phi$;
Compute the average outward flux of $\nabla\phi$:
for *(each point $\mathbf{x}$)* **do**

$\qquad F(\mathbf{x}) = \dfrac{1}{n}\sum_{i=1}^{n} <\mathbf{n_i}, \nabla\phi(\mathbf{x}_i)>$;

$\qquad$(where $\mathbf{x}_i$ is a n-neighbor of $\mathbf{x}$ ($n = 8$ in 2D, $n = 26$ in 3D) and $\mathbf{n_i}$ is the outward normal at $\mathbf{x}_i$ of the unit sphere centered at $\mathbf{x}$)

---

outward flux over a very small neighborhood (a circle in 2D or a sphere in 3D) provides a sufficient approximation to the limiting values. A threshold on the average outward flux yields a close approximation to the medial set, as used in Siddiqi et al. (1999a). However, in general it is impossible to guarantee that the result obtained by simple thresholding is homotopic to the original shape. A high threshold may yield a connected set, but cannot guarantee that it is thin. A low threshold can yield a thin set, but it may be disconnected. The solution, as we shall show in the subsequent section, is to introduce additional constraints to ensure that the resulting medial set is homotopic to the shape. The essential idea is to incorporate a homotopy preserving thinning process, where the removal of points is guided by the average outward flux values. This leads to a robust and efficient algorithm for computing 2D and 3D medial loci.

4.3 Homotopy Preserving Medial Loci

In this section we combine the average outward flux computation with a digital thinning process, where points are removed without altering the object's topology. In digital topology, a point is said to be *simple* if its removal does not change the

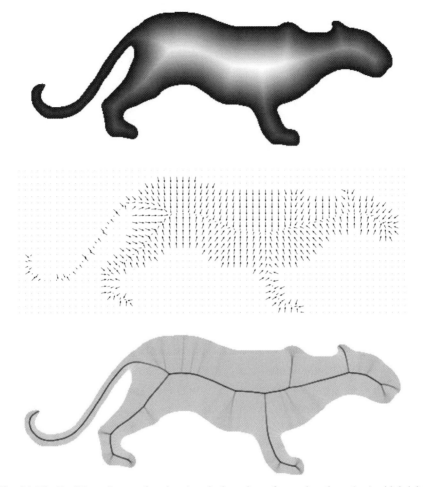

Fig. 4.1 The Euclidean distance function ϕ to the boundary of a panther shape (*top*) with brightness proportional to increasing distance, its gradient vector field $\nabla\phi$ (*center*) and the associated average outward flux (*bottom*). Whereas the smooth regime of the vector field gives zero flux (*medium grey*), strong singularities give large negative values (*dark grey*) in the interior of the object. Adapted from (Dimitrov et al., 2000)

topology of the object. In 2D, we shall consider rectangular lattices, where a point is a unit square with eight neighbors, as shown in Fig. 4.2 (left). Hence, a 2D digital point is simple if its removal does not disconnect the object or create a hole. In 3D, we shall consider cubic lattices, where a point is a unit cube with 6 faces, 12 edges and 8 vertices. Hence, a 3D digital point is simple if its removal does not disconnect the object, create a hole, or create a cavity (Kong and Rosenfeld, 1989). We should note that the 2D version of the algorithm was first developed in Dimitrov et al. (2000), and we review it here for completeness.

4.3.1 2D Simple Points

Consider the 3×3 neighborhood of a 2D digital point $\mathbf{x}$ contained within an object and select those neighbors which are also contained within the object. Now construct a neighborhood graph by placing edges between all pairs of neighbors (not including $\mathbf{x}$) that are 4-adjacent or 8-adjacent to one another. If any of the 3-tuples $\{2,3,4\}$, $\{4,5,6\}$, $\{6,7,8\}$, or $\{8,1,2\}$, are nodes of the graph, remove the corresponding diagonal edges $\{2,4\}$, $\{4,6\}$, $\{6,8\}$, or $\{8,2\}$, respectively. This ensures that there are no degenerate cycles in the neighborhood graph (cycles of length 3). Now, observe that if the removal of $\mathbf{x}$ disconnects the object, or introduces a hole, the neighborhood graph will not be connected, or will have a cycle, respectively. Conversely, a connected graph that has no cycles is a tree. Hence, we have a criterion to decide whether or not $\mathbf{x}$ is simple:

Proposition 4.3.1. *A 2D digital point $\mathbf{x}$ is simple if and only if its 3×3 neighborhood graph, with cycles of length 3 removed, is a tree.*

A straightforward way of determining whether or not a connected graph is a tree is to check that its Euler characteristic $|V| - |E|$ (the number of vertices minus the number of edges) is identical to 1. This check only has to be performed locally, in the 3×3 neighborhood of a point P. Figure 4.2 (right) shows an example neighborhood graph for which P is simple and hence can be removed.

4.3.2 3D Simple Points

In 3D a digital point can have three types of neighbors. Two points are *6-neighbors* if they share a face; two points are *18-neighbors* if they share a face or an edge; and two points are *26-neighbors* if they share a face, an edge or a vertex. This induces three *n-connectivities*, where $n \in \{6,18,26\}$, as well as three *n-neighborhoods* for $\mathbf{x}$, $N_n(\mathbf{x})$. An n-neighborhood without its central point is defined as $N_n^* = N_n(\mathbf{x}) \backslash \{\mathbf{x}\}$. An object A is *n-adjacent* to an object B, if there exist two points $\mathbf{x} \in A$ and $\mathbf{y} \in B$ such that $\mathbf{x}$ is an n-neighbor of $\mathbf{y}$. A *n-path* from $\mathbf{x}_1$ to $\mathbf{x}_k$ is a sequence of points

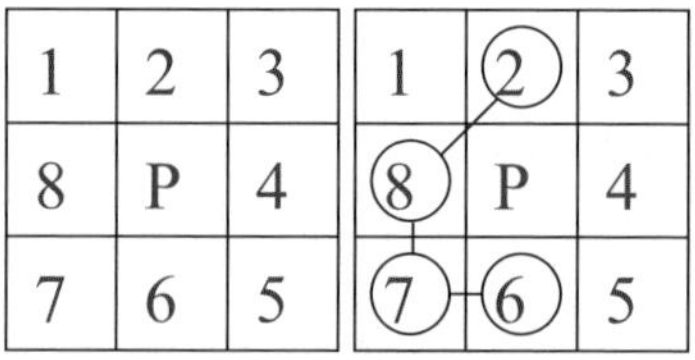

1	2	3
8	P	4
7	6	5

1	②	3
⑧	P	4
⑦	⑥	5

Fig. 4.2 *Left*: A 3×3 neighborhood of a candidate point for removal P. *Right*: An example neighborhood graph for which P is simple. There is no edge between neighbors 6 and 8 (see text)

$\mathbf{x}_1, \mathbf{x}_2, ..., \mathbf{x}_k$, such that for all $\mathbf{x}_i$, $1 < i \leq k$, $\mathbf{x}_{i-1}$ is n-adjacent to $\mathbf{x}_i$. An object represented by a set of points O is *n-connected*, if for every pair of points $(\mathbf{x}_i, \mathbf{x}_j) \in O \times O$, there is a n-path from $\mathbf{x}_i$ to $\mathbf{x}_j$.

Based on these definitions, Malandain et al. (1993) provide a topological classification of a point $\mathbf{x}$ in a cubic lattice by computing two numbers:

- C^*: the number of 26-connected components 26-adjacent to $\mathbf{x}$ in $O \cap N_{26}^*$
- $\bar{C}$: the number of 6-connected components 6-adjacent to $\mathbf{x}$ in $\bar{O} \cap N_{18}$

An important result with respect to our goal of thinning is the following:

Theorem 4.3.1 (Malandain et al., 1993). $\mathbf{x}$ *is* simple *if* $C^*(\mathbf{x}) = 1$ *and* $\bar{C}(\mathbf{x}) = 1$.

We can now determine whether or not the removal of a point will alter the topology of a digital object. When preserving homotopy is the only concern, simple points can be removed sequentially until no more simple points are left. The resulting set will be thin and homotopic to the object. However, without a further criterion the relationship to the medial set will be uncertain since the locus of surviving points depends entirely on the order in which the simple points are removed. In the current context, we have derived a natural criterion for ordering the thinning, based on the average outward flux of the gradient vector field of the Euclidean distance function.

4.3.3 Average Outward Flux Ordered Thinning

Recall from Section 4.2.3, that the average outward flux of the gradient vector field of the Euclidean distance function can be used to distinguish non-medial points from medial ones. This quantity tends to zero for the former but approaches a negative number below a constant times $< \nabla \phi, \mathbf{N} >$ for the latter, where $\mathbf{N}$ is the one-sided normal to the medial axis or surface. Hence, the average outward flux provides a natural measure of the "strength" of a medial point for numerical computations. The essential idea is to order the thinning such that the weakest points are removed first and to stop the process when all surviving points are not simple, or have a total average outward flux below some chosen (negative) value, or both. This will accurately localize the medial set and also ensure homotopy with the original object. Unfortunately the result is not guaranteed to be a thin set, i.e., one without an interior.

One way of satisfying this last constraint is to define an appropriate notion of an end point. Such a point would correspond to the end point of a curve (in 2D or 3D), or a point on the rim of a surface, in 3D. The thinning process would proceed as before, but the threshold criterion for removal would be applied *only to end points*. Hence, all surviving points which were not end points would not be simple and the result would be a thin set.

In 2D, an end point will be viewed as any point that could be the end of a 4-connected or 8-connected digital curve. It is straightforward to see that such a point may be characterized as follows:

Proposition 4.3.2 (2D End Point). *A 2D point* **x** *could be an end point of a 1 pixel thick digital curve if, in a 3 × 3 neighborhood, it has a single neighbor, or it has two neighbors, both of which are 4-adjacent to one another.*

In 3D, the characterization of an end point is more difficult. An end point is either the end of a 26-connected curve, or a corner or point on the rim of a 26-connected surface.

Proposition 4.3.3 (3D End Point). *In* $\mathbf{R}^3$, *if there exists a plane that passes through a point* **x** *such that the intersection of the plane with the object includes an open curve which ends at* **x**, *then* **x** *is an end point of a 3D curve, or is on the rim or corner of a 3D surface.*

This criterion can be discretized easily to 26-connected digital objects by examining nine digital planes in the 26-neighborhood of **x** as in Pudney (1998).

4.3.4 The Algorithm and Its Complexity

The essential idea behind the average outward flux-ordered thinning process is to remove simple points sequentially, ordered by their average outward flux, until a threshold is reached. Subsequently, simple points are removed if they are not end points. The procedure converges when all remaining points are either not simple or are end points. The thinning process can be made very efficient by observing that a point which does not have at least one background point as an immediate neighbor cannot be removed, since this would create a hole or a cavity. Therefore, the only potentially removable points are on the border of the object. This suggests the implementation of the thinning process using a *heap* data structure. A full description of the procedure is given in Algorithm 2. The approach is computationally very efficient. With n the total number of digital points within the original volume and k the number of points within the object, the worst case complexity can be shown to be $\mathcal{O}(n) + \mathcal{O}(k \log(k))$ (Siddiqi et al., 2002).

4.3.5 Labeling the Medial Set

The classification of digital points on the 2D medial set is quite straightforward. Consider a circular path through the digital neighbors of such a point P (see Fig. 4.2) and let n be the number of times this path intersects the medial set. The three generic possibilities are

- $n = 1$, in which case P is an end point.
- $n = 2$, in which case P is an interior (curve) point.
- $n = 3$, in which case P is a branch point.

Algorithm 2: Topology Preserving Thinning

Data   : Object Ω, Average Outward Flux Map.
Result   : (2D or 3D) Skeleton.
for *(each point* **x** *on the boundary of the object)* **do**
   | **if** *(***x** *is simple)* **then**
   | | insert(**x**, maxHeap) with AOF(**x**) as the sorting key for insertion;

while *(maxHeap.size > 0)* **do**
   | **x** = HeapExtractMax(maxHeap);
   | **if** *(***x** *is simple)* **then**
   | | **if** *(***x** *is an end point) and* (AOF(**x**) < *Thresh)* **then**
   | | | mark **x** as a medial surface (end) point;
   | | **else**
   | | | Remove **x**;
   | | | **for** *(all neighbors* **y** *of* **x***)* **do**
   | | | | **if** *(***y** *is simple)* **then**
   | | | | | insert(**y**, maxHeap) with AOF(**y**) as the sorting key for insertion;

Table 4.3 The topological classification of Malandain et al. (1993)

$\bar{C}$	C^*	Type
0	Any	Interior point
Any	0	Isolated point
1	1	Border (simple) point
1	2	Curve point
1	>2	Curves junction
2	1	Surface point
2	>2	Surface-curve(s) junction
>2	1	Surfaces junction
>2	≥ 2	Surfaces-curves junction

The labeling of the 3D medial set is more subtle and relies on the classification of Malandain et al. (1993). Specifically, the numbers C^* and $\bar{C}$, described in Section 4.3.2, can be used to classify curve points, surface points, border points and junction points (see Table 4.3).

However, certain junction points can be misclassified as surface points when special configurations of voxels occur. These points are relabeled in a second step using a new definition for simple surfaces described in Malandain et al. (1993). The procedure is as follows.

Let **x** be a surface point ($\bar{C} = 2$ and $C^* = 1$). Let $A_\mathbf{x}$ and $B_\mathbf{x}$ be the two connected components of $\bar{O} \cap N_{18}$ 6-adjacent to **x**. Two surface points **x** and **y** are in an equivalence relation if there exists a 26-path $\mathbf{x}_0, \mathbf{x}_1, ..., \mathbf{x}_i, ..., \mathbf{x}_n$ with $\mathbf{x}_0 = \mathbf{x}$ and $\mathbf{x}_n = \mathbf{y}$ such that for $i \in [0, ..., n-1]$, $(A_{\mathbf{x}_i} \cap A_{\mathbf{x}_{i+1}} \neq \emptyset$ and $B_{\mathbf{x}_i} \cap B_{\mathbf{x}_{i+1}} \neq \emptyset)$ or $(A_{\mathbf{x}_i} \cap B_{\mathbf{x}_{i+1}} \neq \emptyset$ and $B_{\mathbf{x}_i} \cap C_{\mathbf{x}_{i+1}} \neq \emptyset)$. A *simple surface* is then defined as any equivalence

Fig. 4.3 The topology preserving medial axis (*top*) is extracted from the average outward flux map of Fig. 4.1. The reconstruction as the envelope of the maximal inscribed disks of the medial axis is overlaid in grey on the original shape (*bottom*). Adapted from (Dimitrov et al., 2000)

class of this equivalence relation. The 26-neighborhood of each previously classified surface point **x** is then examined and when it is not a *simple surface*, **x** is relabeled a junction point. Figure 4.4 illustrates the labeling of the 3D medial set of a cylinder as two *simple* sheets connected by a 3D digital curve through two junction points.

The same definition can be used to extract the individual simple surfaces comprising the medial set of a 3D object. The idea is to find an unmarked surface point on a medial surface and use it as a "source" to build its associated simple surface using a depth first search strategy. The next simple surface is built from the next unmarked surface point and so on, until all surface points are marked.

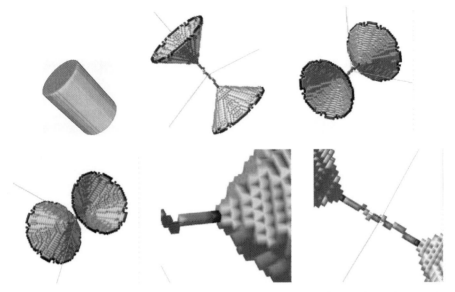

Fig. 4.4 The 3D medial set of a cylinder is labeled as border points (*blue*), surface points (*grey*), curve points (*green*) and junction points (*red*)

4.3.6 Examples

We now illustrate the use of Algorithms 1 and 2 to produce average outward flux-based medial loci with several 2D and 3D examples.

2D Medial Loci

We assume that the input is a 2D binary image where the foreground and background are identified by distinct values. The implementation then uses an exact (signed) distance function to a piecewise circular arc interpolation of the boundary, which allows for sub-pixel computations (details are presented in Dimitrov et al., 2000, 2003). Following this, Algorithms 1 and 2 are used with same average outward flux threshold for each example. Figure 4.3 (middle) shows the sub-pixel medial axis for the panther silhouette. The accuracy of the representation is illustrated in Fig. 4.3 (bottom), where the shape is reconstructed as the envelope of the maximal inscribed discs associated with each medial axis point. Figure 4.5 depicts sub-pixel medial axes for a number of other shapes.

3D Medial Loci

Next we illustrate the algorithm with both synthetic data and graphical models of 3D objects. In both cases we assume that the input is a 3D binary array. We then

Fig. 4.5 Sub-pixel medial sets for a range of 2D shapes, obtained by average outward flux-ordered thinning. We thank Pavel Dimitrov for providing his implementation of the algorithm presented in Dimitrov et al. (2003)

use the D-Euclidean distance function (Borgefors, 1984) which provides a good approximation to the true distance function and apply Algorithms 1 and 2. Once again, the only free parameter is the choice of the average outward flux threshold below which the removal of end points is blocked. For these examples, the value was selected so that approximately 25% of the points within the volume had a lower average outward flux.

Figure 4.6 depicts the 3D medial loci of a rectangular parallelepiped and a cylinder, as well as the reconstructions of the original objects by superimposing the maximal inscribed spheres at the locus of all 3D medial points. As one would expect, the medial set of the parallelepiped is comprised of planes bisecting the adjacent faces and the medial set of the cylinder consists of two cone-like structures which intersect and share a central sheet. Figure 4.7 depicts the medial loci of various graphical models. These models were originally described in the Virtual Reality Modeling Language (VRML) format. Each model was then rescaled to fit in a $128 \times 128 \times 128$ cubic lattice and was then voxelized using a level set based implementation of a surface extraction method on the cloud of 3D (discrete) surface

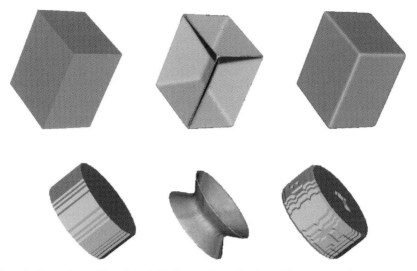

Fig. 4.6 *First column*: The original 3D objects. *Second column*: The corresponding average outward flux-based medial loci. *Third column*: The objects reconstructed from the medial loci in the second column

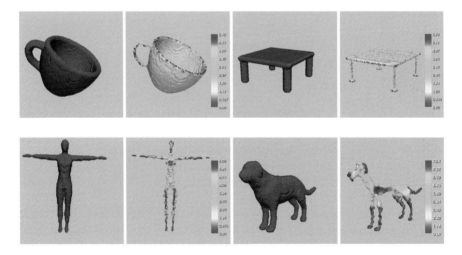

Fig. 4.7 Medial sets for a variety of 3D shapes, obtained by average outward flux-ordered thinning. The color map on the skeletons represents the radius function, with values increasing from red to blue. Results on additional models are presented in Fig. 4.10 (*middle column*)

points (Zhao et al., 2001; Savadjiev et al., 2003). The color map on the skeleton of each binary volume shown in Fig. 4.7 represents the value of the radius function at each medial point, as indicated by the associated color bars. Additional results are presented in Fig. 4.10 (middle column).

4.4 An Object Angle Approach

We now discuss a related approach to detecting medial loci as locations where the gradient of the Euclidean distance function $\nabla\phi$ is multi-valued (Shah, 2005a). This approach exploits the relationship between the gradient vector field and the object angle, using the latter as measure of salience by which to prune an initial coarse representation. It is therefore similar in spirit to the average outward flux-based algorithms discussed in Section 4.3 above, although in its implementation on a discrete lattice it does not incorporate homotopy preserving thinning.

In the following an object is assumed to be a connected open bounded set $\Omega \subset \mathbb{R}^n$ with boundary $\partial\Omega$. Its skeleton is denoted by M and its local geometry is described using the notation in Chapter 1 (Fig. 1.7, Section 2.3). In particular, at a regular medial point $\mathbf{p}$, the object angle is θ and the associated boundary pre-image points are $\mathbf{b}^{\pm 1}$. The approach exploits the property that in the complement of $\partial\Omega \cup M$, the gradient lines of ϕ are straight lines and $\|\nabla\phi\| = 1$. If $\mathbf{x}$ is a point not on $\partial\Omega \cup M$, the gradient line of ϕ passing through $\mathbf{x}$ connects it to its unique closest point on $\partial\Omega$. However, if $\mathbf{x}$ is a regular point $\mathbf{p}$ of M there are exactly two closest boundary points, $\mathbf{b}^{+1}$ and $\mathbf{b}^{-1}$, and the maximal inscribed sphere centered at $\mathbf{p}$ touches $\partial\Omega$ only at these points. We recall from Chapter 1 the relationship between the object angle and the arc-length derivative of Euclidean distance function (the radius) along the skeleton

$$\cos\theta = \frac{d\phi}{ds}. \tag{4.29}$$

We also have

$$\sin\theta = \frac{1}{2}\|D^+\phi - D^-\phi\| \tag{4.30}$$

where $D^{\pm}\phi$ denotes the directional derivatives in the directions $\overrightarrow{\mathbf{b}^{\pm 1}\mathbf{p}}$.

The above formula can be used to extend the notion of $\sin\theta$ to all points in Ω. In particular, for points in the complement of $\partial\Omega \cup M$, because $\nabla\phi = 0$ is continuously differentiable, $D^+\phi = D^-\phi$, so $\sin\theta = 0$.

The result is that the value of $\sin\theta$ can be used to detect the skeleton. This "object angle" extension to all points in Ω/M is referred to as the *grey skeleton* in Shah (2005a). It is defined everywhere except on the set J of points of M where the maximal inscribed disk has more than two points of contact with $\partial\Omega$. Under the assumption that J has co-dimension ≥ 2, the strategy is to determine M/J from $\sin\theta$ in the complement of $\partial\Omega \cup J$ and then extend it to portions of J that lie in the closure of M/J. This strategy still leaves out the non-generic cases, such as discs, circular cylinders and balls, where the closure of M/J does not contain all of J. However, since $\sin\theta$ is used only for the the purpose of extracting M, it is sufficient to use a simple approximation to define it at points in J. At a point $\mathbf{p}$ in J, pick two points, $\mathbf{b}^{+1}$ and $\mathbf{b}^{-1}$, among the set of boundary points nearest to $\mathbf{p}$, such that the object angle θ determined by the vectors $\overrightarrow{\mathbf{b}^{+1}\mathbf{p}}$ and $\overrightarrow{\mathbf{b}^{-1}\mathbf{p}}$ is the largest possible. $\sin\theta$ is then chosen as the object angle definition at $\mathbf{p}$. We note that the association between the object angle and the limiting behavior of the average outward flux of $\nabla\phi$ through a shrinking disc developed in Section 4.2.3 is a more general construction since it applies to all generic cases of medial loci. A possible alternative is to integrate the Laplacian of ϕ in the sense of distributions.

The numerical procedure for estimating θ at each point $\mathbf{x} \in \Omega$ relies on the determination of the gradient directions $\overrightarrow{\mathbf{b}^{+1}\mathbf{x}}$ and $\overrightarrow{\mathbf{b}^{-1}\mathbf{x}}$. These are the directions in which the directional derivative of ϕ is maximum and equals 1. We compute the directional derivatives of ϕ in all *inward* radial directions at every point $\mathbf{x}$ and determine the directions in which it is nearly equal to 1. On a discrete grid there are only finitely many radial directions available depending upon the size of the neighborhood N, and this determines the numerical accuracy of estimating θ. The key point is that since the gradient lines are straight lines and $||\nabla\phi|| = 1$, fairly large neighborhoods can be used. If there are exactly two rays from $\mathbf{x}$ along which the directional derivative of ϕ approximately equals 1, then that point must be on the smooth part of the skeleton and the angle between the two rays may be used to calculate the value of $\sin\theta$ at that point. The presence of more than two such gradient directions at a point $\mathbf{x}$ indicates that it is in J and the value of $\sin\theta$ there is determined as indicated above. Finally, if there is a single gradient direction at a point $\mathbf{x}$, $\sin\theta$ is set to 0 and $\mathbf{x}$ is not in M. This leads to Algorithm 3 for obtaining the object angle extension $\sin\theta$.

Algorithm 3: Object Angle Extension

Data : Object Ω.

Result : (2D or 3D) Object Angle Extension.

1. Compute the Euclidean distance transform ϕ of the object Ω.

2. **foreach** $\mathbf{x} \in \Omega$ **do**
   Scan the boundary $\partial\Gamma$ of a neighborhood Γ of $\mathbf{x}$.
   foreach $\mathbf{y} \in \partial\Gamma$ **do**
   Calculate the derivative of ϕ in the direction $\overrightarrow{\mathbf{yx}}$

$$D_{\mathbf{y}}(\phi) = \frac{\phi(\mathbf{x}) - \phi(\mathbf{y})}{|\overrightarrow{\mathbf{yx}}|}. \tag{4.31}$$

   Determine the local maxima of $D_{\mathbf{y}}(\phi)$ along $\partial\Gamma$. Select those which are approximately equal to 1, within the tolerance determined by the size of Γ. Label them as $D_{\mathbf{y}_i}$, where $i = 1, ..., N$ and N is the number of maxima.

   if $N = 1$ **then**
   | Set $\sin\theta = 0$ at $\mathbf{x}$.
   if $N = 2$ **then**
   | Set $\sin\theta = \frac{1}{2}||D_{\mathbf{y}_1} - D_{\mathbf{y}_2}||$ at $\mathbf{x}$.
   else
   | Set $\sin\theta = \max_{i,j,i\neq j} \frac{1}{2}||D_{\mathbf{y}_i} - D_{\mathbf{y}_j}||$ at $\mathbf{x}$.

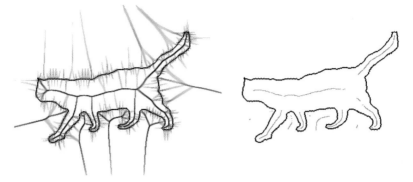

Fig. 4.8 *Left*: The object angle extension $\sin \theta$ with darker shades depicting larger θ. *Right*: The locus of positions with $\theta > 75°$ obtained by thresholding the object angle image

The use of the above object angle extension algorithm is illustrated in Fig. 4.8 (left) for the outline of a cat shape, with darker shades corresponding to higher values for $\sin \theta$, for both the object Ω and its complement. These results were obtained using an adaptive square neighborhood Γ on a discrete lattice. It was chosen to be as large as possible, but no greater than 15×15, provided that all points in Γ were also in the object Ω (or in its complement in the case of the external medial locus). These results can be compared qualitatively with the average outward flux computation of Fig. 4.1 (bottom), although in the latter example the results are shown only within the object. Figure 4.8 (right) depicts those medial loci obtained by thresholding and retaining only points $\mathbf{x}$ with $\theta > 75°$. As with the average outward flux measure a procedure has to be devised to process the object angle extension image to obtain medial loci. The procedure suggested in Shah (2005a) is to carry out a type of reconstruction based on two object angle thresholds, as follows.

1. Choose two thresholds $\overline{\theta}$ and $\underline{\theta}$ for angle θ with $\overline{\theta} > \underline{\theta}$
2. Threshold the object angle image at $\overline{\theta}$ to retain only those loci with large enough object angle
3. Extend the branches of the thresholded skeleton in the direction of increasing ϕ provided that ϕ remains greater than $\underline{\theta}$

These ideas are quite similar in spirit to the topological reconstruction process developed by Malandain and Fernandez-Vidal (1998), where an initial set of fragmented loci are glued together to produce a skeleton that is topologically correct, in that it preserves homotopy type. Both the above approaches face some numerical challenges due to the difficulty of following directions on a discrete lattice. These approaches are distinct in spirit to the homotopy preserving thinning approach of Algorithm 2, where digital points are removed in a manner inversely proportional to object angle, while taking care to preserve homotopy type. An alternative class of methods for obtaining salient medial loci from an object angle image is based on the construction and minimization of appropriate energy functionals in 2D and 3D (Shah, 2005a). This latter approach has also been applied to the segmentation of features such as protrusions and indentations.

4.4.1 Examples

We now present both 2D and 3D examples of the object angle approach developed above, and we compare them to those from the average outward flux method. As long as the threshold $\underline{\theta}$ is sufficiently low, the resulting skeleton is a connected set. The procedure is relatively insensitive to the choice of the thresholds $\overline{\theta}$ and $\underline{\theta}$ except around certain critical values.

2D Medial Loci

Figure 4.9 shows examples of 2D medial loci obtaining using the object angle extension approach of Algorithm 3 followed by the reconstruction process from thresholds, for 3 distinct objects. In these examples the reconstruction thresholds were

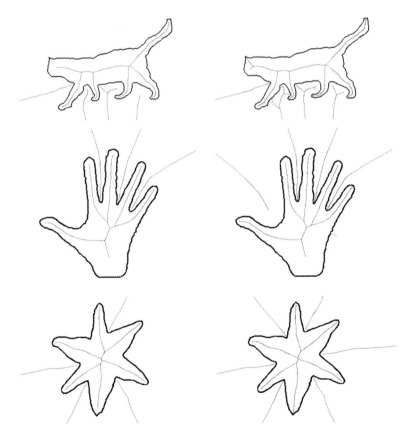

Fig. 4.9 External and internal 2D medial loci obtaining using the object angle extension approach, followed by reconstruction from thresholds. *Left*: $\overline{\theta} = 75°$, $\underline{\theta} = 10°$ *Right*: $\overline{\theta} = 60°$, $\underline{\theta} = 10°$

fixed to be $\underline{\theta} = 10°$ and $\overline{\theta} = 75°, 60°$. Qualitatively these results are comparable to those obtained using the average outward flux based method (Fig. 4.5).

3D Medial Loci

The robust numerical implementation of the reconstruction process from thresholds is a significant challenge in 3D, in part due to the difficulty of following the direction of increasing ϕ on a discrete lattice. The object angle extension measure however,

Fig. 4.10 A comparison of medial sets for various 3D shapes (*left column*), obtained by average outward flux-ordered thinning (*middle column*), and object angle extension order-ordered thinning (*right column*). The results are in qualitative agreement

as provided by Algorithm 3, can be used in much the same fashion as the average outward flux measure to guide a homotopy preserving thinning process. In the right column of Fig. 4.10 we show examples of 3D medial loci obtained using this idea, with those obtained using average outward flux-ordered thinning shown in the middle column for comparison. The results are in qualitative agreement.

4.5 Discussion and Conclusion

In this chapter we have shown how considerations from classical mechanics and geometric optics lead to fresh insights into the computation of medial loci. We have considered the Hamiltonian system associated with the vector field underlying Blum's grassfire flow, i.e., the gradient, $\nabla \phi$ of the Euclidean distance function ϕ to the object boundary $\partial \Omega$. We have shown that an average outward flux measure computed over shrinking neighborhoods provides an effective way for detecting medial loci, which is a special case of the medial integrals developed in Chapter 3. For the case of shrinking disks this measure also reveals the object angle, so an explicit mapping from medial points back to the object boundary is provided. We have also discussed a related approach by which medial loci are detected via an extension of the object angle to Ω, along with a procedure to determine locations where $\nabla \phi$ is multi-valued.

The approaches to detecting singularities of the grassfire flow developed in this chapter have the advantage that their development is rooted in principles from classical mechanics. Furthermore, the numerical methods based on average outward flux and the object angle are well supported an by extended divergence theorem that applies to vector fields having a discontinuity along a locus of points and also have the property that the algorithms are essentially the same in 2D and 3D. In their numerical implementation the techniques are further refined to obtain topologically correct medial loci on a discrete lattice.

A potential weakness faced by the approaches in this chapter that numerical errors due to approximations on a discrete lattice do occur. For example, in their current implementation in 3D, average outward flux based skeletons use an approximate digital distance transform, as well as a discrete lattice for thinning. Thus ϕ is imprecise, and the detection of medial loci is only as accurate as the grid sampling. In 2D some of these limitations have been overcome by developing a notion of subpixel medial loci, where the boundary is represented in a continuous fashion and an exact Euclidean distance function is used (Dimitrov et al., 2003; Dimitrov, 2003). In 3D more recent work has extended the average outward flux computation by using computational geometry techniques to work with a polyhedral mesh as $\partial \Omega$ and to use "exact" Euclidean distance functions to sample points on a shrinking sphere densely and continuously. This in turn leads to a coarse to fine algorithm for computation where the grid is iteratively refined at regions (voxels) through which the medial locus passes (Stolpner and Siddiqi, 2006). Preliminary evidence suggests that the medial loci thus obtained have the regularity properties to be expected of the Blum skeleton, which are discussed in Chapter 2. Thus, these refined algorithms have the potential to lead to a computational instantiation of the medial to boundary theory developed in Chapter 3.

Acknowledgements We are grateful to James Damon, Pavel Dimitrov, Carlos Phillips, Allen Tannenbaum and Steven Zucker for collaborations towards the development of the skeletonization techniques reviewed in this article. We thank Peter Savadjiev and Juan Zhang for their help with the numerical examples. This work was supported by grants from the Natural Sciences and Engineering Research Council of Canada, the Canadian Foundation for Innovation and FQRNT Québec.

Chapter 5
Discrete Skeletons from Distance Transforms in 2D and 3D

Gunilla Borgefors, Ingela Nyström, and Gabriella Sanniti di Baja

Abstract In this chapter we present discrete methods to compute the digital skeleton of shapes in 2D and 3D images. In 2D, the skeleton is a set of curves, while in 3D it is a set of surfaces and curves, the surface skeleton, or a set of curves, the curve skeleton. A general scheme could, in principle, be followed for both 2D and 3D discrete skeletonization. However, we will describe one approach for 2D skeletonization, mainly based on marking, in the distance transform, the shape elements that should be assigned to the skeleton, and another approach for 3D skeletonization, mainly based on iterated element removal. In both cases, the distance transform of the image will play a key role to obtain skeletons reflecting important shape features such as symmetry, elongation, and width.

5.1 Introduction

We understand the MAT to begin from a binary image in $\mathbb{R}^2$ or $\mathbb{R}^3$ and transform it into the Blum medial axis of that region, including the connections within simple regions of the axis and at branch points as well as the radius value at each point. The binary image is provided discretely as a set of pixels (in 2D) or voxels (in 3D), which we will call "image elements" or just "elements". Thus it is natural to consider the discrete MAT as a transformation from that discrete binary image to a

G. Borgefors
Centre for Image Analysis, Swedish University of Agricultural Sciences, Sweden
e-mail: gunilla@cb.uu.se

I. Nyström
Centre for Image Analysis, Uppsala University, Sweden
e-mail: ingela@cb.uu.se

G. Sanniti di Baja
Institute of Cybernetics "E. Caianiello," C.N.R., Pozzuoli, Naples, Italy
e-mail: gsdb@cib.na.cnr.it

K. Siddiqi and S. Pizer (eds.) *Medial Representations – Mathematics, Algorithms and Applications*.
© Springer Science + Business Media B.V. 2008

discrete list of image elements on the medial locus, the inter-element connections on the medial locus, and the radius values at these elements. The literature is replete with methods that try to compute this discrete MAT. For these approaches there is no need to convert discrete images to continuous structures, or to convert continuous results back to discrete data.

These methods divide into two categories. In the first category one uses techniques of discrete erosion or the fitting of discrete disks or balls into the binary image. The methods of mathematical morphology (Serra, 1982) are especially apt for these operations. The difficulty is that the connective structure of the medial locus may be hard to compute in certain situations.

In the second category one computes a discrete distance function from the discrete boundary of the binary image, and then one computes the medial axis from the ridge of this distance function. This approach, making use of ideas of discrete geometry and topology, is the focus of this chapter. Computing skeletons from distance transforms is convenient since it is easy to fulfill two of the important properties that a skeleton is often expected to satisfy, namely reversibility and centrality. Furthermore, distance based methods are computationally convenient since distance labeling makes it possible to simultaneously discriminate all the borders that would successively characterize the object during iterated border identification and removal.

An example illustrating the discrete computation of skeletons in digital images is given in Fig. 5.1. One option is to work within the pixels or voxels and produce skeletons formed by a list of such elements. In this approach we face problems specific to the digital space. Some of these problems are relevant for skeletonization. A well known example is the need of using a different connectivity type for the object and for its complement (the background). Using the same connectivity type would create a topological paradox: a closed curve/surface would not divide the background into disjoint parts, or the background would be divided into disjoint parts by an open curve/surface. Another problem relevant for digital skeletonization is the impossibility of giving a precise solution to apparently easy tasks, such as identifying the middle point in a segment. If the segment consists of an even number of elements, any of two elements—or both!—will be found as the middle point if we want a discrete solution, i.e., a solution consisting of image elements.

Depending on problem domain, the discrete skeleton is expected to reflect a number of features characterizing the represented shape, such as symmetry, elongation, width, and curvature. We regard a skeleton as satisfactory if it is thin (to actually have lower dimensionality) and centered within the shape (to reflect shape symmetry and curvature) and if its elements are labeled with their distance from the

Fig. 5.1 Discrete skeletons. *Left*: a rectangle and its skeleton. *Middle*: the surface skeleton of a box. *Right*: the curve skeleton of a box

initial background (medial radius). Moreover, we aim for skeletons homotopic to the represented shapes and such that each perceptually meaningful portion of the shape has a corresponding entity in the skeleton. For example, a branch should be found in correspondence with any elongated part or shape protrusion. Last, but not least, skeletonization should be reversible, to consider the skeleton as a really faithful representation of the shape. Generally, both in the digital and in the continuous space, reversibility is not possible when shape dimensionality is reduced more than one dimension, e.g., when a 3D shape is reduced to a set of curves. In this case, reversibility should be interpreted as similarity: the skeleton should resemble the shape enough so that it is possible to use it in place of the shape itself. Unfortunately, in digital space, reversibility is not compatible with the skeleton being thin, wherever the shape is an even number of elements wide. This problem can be solved in two possible ways. One possibility is to have a skeleton that is partly two elements thick but such that full recovery is possible. Alternatively, the skeleton can be thin, but then some elements along the border of the shape cannot be recovered. Which of the two solutions to adopt depends on the specific task. Here, we favor the latter. With all the above features, the discrete skeleton is a promising tool for an increasing number of applications.

Discrete skeletonization algorithms can be based on sequential or parallel operations and can be implemented on different architectures. This creates a huge variety of possible ways to describe skeletonization. The most common and intuitive approach to skeletonization is based on iterated element removal. Given an input image, skeletonization changes shape elements to background elements, until a subset of the shape with the desired properties is obtained. Of course, topology preserving removal operations should be applied only to border elements, border after border, to obtain a homotopic skeleton, centered within the shape. This approach is time-consuming if it is implemented on sequential conventional computers by repeatedly scanning the image. However, special architectures could be used, or smart implementations could be written to avoid repeated, time-consuming scans of the image. In our opinion, the main problem with iterative methods is the need to find good criteria allowing the skeleton to keep elements necessary to represent perceptually meaningful parts of the shape. For example, once they are found, skeleton branches should not be shortened, as this may cause diminished representativeness of the skeleton. Most of the algorithms in the literature use an end point detection criterion to prevent shortening of skeleton branches, but the quality of the results is questionable. This detection criterion is based on a property (number of neighbors in the skeleton) that characterizes end points only *after* the skeleton has been obtained, but is not necessarily fulfilled during skeletonization. The solution that we offer here, when describing iterative skeletonization, is to find alternative ways, based on distance information, to identify those elements that will be end points in the final skeleton.

Another approach to skeletonization is based on marking—directly on the shape or, better, on the distance transform of the image—the elements that definitely should be part of the skeleton, i.e., the most centrally located elements. After this marking, it is possible to identify and mark further elements which, per se, would have no reason to be assigned to the skeleton but whose assignment to the skeleton becomes evident once elements in their proximity have been marked as skeleton elements. Distance transform based skeletonization has the advantage of producing symmetric, at most two-elements thick, fully reversible skeletons. Distance transform based skeletonization should be followed by iterative thinning to pass from a two-element thick skeleton to a thin skeleton. Different distance transforms can be used, including the Euclidean distance transform. Our preference, however, is weighted distance transforms, where local distances (between neighbors) are integer (and generally small) numbers. A weighted distance transform, where distance values are not very dependent on the rotation of the shape, is preferred when the skeleton is used for quantitative analysis and to make the skeleton itself as stable as possible under rotation. For specific applications and, in general when working with elongated thin shapes, simple distances, like the well known city-block distance in 2D or its equivalent in 3D, can be used.

The literature on discrete skeletonization is very rich, and a review would require a whole book. In this chapter, we limit ourselves to describe some of the work on 2D and 3D skeletonization where we ourselves have been involved. Even so, a review of all our work would be very long. Thus, we describe only one algorithm for 2D skeletonization and one for 3D skeletonization. Both approaches (iterative skeletonization and distance transform based skeletonization) will be used. In the 2D case, we will mainly use distance transform based skeletonization and reserve iterative skeletonization only for final thinning. In the 3D case, we will mainly follow the iterative approach but will use distance information to detect the voxels most centrally located within the shape, as these should not be removed. Thus, in both 2D and 3D, distance transforms will play a key role for skeletonization. We describe the algorithms in detail that we deem enough for interested readers to implement them. The descriptions are intended to be easily understandable; for this reason, the given algorithms are far from being optimized.

5.2 Definitions and Notions

In this section, we will list the definitions and notions used in this chapter. References are given where the definitions are used.

Consider a digital image I, whose elements $\mathbf{x} \in \mathbb{Z}^n$, $n = 2, 3$. Elements are pixels for $n = 2$ and voxels for $n = 3$. The original images, from which skeletons are computed, are binary, i.e., $I(\mathbf{x}) \in \{0, 1\}$.

Definition 5.2.1. A *shape* S_o is what is called a binary image in Chapter 1: the subset of the elements with value $I(\mathbf{x}) = 1$. The complement of S_o, $\overline{S}_o$, is called the *background*. In 2D, S_o is 8-connected and $\overline{S}_o$ 4-connected. In 3D, S_o is 26-connected and $\overline{S}_o$ 6-connected.

There are no limitations on the topology of S_o or $\overline{S}_o$, i.e., on the number of components, holes, etc. For the sake of simplicity, we will assume that S_o consists of a single component and, in the 3D case, that S_o is a solid object, i.e., it can include tunnels, but not cavities.

Definition 5.2.2. As defined in Chapter 4, a *simple-point* is an element of S_o that can be set to $\overline{S}_o$ without changing the topology of S_o.

Definition 5.2.3. In 3D, the *face-neighbors* of $\mathbf{x}$ are denoted $N_f(\mathbf{x})$. In 2D and 3D, the *edge-neighbors* are denoted $N_e(\mathbf{x})$, and the *vertex-neighbors* are denoted $N_v(\mathbf{x})$. In 2D, the *"knight-neighbors,"* i.e., those reached by the knight's step in chess, are denoted $N_k(\mathbf{x})$. Both in 2D and in 3D, $N(\mathbf{x})$ denotes all elements that are neighbors of $\mathbf{x}$ in the specific metric.

Definition 5.2.4. The digital *distance* between two points $\mathbf{x}$ and $\mathbf{y}$ in $\mathbb{Z}^n$ is the length of the shortest path connecting $\mathbf{x}$ to $\mathbf{y}$, where the path consists of steps between close neighbors.

The distance thus depends on the chosen neighborhood relation and the definition of path length.

Definition 5.2.5. As discussed in Chapter 1, the *distance transform* (DT) is an image where each element in S_o has the value that is the shortest distance from that element to $\overline{S}_o$. The distance can be measured in any metric.

From the definition, it follows that, in the DT, $I(\mathbf{x}) > 0$ if $\mathbf{x} \in S_o$ and $I(\mathbf{x}) = 0$ if $\mathbf{x} \in \overline{S}_o$. We will only use integer valued DTs.

Definition 5.2.6. The *reverse* distance transform (rDT) is an image computed from a set of seed elements with seed values s_i. Each element in the rDT has the value $\max_i[0, s_i - d]$, where d is the distance between the seed element and the rDT element.

Definition 5.2.7. In 2D a *center of maximal disk* (CMD) is the center element in a maximal disk (see Section 2), labeled with the radius of the disk. In 3D a *center of maximal ball* (CMB) is defined equivalently. If the dimension is arbitrary, we will refer to the *maximal object* and *center of maximal object* (CMO), respectively.

This definition is valid for any metric, any dimension, and in both $\mathbb{Z}$ and $\mathbb{R}$.

Remark 5.2.1. A maximal object in $\mathbb{Z}^n$ does not have all the properties of a maximal object in $\mathbb{R}^n$. One important difference is that, due to digitization effects, in $\mathbb{Z}^n$ the maximal object is not guaranteed to touch the border of S_o in more than one place, which is always true in $\mathbb{R}^n$ except at medial end points. This, in turn, means that defining medial axis elements in $\mathbb{Z}^n$ cannot be done by finding elements equidistant to more than one point/segment of the border of S_o, which is a common approach in $\mathbb{R}^n$.

Remark 5.2.2. The union of the maximal objects is equivalent to the original shape. This means that if the seed elements consist of the set of CMOs, then S_o is equal to the set of rDT elements with values larger than zero.

The skeleton of a shape has been defined and named in different ways in the literature. The various definitions are based on a number of properties. We use

Definition 5.2.8. In 2D and 3D, the *skeleton* S_s of S_o is a subset of S_o with the following properties.

(a) S_s has the same topology as S_o.
(b) S_s is centered in S_o.
(c) S_s is one-element thick.
(d) S_s allows recovery of S_o.

Two of these properties are contradictory. A one-element thick skeleton does not usually allow complete recovery, meaning that some border elements of S_o will be lost if a thin S_s is used as the set of seeds elements for the rDT. If S_o is recovered, except for some non-significant (with respect to the application) border elements, we will denote this *recovery*, and reserve *complete recovery* for the case where exactly all elements are recovered.

In 2D, S_s is a set of curves. In 3D, S_s (the *surface skeleton*) consists of surfaces and curves; if S_o is a solid object, we can reduce it to a set of curves S_c (the *curve skeleton*).

Definition 5.2.9. In 3D, the *curve skeleton* S_c of S_o is a subset of S_o with the following properties.

(a) S_c has the same topology as S_o.
(b) S_c is centered in S_o.
(c) S_c is one-element thick.

Differently from S_s, the 3D curve skeleton does not have the recovery property. However, it is generally required that S_c is "similar" to S_o.

Similarity can be interpreted in many different ways, but some criterion is necessary; otherwise a set S_o without holes or tunnels can be reduced to a single element and that element still be called a curve skeleton. We interpret similarity in the sense that enough shape information must be retained by the curve skeleton so that S_c adequately represents S_o. In this chapter pruning a skeleton, which is defined in Chapter 1 as removing elements with low significance according to some significance criterion, should fulfill the criteria in Definition 5.2.8 (respectively, Definition 5.2.9), except that non-significant parts of the shape will not be recovered (represented).

Elements in $\overline{S}_o$, $\overline{S}_s$, and $\overline{S}_c$ have value zero. Elements in S_o, S_s, and S_c, assume different integer values depending on the stage of the computations.

5.3 Distance Transforms

In a distance transform, the elements of a shape S_o are marked by their distance in such a way that the results of all steps of an iterative skeletonization algorithm are visible simultaneously. The locus of "ridges" in the DT forms the medial axis, which is analogous to the locus where wavefronts meet in the grassfire transformation. Hence, the DT is a very useful structure for skeleton extraction. Distance transforms are computed efficiently using marching approaches that propagate the boundary, which is set to a zero distance. Each step of the march uses a neighborhood of a particular size.

Besides computing skeletons, DTs are useful for many other shape related tasks, e.g., segmentation, watershed, Voronoi tessellation, and compression, as described in Borgefors (1994). By also propagating other information together with the distances, we get the *salience* DTs (Rosin and West, 1995). Salience DTs are an earlier and digital version of the Fast Marching method (Sethian, 1996). A good underlying concept for all DTs is the one proposed in Definition 5.2.4, from Yamashita and Ibaraki (1986). Much has been written since the concept was first defined in Rosenfeld and Pfaltz (1966). For DTs in 2D, see, e.g., (Rosenfeld and Pfaltz, 1968; Borgefors, 1986; Vossepoel, 1988; Borgefors, 1991; Das and Chatterji, 1990; Thiel and Montanvert, 1992). For DTs in 3D, see, e.g. (Verwer, 1991; Borgefors, 1996; Svensson and Borgefors, 2002a,b).

5.3.1 2D Distance Transforms

In 2D, we will consider neighborhoods up to 5×5. As we should use isotropic distances, we have three neighborhood relations, N_e, N_v, and N_k, and thus three different steps in the graph. The distance between edge-neighbors is a, between vertex-neighbors b, and between knight-neighbors c. These step lengths are called *local distances*, and the resulting DTs are called *weighted* distance transforms.

All combinations of local distances a, b, and c result in metrics, but not all choices of a, b, and c will result in suitable DTs. DTs should have the following property.

Definition 5.3.1. Consider two image elements that can be connected by a straight line, i.e., by using only one type and direction of local step. If that line defines the distance between the elements, i.e., is a minimal path, then the resulting DT is *semi-regular*. If there are no other minimal paths, then the DT is *regular*.

In a 3×3 neighborhood, a DT is regular iff $a < b < 2a$. In a 5×5 neighborhood, a DT is regular iff $2a < c < a + b$ and $3b < 2c$.

The local distances can be optimized using different criteria. The purpose is always to achieve as rotation independent DTs as possible. This is usually expressed as being as close as possible to the digital Euclidean distance, with the "error" being minimized according to some norm over some structure. Independently of the norm used, the practical results become very similar. Good regular integer local distances are, in a 3×3 neighborhood, $a = 3$, $b = 4$, denoted $\langle 3,4 \rangle$, with a rotation dependence of $\pm 6.8\%$ and, in a 5×5 neighborhood, $a = 5$, $b = 7$, $c = 11$ ($\langle 5,7,11 \rangle$) with a rotation dependence of $\pm 1.9\%$. Rotation dependence is the maximum difference of the distance values for the same shape when it is placed in different orientations. The difference from the Euclidean DT can be larger, as the DT can systematically

Fig. 5.2 A common traffic sign in Sweden: "Moose warning"

Fig. 5.3 The $\langle 3,4 \rangle$ distance transform of the moose

over- or underestimate the Euclidean distance. The circle (i.e., all pixels equidistant from a center pixel) of $\langle 3,4 \rangle$ is an octagon and of $\langle 5,7,11 \rangle$ a hexadecagon (16 sides). We use $\langle 3,4 \rangle$ in our examples.

As a 2D running(!) example, we will use the moose in Fig. 5.2. The $\langle 3,4 \rangle$ DT of our moose is found in Fig. 5.3.

Distance transforms that have been popular since the 1960s are city-block (or D^4) and chessboard (or D^8) DTs. In city-block, $a = 1$ and only edge-neighbors are considered. In chessboard, $a = b = 1$. These DTs are a "degenerate" case of weighted DTs and are both semi-regular. To be consistent, we will denote them $\langle 1, \infty \rangle$ and $\langle 1, 1 \rangle$, respectively. Their circles are diamonds and squares, respectively, and the rotation dependencies are 20.0% and 20.7%, respectively. These DTs are the least complex to compute, and they need the simplest algorithms when using them in various contexts. For elongated thin shapes, $\langle 1, \infty \rangle$ is often a good choice.

5.3.2 3D Distance Transforms

In 3D we will only consider the $3 \times 3 \times 3$ neighborhood with the distance between face-neighbors denoted a, between edge-neighbors b, and between vertex-neighbors c. A weighted DT in 3D is regular iff $a < b < 2a$ and $b < c < 2b/3$. Good regular integer local distances are $a = 3$, $b = 4$, $c = 5$ ($\langle 3,4,5 \rangle$) with a rotation dependence of $\pm 10.0\%$. The sphere of $\langle 3,4,5 \rangle$ is a 24-sided polyhedron (one of the Catalan solids).

The generalization of the city-block DT to 3D is D^6, i.e., where only face-neighbors are considered, $a = 1$ or $\langle 1, \infty, \infty \rangle$. The generalization of the chessboard DT is D^{26} where $a = b = c = 1$ or $\langle 1,1,1 \rangle$. Both $\langle 1, \infty, \infty \rangle$ and $\langle 1,1,1 \rangle$ are semi-regular. They have rotation dependencies of $\pm 63.4\%$ and $\pm 36.6\%$, respectively. Their spheres are the octahedron and the cube, respectively. Here, we will only discuss skeletons based on $\langle 1, \infty, \infty \rangle$ and $\langle 1,1,1 \rangle$. Skeletons based on $\langle 1, \infty, \infty \rangle$ are suited for thin elongated shapes.

```
Forward pass:
    for i = 1, 2, ..., lines do
        for j = 1, 2, ..., columns do
            If I₀(x) = 1
                I₁(x) = min_{z∈f_mask}{I₀(x+z)+w(z)}

Backward pass:
    for i = lines, lines−1, ..., 1 do
        for j = columns, columns-1, ..., 1 do
            I₂(x) = min_{z∈b_mask}{I₁(x+z)+w(z)}
```

Fig. 5.4 Pseudo-code for computing weighted DTs in 2D. The original binary image is I_0. The co-ordinates of $\mathbf{z}$ are relative to $\mathbf{x}$. The f_mask consists of all pixels $\mathbf{z} \in N(\mathbf{x})$ that have already been visited in the forward scan, but not $\mathbf{x}$ itself. The b_mask consists of all pixels $\mathbf{z} \in N(\mathbf{x})$ that have already been visited in the backward scan, together with $\mathbf{x}$ itself (i.e., including $\mathbf{z} = \mathbf{0}$). The weight w is the appropriate value from $\langle a,b,c \rangle$, depending on the neighborhood relation between $\mathbf{x}$ and $\mathbf{z}$, and $w(\mathbf{x}) = 0$

In both 2D and 3D, weighted DTs are computed in $O(n)$ time using two raster scans through the image, where n is the number of elements in I. See Fig. 5.4 for the 2D pseudo-code. The 3D code is the same, except scanning is done in one more dimension.

5.3.3 Euclidean Distance Transforms

In addition to weighted distance transforms, much effort has been spent on constructing efficient algorithms producing the error-free digital Euclidean distance transform (EDT). The element values are vectors, but after computation, they are usually set to the square of the distance values. The "neighborhood" that must be considered is, in principle, equal to the whole image, as there are no limitations on the length and the direction of the steps in the minimal paths, refer to Definition 5.2.4. See, e.g. (Danielsson, 1980; Ye, 1988; Ragnemalm, 1993; Maurer et al., 2003).

Note that the EDT is as sensitive to digitization effects as all other digital DTs. Since digital shape cannot be perfectly preserved during rotation, the EDT is not completely rotation independent. This, and noise, will generate spurious effects that can easily be of the same severity as similar imperfections generated by using more rotation dependent metrics.

5.4 Centers of Maximal Disks/Balls

When computing the skeleton S_s of a shape S_o, per Definition 5.2.8d we demand that S_o can be recovered from the skeleton together with the value of the radius r, which comes from the value of the distance function at the CMO. As S_o is the union of its maximal objects, this property is guaranteed if the CMOs are enclosed in S_s. In addition to ensuring complete recovery, the set of CMOs also fulfills Definition 5.2.8b. Since the set of CMOs is likely to be two elements thick at parts, inclusion of all CMOs in the skeleton would prevent the skeleton from having property (c). The skeleton cannot have the two properties (c) and (d) at the same time. In this section we will describe how the CMOs are computed for different distance transforms.

5.4.1 Centers of Maximal Disks

For $\langle 1, \infty \rangle$ and $\langle 1, 1 \rangle$, the CMDs are equivalent to the local maxima in the DT, i.e., $\mathbf{x}$ is a CMD if the following condition is satisfied

$$I(\mathbf{x}) \geq I(\mathbf{y}), \ \mathbf{y} \in N(\mathbf{x}) \tag{5.1}$$

where $N(\mathbf{x})$ is the appropriate neighborhood of $\mathbf{x}$, i.e., the same as that used when computing the DT. For $\langle 1, \infty \rangle$, only N_e is thus considered.

For weighted DTs $\langle a, b \rangle$ in the 3×3 neighborhood, the smallest distance between neighbors is larger than one, since DT values are linear combinations with positive integer coefficients of a and b. In weighted DTs, a pixel can be a CMD even if it has neighbors with higher values. This must be taken into consideration; otherwise some CMDs will be missed (Arcelli and Sanniti di Baja, 1988a,b; Borgefors et al., 1990). CMDs can be detected correctly by adding a or b to $I(\mathbf{x})$ when comparing $I(\mathbf{x})$ with the neighboring DT values, at least for DT values over a certain threshold T, with T monotonic with a. The condition used is

$$\begin{aligned} I(\mathbf{x}) + a > I(\mathbf{y}), \ \mathbf{y} \in N_e(\mathbf{x}) \ \text{and} \\ I(\mathbf{x}) + b > I(\mathbf{y}), \ \mathbf{y} \in N_v(\mathbf{x}). \end{aligned} \tag{5.2}$$

For $I(\mathbf{x}) \geq T$, $\mathbf{x}$ is a CMD if condition (5.2) is satisfied. For $I(\mathbf{x}) < T$, condition (5.2) can be used, provided that some DT values are replaced by "equivalent labels." For $\langle 3, 4 \rangle, T = 7$ and to make condition (5.2) regular, the DT values 3 and 6 must be substituted by 1 and 5, respectively. Thereafter, condition (5.2) can be used for *all* DT values $I(\mathbf{x})$. In general, the equivalent labels are linked to non-occurring values in the DT. In $\langle 3, 4 \rangle$, the values 1, 2, and 5 do not occur. A rule that can be used for all $\langle a, b \rangle$ is: Replace each value by the smallest non-occurring lower value before checking condition (5.2). To exemplify further, for $\langle 5, 7 \rangle$ we get $T = 25$. Values $1, 2, 3, 4, 6, 8, 9, 11, 13, 16, 18, 23$ do not occur, so $5 := 1, 7 := 6, 10 := 8, 12 := 11, 14 := 13, 17 := 16, 19 := 18$, and $24 := 23$.

The $\langle 3, 4 \rangle$ CMDs of the moose are found in Fig. 5.5.

For weighted DTs $\langle a, b, c \rangle$ in the 5×5 neighborhood, computing CMDs is more complex (Borgefors, 1993). Also knight-neighbors should be taken into account. For large enough values, condition (5.2) can be used if adding

$$I(\mathbf{x}) + c > I(\mathbf{y}), \ \mathbf{y} \in N_k(\mathbf{x}),$$

Fig. 5.5 The set of $\langle 3,4 \rangle$ CMDs (*black*) in the moose (*grey*)

Fig. 5.6 The reduced set of $\langle 3,4 \rangle$ CMDs (*black*) in the moose (*grey*)

but the threshold T is usually large. In fact, $T = 61$ for $\langle 5,7,11 \rangle$. Worse is the fact that for values $I(\mathbf{x}) < T$, the simple replacement rules are not sufficient. Look-up tables are necessary.

For the EDT, the CMD computation is even more complex (Borgefors et al., 1991; Remy and Thiel, 2003). Look-up tables are the only reasonable possibility. A further complication is that checking neighbors in the 3×3 neighborhood is not enough. The larger the distance values that occur are, the larger the neighborhood that must be considered to correctly identify CMDs becomes, as proven in Remy and Thiel (2003).

5.4.2 Centers of Maximal Balls

As in 2D, for $\langle 1,\infty,\infty \rangle$ and $\langle 1,1,1 \rangle$ the CMBs are equivalent to the local maxima in the DT, using the neighborhoods satisfying condition (5.1). For weighted DTs $\langle a,b,c \rangle$ in the $3 \times 3 \times 3$ neighborhood, similar rules as those in condition (5.2) are valid if equivalent labels are used. Replacements are needed similarly to the 2D case discussed in Section 5.4.1. In particular, for $\langle 3,4,5 \rangle$, once the DT is computed the value 3 should be replaced by the equivalent label 1 (i.e., $T = 4$ since only values 1 and 2 do not occur in the DT). After replacement, $\mathbf{x}$ is a CMB if the following condition is satisfied

$$
\begin{aligned}
I(\mathbf{x}) + a &> I(\mathbf{y}),\ \mathbf{y} \in N_f(\mathbf{x})\ \text{ and} \\
I(\mathbf{x}) + b &> I(\mathbf{y}),\ \mathbf{y} \in N_e(\mathbf{x})\ \text{ and} \\
I(\mathbf{x}) + c &> I(\mathbf{y}),\ \mathbf{y} \in N_v(\mathbf{x}).
\end{aligned}
\tag{5.3}
$$

We are not aware of any exhaustive investigation into CMBs for the 3D EDT. Of course, one can always build a ball for each possible EDT value and then, for all voxels in S_o, test if the balls centered on voxels in $N(\mathbf{x})$ cover the ball centered on $\mathbf{x}$, but this is not a very efficient approach.

5.4.3 Reduced Set of Centers of Maximal Objects

The definition of CMO requires that a maximal object is not covered by any other *single* maximal object in the shape. However, it often happens that a maximal object is completely covered by the union of two or more other maximal objects. Thus, if recovery of the shape is the only goal, the set of CMOs can often be considerably reduced (Borgefors and Nyström, 1997). The reduced set of CMDs for our moose is found in Fig. 5.6.

In general, the number of elements in the reduced set is not much smaller than the number of CMOs for the city-block and D^6 metrics, but for other metrics, including the Euclidean, the number of elements in the reduced set is typically about 20% of that in the set of CMOs. If a minimal description of the shape is desired, e.g., for compression purposes, computing the reduced set of CMOs is worth the effort. When computing skeletons, we prefer starting from the full sets of CMOs. One important reason is that for a symmetric shape, the set of CMOs is symmetric whereas the reduced set is usually not. Another reason is that fewer linking elements have to be added to obtain a skeleton with the same topological properties as the original shape.

5.4.4 Reverse Distance Transforms

The reverse distance transform is computed in an image with seed values that can be interpreted as radii, with all other elements having the value zero. Even though much less used than the DTs, the rDT was also defined in Rosenfeld and Pfaltz (1966). In our context, the seed elements are the CMOs, and the rDT is used to recover the original shape S_o.

For weighted DTs, the rDT can be computed by an algorithm very similar to that in Fig. 5.4. The differences are that *all* image elements are checked in the forward pass, **x** is included in the f_mask, weights w are *subtracted* instead of added, and minimization is replaced by *maximization* (see, e.g., Arcelli and Sanniti di Baja, 1988a).

For Euclidean DTs, computing the rDT is less straightforward, due to the unlimited length of the steps in minimal paths. An early and a late solution are Borgefors et al. (1991) and Coeurjolly (2003), respectively.

5.4.5 Role of Centers of Maximal Objects in Skeletons

As noted before, the set of CMOs is central in the shape S_o and allows recovery. However, since the skeleton is required to be one-element thick, some CMOs should be removed from the set of CMOs. In turn, since the skeleton is required to be homotopic to S_o, some elements should be added to the set of CMOs. Not many elements

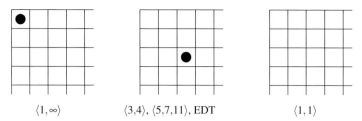

Fig. 5.7 Outermost CMD (*black dot*) generated by different DTs in a 90° corner

have to be removed or added, but the difference these make is very important for shape manipulation and analysis. In fact, it is preferable to use traceable, i.e., one-element thick and connected, skeletons whose branches correspond to identifiable parts of S_o, compared to working with the non-connected, thick set of CMOs.

The set of CMOs is more than one-element thick where the local thickness of S_o is an even number of elements. If the set is thinned to one-element thickness, then complete recovery is not possible by computing rDT from this thinned set, as some border elements in S_o will be missed. Recovery in the less strict sense is still possible.

The set of CMOs is not guaranteed to have the same topology as S_o. Readily apparent is that the set is not properly connected. Thus, a number of elements have to be added to form a skeleton. This can be done is several ways. One approach is to perform iterative skeletonization in the DT instead of in I, so that not only the CMOs but also other elements necessary to preserve topology can be kept in the skeleton. The iterations are guided by the DT values. Another approach is to examine the DT to detect both CMOs and saddle-points (defined for the 2D case in Section 5.5.1.2 and for the 3D case in Section 5.6.1) and then connect them by path growing along the DT gradient. We follow the second approach in 2D and the first in 3D, and we describe the details in Sections 5.5 and 5.6, respectively.

The set of CMOs, in addition to allowing recovery, also has the very important function of defining the "peripheral" parts of the skeleton. All maximal objects share at least one element of their border with the border of S_o. The maximal objects corresponding to the "outermost" CMOs share a significant connected portion of their border with the border of S_o. In a 2D 90° corner with edges oriented along the coordinate axes, the outermost $\langle 3,4 \rangle$ CMD is the third pixel diagonally from the corner, with radius 9. The $\langle 3,4 \rangle$ circle with radius 9 is a 5×5 square fitting the corner perfectly. The outermost CMD is in the same place also in the $\langle 5,7,11 \rangle$ DT and in the EDT; its radius is 15 and 3, respectively. In $\langle 1,\infty \rangle$, the corner pixel itself is the outermost CMD, whereas in $\langle 1,1 \rangle$ this corner does not generate any CMD at all. See Fig. 5.7.

Thus, insisting that the CMOs are part of the skeleton, there is no necessity for any end point detection criterion. As noted in the Introduction, defining an end point criterion is a difficult problem, and many solutions have been suggested. Not having to consider this simplifies skeletonization algorithms considerably.

5.5 Skeletons of 2D Shapes

The 2D skeletonization procedure followed here is found in Sanniti di Baja and Thiel (1996). It consists of finding CMDs and saddle-points in the DT and then linking them to produce a topologically correct skeleton S_s. Any regular $\langle a,b \rangle$ weighted DT can be used, or the semi-regular $\langle 1,\infty \rangle$ or $\langle 1,1 \rangle$. We first compute a nearly-thin skeleton, which is a set of pixels having all the skeleton properties except one-pixel thickness. The nearly-thin skeleton may be partly two pixels thick. It also contains branches originating from noise pixels. The nearly-thin skeleton is post-processed to make a thin and more stable skeleton.

5.5.1 Computing the Nearly-Thin 2D Skeleton

5.5.1.1 Neighborhood Operations

As we use 3×3 neighborhood DTs, we will only need to use 3×3 neighborhood operations throughout.

Many different criteria to preserve the topology while thinning are described in the literature. Common criteria are either to use a (large) number of masks or to use arithmetic checks, computed by investigating the neighborhood of a pixel $\mathbf{x}$. The eight neighbors $\mathbf{y} \in N(\mathbf{x})$ are numbered consecutively, modulo 8. Odd numbers are used for edge-neighbors and even numbers for vertex-neighbors. The sub-image $I(\mathbf{x})$, $\mathbf{x} \in N(\mathbf{x})$ is binary and $\bar{I}(\mathbf{y}_i) = 1 - I(\mathbf{y}_i)$.

We use the *crossing number* X_4, introduced by Rutovitz (1966), and the *connectivity number* C_8, introduced by Hilditch (1969) and reformulated by Yokoi et al. (1975).

Definition 5.5.1.

$$X_4(\mathbf{x}) = \frac{1}{2} \sum_{i=1}^{8} |I(\mathbf{y}_{i+1}) - I(\mathbf{y}_i)|$$

X_4 is equal to the number of 4-connected components (of object or background) in $N(\mathbf{x})$.

Definition 5.5.2.

$$C_8(\mathbf{x}) = \sum_{i=1}^{4} \bar{I}(\mathbf{y}_{2i-1}) - \bar{I}(\mathbf{y}_{2i-1}) \cdot \bar{I}(\mathbf{y}_{2i}) \cdot \bar{I}(\mathbf{y}_{2i+1})$$

C_8 is equal to the number of 8-connected components of S_o in $N(\mathbf{x})$, except if all four edge-neighbors of $\mathbf{x}$ are in S_o, when $C_8 = 0$.

For an isolated pixel crossing number and connectivity number give the result zero. Moreover, $X_4 = C_8 = 0$ for a pixel whose eight neighbors $\in S_o$.

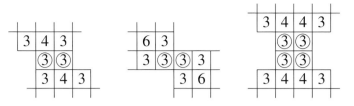

Fig. 5.8 Saddle-points (*circled*) detected by the three criteria. *Left*: Criterion 1. *Middle*: Criterion 2. *Right*: Criterion 3

X_4 and C_8 are used to check the local topology around a pixel $\mathbf{x}$. If $I(\mathbf{x}) > 1$, e.g., in the DT, a suitable binarization of $N(\mathbf{x})$ is necessary. The definitions of "shape" and "background" pixels in $N(\mathbf{x})$ will depend on the context; these definitions will be given in the appropriate places.

5.5.1.2 Saddle-Points

With this approach to skeletonization, we do not only need to identify the CMDs but also the *saddle-points* (SP) in the DT, i.e., pixels located in saddles of the DT. SPs occur in "necks" of S_o, connecting areas of larger local thickness. Most SPs are also CMDs, but not all of them. A pixel $\mathbf{x}$ in S_o is an SP if at least one of the following three criteria is satisfied:

1. $C_8 = 2$, where $\mathbf{y}$ is considered as a shape pixel if $I(\mathbf{y}) > I(\mathbf{x})$, and as a background pixel otherwise.
2. $X_4 = 2$, where $\mathbf{y}$ is considered as a background pixel if $I(\mathbf{y}) < I(\mathbf{x})$, and as a shape pixel otherwise.
3. There exists a 2×2 block of pixels in $N(\mathbf{x}) + \mathbf{x}$ with the same value as $\mathbf{x}$. In regular DTs (i.e., if $a > 1$) only the smallest distance value a has to be checked, but in semi-regular DTs (i.e., if $a = 1$) all values have to be checked.

Examples of saddle-points found according to the three criteria are found in Fig. 5.8. Both CMDs and SPs are found during one inspection of the DT.

5.5.1.3 Linking Pixels

Linking pixels connect CMDs and SPs. Linking is not straightforward, as there is a risk of producing thick "fans" of linking pixels in some configurations. By using directional derivatives of the DT, this is avoided.

The DT is scanned once. When a CMD or an SP $\mathbf{x}$ is detected, the scan is interrupted and $N(\mathbf{x})$ is inspected. The number of 8-connected components of pixels $\mathbf{y} \in N(\mathbf{x})$ with $I(\mathbf{y}) > I(\mathbf{x})$ is counted. There are at most two such components, H_i, $i = 1, 2$. An ascending path is started from any H_i. For each $\mathbf{y} \in H_i$, compute the directional derivative $d^{\mathbf{y}}I$

$$d^{\mathbf{y}}I = \tfrac{1}{\alpha}(I(\mathbf{y}) - I(\mathbf{x})), \quad \text{when } \mathbf{y} \in N_e(\mathbf{x}),$$
$$d^{\mathbf{y}}I = \tfrac{1}{\beta}(I(\mathbf{y}) - I(\mathbf{x})), \quad \text{when } \mathbf{y} \in N_v(\mathbf{x}),$$

where $\alpha = a$, $\beta = b$ for regular DTs and $\alpha = 2$, $\beta = 3$ for the semi-regular $\langle 1, \infty \rangle$ and $\langle 1, 1 \rangle$.

In *each* component H_i the $\mathbf{y}$ with the highest directional derivative is chosen as the linking pixel. At most two $\mathbf{y}$ in the same H_i have the same highest directional derivative. If this is the case, the edge-neighbor is chosen. Path growing then continues from the newly identified linking pixels until no positive directional derivative is found in $N(\mathbf{x})$.

5.5.1.4 Spurious Hole Filling

To preserve topology, the skeleton must have a loop around any hole in S_o but not in any other place. However, spurious loops can be created while growing linking paths through successive vertex-neighbors, when the paths are very close and parallel. Pixels in one path can then have vertex-neighbors in the other path, creating small holes in S_s. For semi-regular $\langle a, b \rangle$ DTs, the size of these holes is always a single pixel. They must be detected and filled.

A pixel $\mathbf{x} \in \bar{S}_s$ is a spurious hole if $\mathbf{x}$ was initially in S_o and for all $\mathbf{y} \in N_e(\mathbf{x})$, $\mathbf{y} \in S_s$. If $\mathbf{x}$ is changed to S_s, the hole disappears. It is required that $\mathbf{x}$ was initially in S_o; otherwise we would fill real small holes.

5.5.1.5 Algorithm

The algorithm to compute the nearly-thin skeleton can be summarized as follows, with indication of the computing complexity. The number of pixels in S_s is l, in S_o is m, and in I is n.

1. Compute the chosen DT (2 scans, $O(n)$)
2. Detect CMDs, SPs and linking pixels (1 scan $O(m)$, some path-growing $O(p+1)$, where p is the number of pixels in the paths)
3. Detect and fill spurious holes (1 scan, $O(l)$)

A very similar procedure can be used to produce skeletons from the $\langle 5, 7, 11 \rangle$, but some extra steps and checks are necessary, and the 5×5 neighborhood must be used (Sanniti di Baja and Thiel, 1996).

The $\langle 3, 4 \rangle$ nearly-thin skeleton of the moose and a close-up of the moose head are found in Figs. 5.9 and 5.10, respectively. The set of CMDs are presented in black, while saddle-points and linking pixels are in dark grey. We will illustrate the subsequent post-processing steps using this head, to make the pixel-level effects clearer.

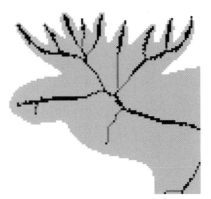

Fig. 5.9 The $\langle 3,4 \rangle$ nearly-thin skeleton of the moose

Fig. 5.10 Close-up of the $\langle 3,4 \rangle$ nearly-thin skeleton of the moose head

5.5.2 Post-Processing, 2D Case

Two types of pixels in a 2D skeleton play a very important role:

1. *End Points.* Skeleton pixels with $X_4 = 1$
2. *Branch points.* Skeleton pixels with $X_4 > 2$ or with more than two neighbors belonging to the skeleton

All remaining skeleton pixels are called *internal points.* In a one-pixel thick skeleton, end points, internal points and branch points have one, two, and more than two neighbors, respectively, in the skeleton.

5.5.2.1 Final Thinning

The first post-processing step is to thin the nearly-thin skeleton to one-pixel thickness. Standard topology preserving removal operations must be used with caution, as these may remove end points, thus shortening the branches.

A set of four 3×3 masks around the tested pixel $\mathbf{x}$ can be used (Sanniti di Baja, 1994). Figure 5.11 shows one of them; the other three are 90° rotations of this mask. These masks allow the removal of any pixel $\mathbf{x}$ of S_s (i.e., setting the pixel to $\overline{S}_s$) without altering the topology of S_s, and without causing branch shortening. The center pixel of the mask can be removed from the skeleton if the two pixels marked with closed circles are in S_s and the pixel marked with an open square and at least one of the pixels marked with open circles are in $\overline{S}_s$. Unmarked pixels are "don't care." With two S_s pixels in $N_e(\mathbf{x})$ (closed circles) that are vertex-neighbors to each other and with a $\overline{S}_s$ pixel (open square) in the opposite vertex-neighbor position, removing $\mathbf{x}$ will not change the 8-connectedness of S_s. Since (at least) one edge-neighbor (open circle) is in $\overline{S}_s$, no hole will be created in the skeleton. Since at least two neighbors (closed circles) are in S_s, end points will not be removed.

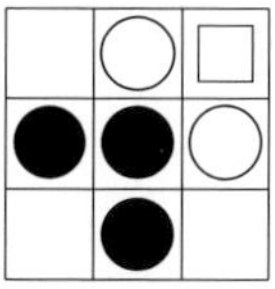

Fig. 5.11 One of four rotations of the mask for final thinning. See text

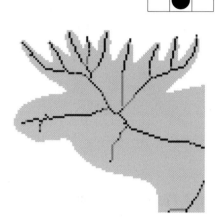

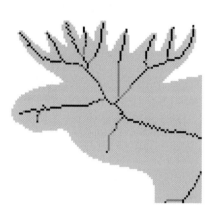

Fig. 5.12 The skeleton of the moose head

Fig. 5.13 The beautified skeleton of the moose head

Final thinning is achieved by applying the four masks to each S_s pixel in one raster scan. A few superfluous pixels may still remain after one application of the four masks, as a pixel may become removable only after its neighbors have been visited. If desired, mask application can be repeated a second time to remove these superfluous pixels. The (one-pixel thick) skeleton of the moose head is found in Fig. 5.12.

Final thinning has two undesired effects on the skeleton: some border pixels of S_o can no longer be recovered from S_s, and S_s is somewhat jagged where the nearly-thin skeleton was two pixels thick. The first effect can be minimized by performing final thinning in two raster scans. During the first scan, *only* non-CMD pixels can be removed from S_s. During the second scan, all pixels of S_s are checked against the four masks. Thus, maintaining CMDs in S_s is favoured. The second effect can be somewhat ameliorated but never completely solved by the next post-processing step.

5.5.2.2 Beautifying

Thinning the skeleton with the masks in Fig. 5.11 will unavoidably make it jagged in places, see Fig. 5.12. This occurs where the nearly-thin skeleton is actually two pixels thick and because the masks are applied sequentially rather than in parallel. A simple (but perhaps whimsically named) beautifying algorithm removes this induced jaggedness (Sanniti di Baja, 1994).

Four masks are used, where the center pixel and the two vertex-neighbors not aligned with it belong to S_s, and the remaining neighbors belong to $\overline{S}_s$. See Fig. 5.14,

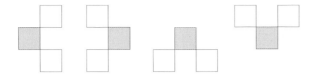

Fig. 5.14 Jaggedness is reduced if the grey pixel is shifted into alignment with the white pixels

where the four configurations for the three pixels belonging to S_s are shown. If any of the four masks fits, during one raster scan over the image, the center pixel (grey) is "shifted" into edge-connected alignment between the vertex-neighbors (white). Of course, the center pixel can only be shifted if the pixel in that position was initially in the original shape S_o. The beautified skeleton of the moose head is found in Fig. 5.13.

The beautifying process can be more strict, by only allowing shifting to positions where there was a skeleton pixel before final thinning. More jaggedness will remain, but fewer border pixels will be lost when reconstructing the shape.

5.5.2.3 Pruning

The skeleton computed so far is correct in the sense that it represents *all* parts of the shape S_o. If all parts are significant, nothing more should be done. However, even if image capture is perfect, which seldom happens, digitization and rotation cause "noise" on the shape border. This noise will in turn generate S_s branches that are not significant in the application. These branches can be removed in (at least) three ways: (1) S_o can be smoothed prior to skeletonization, e.g., by morphological filtering (Serra, 1982); (2) the skeletonization algorithm itself can avoid creation of non-significant branches; or (3) S_s can be pruned as a post-processing procedure. Smoothing certainly helps, but to yield robust skeletons, some pruning is usually necessary. We prefer a post-processing step, mostly because it gives more control over the results.

Numerous articles have been written on skeleton pruning, both in the continuous and the digital domains, using different significance measures. Pruning is discussed in Chapters 1, 6, and 7, and a systematization of pruning in the continuous domain is found in (Shaked and Bruckstein, 1998). There, *acceptable* pruning methods are defined as preservation of topology, being well-conditioned, and locality of significance (see Section 1).

Meeting these pruning criteria means that pruning can only be done by removing, along a skeleton branch, pixels from the end point inwards, and that the decision whether a pixel will be removed can be taken by local operations. As discussed in Chapter 6, significance measures are frequently related to *area* or *elongation* of the portion of S_o lost in recovery. Another measure, used mostly in the continuous case, is the *boundary length* corresponding to a skeleton branch.

We use elongation as the significance measure (Sanniti di Baja, 1994; Attali et al., 1995). We do not perform pruning if the branch corresponds to a region decreasing in width (i.e., if $I(\mathbf{x})$ decreases along the branch). The significance of an S_s pixel $\mathbf{x}$ is the length, in pixels, of the part of the protrusion that is lost if $\mathbf{x}$ is removed. The significance can be computed as follows. Let $\mathbf{x}$ be an end point and $\mathbf{y}$ an internal point on the same skeleton branch. Let d be the distance between $\mathbf{x}$ and $\mathbf{y}$ (in the DT metric used), and let ε be the pruning threshold used. Then, all pixels in the skeleton branch from $\mathbf{x}$ up to $\mathbf{y}$ can be removed if

$$I(\mathbf{x}) - I(\mathbf{y}) + d \leq \varepsilon \tag{5.4}$$

This criterion will correctly measure the elongation of the protrusion lost if the skeleton branch has low curvature between $\mathbf{x}$ and $\mathbf{y}$. If this is not the case, the length d should be geodesic, i.e., measured along the skeleton curve.

Equation (5.4) will not take protrusion sharpness into account. It can be argued that a sharp protrusion is more significant than a blunt one. A protrusion is sharp if the difference in radii, $I(\mathbf{y}) - I(\mathbf{x})$, is large compared to d. By introducing a correction factor in (5.4), sharp protrusions do not have their elongation reduced by pruning:

$$I(\mathbf{x}) - I(\mathbf{y}) + d \leq \varepsilon \left(\frac{I(\mathbf{y}) - I(\mathbf{x}) + 1}{d} \right) \tag{5.5}$$

The pruning algorithm is thus

1. List all end points.
2. For each end point $\mathbf{x}$, mark as non-significant each internal point $\mathbf{y} \in S_s$ inwards from $\mathbf{x}$, until (5.4) (or (5.5)) is no longer valid, a branch point is reached, or $I(\mathbf{y}) < I(\mathbf{x})$.
3. For each branch including non-significant pixels, remove the end point and all successive non-significant pixels in the branch.

The pruned skeleton of the moose head, using (5.4) with pruning threshold $\varepsilon = 5$, is found in Fig. 5.17. The beautified skeleton computed for a rotated head is shown in Fig. 5.15, and the pruned version of this skeleton is shown in Fig. 5.16. This was done to show that mildly pruned $\langle 3, 4 \rangle$ skeletons are quite stable under rotation.

Slightly different results could be obtained if pruning is performed before beautifying. The differences, however, can generally be disregarded.

In Fig. 5.18, we show how much of the moose head is recovered (by the rDT) from the pruned skeleton, $\varepsilon = 5$. The protrusions have been shortened by five pixels (as expected). Otherwise there are few differences.

Using lost area as a significance measure for pruning is more difficult in the digital domain, as it is not straightforward to compute the area difference between overlapping digital disks (Sanniti di Baja, 1994; Attali et al., 1995). The area must be computed by pixel counting, rather than by mathematical formulae, as can be done in the continuous domain. The procedure for $\langle 1, 1 \rangle$ is found in Cordella and Sanniti di Baja (1989).

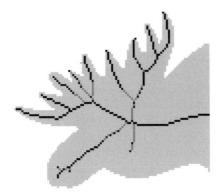

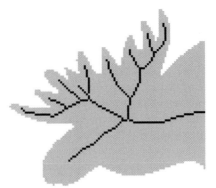

Fig. 5.15 The beautified skeleton of the moose head rotated 30 degrees

Fig. 5.16 The pruned skeleton from Fig. 5.15 (pruning threshold 5)

Fig. 5.17 The pruned skeleton from Fig. 5.13 (pruning threshold 5)

Fig. 5.18 Recovered moose head (black) from the skeleton in Fig. 5.17. Original moose head in grey

Fig. 5.19 The $\langle 3,4 \rangle$ beautified skeleton of the moose (no pruning)

Fig. 5.20 Recovered moose from the skeleton in Fig. 5.19

Finally, the beautified non-pruned $\langle 3,4 \rangle$ skeleton of the whole moose is presented in Fig. 5.19, and the shape recovered from this skeleton in Fig. 5.20. The recovered shape can hardly be distinguished from the original.

The described final thinning and beautifying methods can be applied to any digital skeleton, and the pruning method to any DT based skeleton. The post-processing methods are thus not dependent on the skeletonization algorithm in Section 5.5.1.

5.6 Skeletons of 3D Shapes

For 3D images the surface skeletons S_s are thin surfaces, possibly including curves. As already said, the centers of maximal balls of a shape generally do not constitute a connected set, and finding saddle-points and linking voxels in the DT to compute S_s is not straightforward. Also the iterative approach to skeletonization presents more difficulties in 3D. Removing voxels from the shape by using topology preserving operations, while keeping the CMBs in the skeleton, is not enough to reach the desired goal. The result tends to be a skeleton that consists of many curves, running close to each other, rather than actual surfaces. This is not what is generally wanted, as these bunches of curves are difficult to analyze and manipulate. Therefore extra, surface preserving rules are necessary that depend on the connectedness type considered for the (surface) border of the shape. This is somewhat analogous to what happens in 2D, where the border of a shape can be defined as an 8-connected contour (i.e., the set of shape pixels having an edge-neighbor in the background) or as a 4-connected contour (i.e., the set of shape pixels having an edge- or a vertex-neighbor in the background). Selection of the contour connectedness type conditions the resulting skeleton. It also partly conditions the connectedness type of the skeleton, which should be a set of 8-connected or 4-connected curves, respectively. Actually, 8-connectedness is preferred even when 4-connected contours are chosen, because this makes the number of skeleton pixels as small as possible.

If a DT is employed to define successive shape contours, the city-block and the chessboard distances should be used to originate 8-connected and 4-connected contours, respectively. In 3D, if D^6 and D^{26} are used, the border of the shape is the set of shape voxels having a face-neighbor in the background (18-connected surface) and the set of shape voxels having a face-, or an edge-, or a vertex-neighbor in the background (6-connected surface), respectively. In both cases, 18-connected surface skeletons are preferred, but the surface preserving rules to be used will unfortunately be different for the two different DTs, originating rather different borders.

Curve skeletons, S_c, are also of interest in 3D, but there is no complete agreement on how S_c should be defined. There are two classes of approaches for curve skeletonization: either the curve skeleton is extracted directly from the shape S_o, or the curve skeleton is extracted from the surface skeleton S_s. We use the latter approach. In any case, a 3D curve skeleton cannot have the recovery property, as too much information is lost. S_c can still be a very useful representative structure, especially when analyzing images mainly containing thin elongated shapes, e.g., blood vessels and fibres.

As mentioned before, in 3D we will follow the iterative approach to skeletonization, with the CMBs of $\langle 1, \infty, \infty \rangle$ and $\langle 1, 1, 1 \rangle$ DTs as anchor points, and the DTs themselves used to guide the iterations. The 3D version of the 2D method described in Section 5.5 would be more complex (see (Sanniti di Baja and Svensson, 2000) for the $\langle 1, \infty, \infty \rangle$ version). We will also present post-processing algorithms (final thinning, beautifying, and pruning) similar to those in 2D. Our 3D running example is a human hand. See Fig. 5.21 (image copyright J. Toriwaki and K. Katada, Nagoya University, Japan).

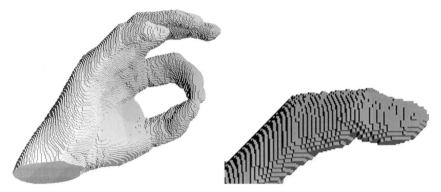

Fig. 5.21 Running example for 3D skeletonization: a hand, left, and close-up of one of its fingers, right

5.6.1 Computing the Nearly-Thin Surface Skeleton

The methods presented here have been developed over a number of years (Borgefors et al., 1996, 1999; Svensson et al., 1999; Svensson, 2002).

5.6.1.1 Neighborhood Operations

We need a number of neighborhood operations to determine the local topology in 3D (see, e.g., Malandain et al., 1993; Saha and Chaudhuri, 1994). We need to be able to prevent creation of disconnections, cavities, and tunnels in $N(\mathbf{x})$.

Let $\mathbf{x} \in S$, where S is the current set of shape voxels.

Definition 5.6.1. N^{26} is the number of (26-connected) components of S in $N(\mathbf{x})$.

A voxel $\mathbf{x}$ having $N^{26} = 1$ can be removed without altering the number of components of S in $N(\mathbf{x})$.

Definition 5.6.2. $\overline{N}_f^{18}$ is the number of (6-connected) components of $\overline{S}$ in $N_f(\mathbf{x}) \cup N_e(\mathbf{x})$ that are face-adjacent to $\mathbf{x}$.

A voxel $\mathbf{x}$ having $\overline{N}_f^{18} = 1$ can be removed without creating cavities or tunnels in $N(\mathbf{x})$. In fact, for a voxel $\mathbf{x}$ having no face-neighbors in $\overline{S}$, whose removal would create a cavity, $\overline{N}_f^{18} = 0$. In turn, for a voxel whose removal would create a tunnel, $\overline{N}_f^{18} \geq 2$.

Thus, N^{26} and $\overline{N}_f^{18}$ are used to check whether removal of $\mathbf{x}$ can be done, without creating disconnections (or even disappearance), cavities, and tunnels in $N(\mathbf{x})$.

Analogously to the 2D case, saddle-points (SPs) are located in saddles of the DT, connecting parts of the shape with larger local thickness. Criteria similar to those in Section 5.5.1.2 can be suggested by using the appropriate connectedness types

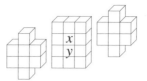

Fig. 5.22 The extended neighborhood $N_{ext}(\mathbf{x})$ used to compute $\overline{N}_f^{27}$

(26-connectedness for shape and 6-connectedness for background). In particular, the third criterion, dealing with two-element thick saddles, can be checked as follows.

Definition 5.6.3. The voxel $\mathbf{x} \in S$ is in a *two-voxel saddle configuration* with $\mathbf{y} \in S$ if $\mathbf{x}$ and $\mathbf{y}$ are the inner voxels in a $1 \times 1 \times 4$ neighborhood, where the two outer voxels are in $\overline{S}$.

Of course, a $1 \times 4 \times 1$ or a $4 \times 1 \times 1$ neighborhood is used, depending on the position of $\mathbf{y}$ with respect to $\mathbf{x}$.

For a shape voxel $\mathbf{x}$ in a two-voxel saddle configuration with $\mathbf{y}$, an extended neighborhood, $N_{ext}(\mathbf{x})$, is used. This neighborhood is $3 \times 3 \times 4$ voxels with the cardinal direction where $N_{ext}(\mathbf{x})$ is four voxels thick determined by the position of $\mathbf{y}$, see Fig. 5.22. We will use

Definition 5.6.4. $\overline{N}_f^{27}$ is the number of (6-connected) components of $\overline{S}$ in $N_{ext}(\mathbf{x})$ that are face-adjacent to $\mathbf{x}$ or to its inner face-neighbor $\mathbf{y}$.

Efficient algorithms for computing N^{26} and $\overline{N}_f^{18}$ are found in (Borgefors et al., 1997). $\overline{N}_f^{27}$ can be computed efficiently by an obvious extension of these algorithms.

5.6.1.2 Iterative Voxel Removal

Once the DT has been computed and the CMBs identified, iterative voxel removal starts. In each iteration, all border voxels (i.e., voxels with the same distance value d) are considered. First, among non-CMBs, we mark all simple voxels, i.e., voxels with $N^{26} = \overline{N}_f^{18} = 1$. This is a parallel step, as all voxels can be marked simultaneously.

If a simple voxel is removed, the remaining set of shape voxels will still have the same topology as S_o, but surfaces are not preserved. As remarked before, we need an extra condition, which is DT-dependent. Also, all marked simple voxels cannot be removed simultaneously, as removing one voxel can make a previously simple neighbor, non-simple. Therefore, the marked voxels are checked sequentially, and those that are still simple and fulfill the extra condition are removed.

Definition 5.6.5. In $\langle 1, \infty, \infty \rangle$, a voxel $\mathbf{x}$ fulfills the *surface condition* SC_6 if there exist two opposite neighbors $\mathbf{y}$ and $\mathbf{z}$ in $N_f(\mathbf{x})$ such that $\mathbf{y}$ is a background voxel and $\mathbf{z}$ is an internal voxel (a shape voxel but not a border voxel).

This means that a simple voxel is included in S_s if in the DT it has no pair of opposite face-neighbors, where one has smaller DT value and has not been assigned to the skeleton (i.e., is interpreted as a background voxel) and the other has a higher DT value (i.e., is an internal voxel). SC_6 prevents removal of CMBs. In fact, if $\mathbf{x}$ is a CMB, all its face-neighbors have values smaller than $I(\mathbf{x})$. SC_6 also prevents removal of SPs. In fact, the two opposite face-neighbors of an SP $\mathbf{x}$ in each pair have values either both smaller or both larger than $I(\mathbf{x})$, whichever the pair of opposite face-neighbors is. Therefore, SPs do not need to be explicitly identified in 3D.

Definition 5.6.6. In $\langle 1,1,1 \rangle$, a voxel $\mathbf{x}$ fulfills the *surface condition* SC_{26}, if there does not exist two opposite neighbors $\mathbf{y}$ and $\mathbf{z}$ in $N_f(\mathbf{x})$ such that both $\mathbf{y}$ and $\mathbf{z}$ are either internal voxels or are border voxels not marked as removable.

This means that a simple voxel is not removed if in the DT, in some direction, it is placed between face-neighbors with either have higher DT value(s) or have equal DT value(s) and are not marked.

In $\langle 1,\infty,\infty \rangle$, no voxel with value d has an internal face-neighbor with the same value d. Therefore, it is enough to perform the thinning operation only once for each distance value d. In $\langle 1,1,1 \rangle$, voxels with distance value d may have internal face-neighbors with the same distance value d. In this case, the thinning operation has to be repeated thrice for each distance value d, so that each d-valued voxel actually has a possibility to be removed.

5.6.1.3 Algorithm

The algorithm to compute the nearly-thin surface skeleton can be summarized as follows, with indication of the computing complexity. The number of voxels in S_o is m and in I is n.

1. Compute the DT, $\langle 1,\infty,\infty \rangle$ or $\langle 1,1,1 \rangle$ (2 scans, $O(n)$).
2. Detect CMBs (1 scan, $O(m)$).
3. For increasing distance values d:

   3a. Mark border voxels valued d that are simple but not CMBs (1 scan, $O(v)$, where v is the current number of border voxels).
   3b. Remove sequentially marked voxels that are simple and fulfill the surface condition SC (1 scan, $O(p)$, where p is the number of marked voxels).

   If $\langle 1,1,1 \rangle$ is used, repeat 3a and 3b three times. (2Δ scans, $O(m)$ for $\langle 1,\infty,\infty \rangle$ and 6Δ scans, $O(m)$ for $\langle 1,1,1 \rangle$, where Δ is the maximum value in the DT).

In principle, the same algorithm can run for other 3D DTs, but we have not yet developed reasonable surface conditions SC for these metrics.

The resulting nearly-thin surface skeletons of the finger of the hand are found in Fig. 5.23.

Fig. 5.23 The $\langle 1, \infty, \infty \rangle$ *(left)* and $\langle 1, 1, 1 \rangle$ *(right)* nearly-thin surface skeletons of the finger in Fig. 5.21

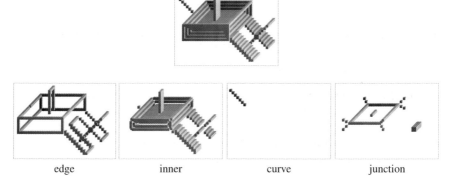

| edge | inner | curve | junction |

Fig. 5.24 A surface-like object, top, with its voxel classification, bottom

5.6.2 Post-Processing, Surface Skeleton

5.6.2.1 Voxel Classification

We classify the voxels of S_s into one of four classes: curve, edge, inner, or junction voxels (Svensson et al., 2002b). A small example is found in Fig. 5.24, where a surface-like object, i.e., an object at most two voxels thick, is shown with its classes of voxels.

The classification of a voxel $\mathbf{x} \in S_s$ is performed using the following scheme. We first classify voxels as junction, inner, curve, or border voxels. Border voxels are then classified as edge or curve voxels. Our classification correctly treats two-voxel thick surfaces and curves.

1. If $N^{26} \geq 2$ and $\overline{N}_f^{18} \geq 1$, then $\mathbf{x}$ is a *curve* voxel.
2. If $N^{26} = 1$ and $\overline{N}_f^{18} > 2$, then $\mathbf{x}$ is a *junction* voxel.
3. If $N^{26} = 1$ and $\overline{N}_f^{18} = 0$, then $\mathbf{x}$ is a *junction* voxel.
4. If $N^{26} = 1$ and $\overline{N}_f^{18} = 2$, then $\mathbf{x}$ can be an *inner* voxel in a one-voxel thick area, or a *junction* voxel in a two-voxel thick area.
5. If $N^{26} = 1$ and $\overline{N}_f^{18} = 1$, then $\mathbf{x}$ can be a *border* voxel, an *inner* voxel in a two-voxel thick area, or a *junction* voxel in a two-voxel thick area.

For any voxel $\mathbf{x}$ satisfying the above condition 4 (5), and located in a two-voxel thick saddle with a voxel $\mathbf{y}$ that also satisfies condition 4 (5), the following is done:

6. If $\overline{N}_f^{27} = 2$, then $\mathbf{x}$ is an *inner* voxel.
7. If $\overline{N}_f^{27} > 2$, then $\mathbf{x}$ is a *junction* voxel.
8. If $\overline{N}_f^{27} = 1$, then $\mathbf{x}$ is a *border* voxel.

Any voxel $\mathbf{x}$ satisfying the above condition 4 (5), but not located in a two-voxel thick saddle is an *inner* voxel (*border* voxel). Finally, for all voxels already classified as border voxels, the classification into *edge* or *curve* voxel is done as follows:

9. If all voxels in $N(\mathbf{x})$ that belong to S_s are border voxels, then $\mathbf{x}$ is a *curve* voxel.
10. If a voxel $\mathbf{y}$ in $N(\mathbf{x})$ is a curve voxel with no inner voxel in $N(\mathbf{y})$, then $\mathbf{x}$ is a *curve* voxel.
11. If neither condition 9 nor condition 10 are valid, then $\mathbf{x}$ is an *edge* voxel.

This classification is not quite exhaustive, since in some (not frequent!) cases, junctions are locally thicker than two voxels. Using the above classification can then yield some voxel classifications that are not as they would be in the "ideal" case, but the classification is usually good enough. An exhaustive classification can be found in Sanniti di Baja and Svensson (2000).

5.6.2.2 Final Thinning

The 3D nearly-thin skeleton can be thinned to one voxel by using a simple $1 \times 1 \times 4$ neighborhood operation in six directions (Borgefors et al., 1999). The thinning method is applicable to both nearly-thin surfaces and nearly-thin curves.

A voxel $\mathbf{x} \in S_s$ is removed from S_s if $\mathbf{x}$ is simple, and is in a two-voxel saddle configuration with a face-neighbor $\mathbf{y} \in S_s$.

Thinning is accomplished by sequentially removing voxels in the "up" and "down" directions, the "left" and "right" directions, and the "front" and "back" directions, i.e., in six scans.

If the surface skeleton is going to be compressed into a curve skeleton, the nearly-thin surface skeleton does *not* need to be thinned before curve skeleton computation.

5.6.2.3 Pruning the Nearly-Thin Surface Skeleton

As can be seen in Fig. 5.23, the surface skeletons tend to be "furry," i.e., despite the surface condition SC, at the borders of the surfaces there are a number of short curves that are noise rather than significant parts of S_s. They should thus be pruned (Borgefors et al., 2000). The definition of acceptable pruning given in Section 5.2 is used, here in 3D. Here our significance criterion is the curve length. Our pruning can handle nearly-thin surface skeletons.

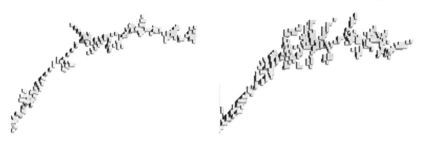

Fig. 5.26 The $\langle 1, \infty, \infty \rangle$ (*left*) and $\langle 1, 1, 1 \rangle$ (*right*) nearly-thin curve skeletons of the finger in Fig. 5.23

  2b. Mark as candidate for removal non-significant voxels that are or have become edge-voxels.
  2c. Sequentially remove simple marked voxels.

3. Iterate until no more voxels are removed:

  3a. Classify the voxels in the current surface (curve, edge, junction, and inner).
  3b. Mark as candidate for removal voxels that are or have become edge-voxels.
  3c. Sequentially remove simple marked voxels.

The resulting $\langle 1, \infty, \infty \rangle$ and $\langle 1, 1, 1 \rangle$ nearly-thin curve skeletons of the finger of the hand are found in Fig. 5.26.

5.6.4 Post-Processing, Curve Skeleton

5.6.4.1 Final Thinning

Thinning of the nearly-thin curve skeleton could be performed by exactly the same algorithm used for thinning the nearly-thin surface skeleton. However, there are a number of configurations that would remain thick. Also, some end points could be removed and the relative branches shortened. This is the case for any method that removes simple-points without considering if they are end points. This will cause branch shortening and thus loss of information. Better results are achieved by the method presented here (Svensson and Sanniti di Baja, 2003).

Tips of branches in S_c can consist of a set of up to four voxels, depending on curve thickness and curve orientation. One of the tip voxels should become an end point and be preserved; the others should be removed. Two-voxel thick curves and junctions should be thinned to one-voxel thickness, if possible. Not all junctions *can* be thinned to one voxel while preserving topology, e.g., in the configuration where eight vertex-connected curves meet in a block of $2 \times 2 \times 2$ voxels (all voxels in the block are branch points, and the block itself is a *cluster* of branch points).

To identify actual and possible end points, we need to count the number of neighbors of a voxel.

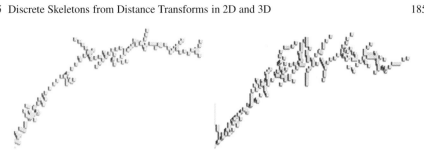

Fig. 5.27 The $\langle 1, \infty, \infty \rangle$ (*left*) and $\langle 1, 1, 1 \rangle$ (*right*) curve skeletons from Fig. 5.26

Definition 5.6.7. Let $\mathbf{x} \in S_c$. $\#N_s^{26}$ is the number of voxels in $N(\mathbf{x})$ belonging to S_c.

In the first part of the process, we remove voxels that are placed in two-voxel thick saddle configurations, per Definition 5.6.3. These voxels occur where the curve is two voxels thick in one of the three cardinal directions. First, we remove only non-significant voxels (i.e., voxels that were not classified as curve or junction voxels in S_s), and then we repeat the process for all voxels. Here, we need no end point criterion, except the obvious one: a voxel is an end point if $\#N_s^{26} = 1$. In thick tips aligned along the cardinal axes, the voxel that remains as end point depends on the scanning order.

There will still remain thick tips and thick junctions, as those not aligned along the cardinal axes are not thinned by the first procedure, e.g., curve segments consisting of pairs of edge-neighbors. Here, we will resort to the usual parallel marking and sequential removal method. First, definite end points, those with $\#N_s^{26} = 1$, and candidate end points are marked. Candidate end points are defined as voxels with $N_s^{26} = 1$ and $2 \leq \#N_s^{26} \leq 3$. Some of these candidates are *isolated* end points, i.e., have no marked neighbor, some are placed in thick tips, and some are placed in T- and L-junctions. Before thinning, candidate end points are checked, and only those that are isolated or are simple with respect to the set of end points (definite and candidate) are kept. Thus, only one marked voxel is left at each branch tip. Unmarked simple voxels are then removed, again in two steps, where the first only removes non-significant voxels. Finally, marked voxels that are simple and have $\#N_s^{26} > 1$ are removed. These are voxels in L- and T-junctions and should be removed to make the curve as thin as possible.

The curve thinning algorithm can be viewed in detail in Svensson and Sanniti di Baja (2003). When implementing the curve thinning algorithm, the complete image should not be scanned over and over. Instead, "interesting" voxels should be put in lists, and in most cases only the voxels in the lists should be inspected. For example, voxels in two-voxel thick saddle configurations can be placed in a saddle-list, and candidate end points in a candidate-end-point-list.

The $\langle 1, \infty, \infty \rangle$ and $\langle 1, 1, 1 \rangle$ curve skeletons of the finger are found in Fig. 5.27.

Fig. 5.28 Basic skeleton configurations for beautification
in 3D. If any of these configurations, or any of their twelve
rotations, occurs the grey
voxel is shifted into alignment
with the white voxels

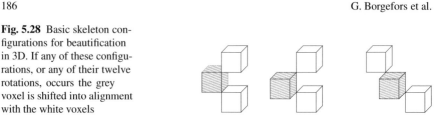

5.6.4.2 Beautifying

As in the 2D case, the final thinning of S_c will unavoidably make the set jagged
in places (see Fig. 5.27). Therefore, we apply a beautifying procedure. We use the
same idea as in 2D: if a piece of the curve is in any of a specified set of three-voxel
configurations, the center voxel is shifted into alignment with the other two voxels.
In 2D we needed only one configuration in four rotations. In 3D we need three
configurations, each in twelve rotations. The three basic skeleton configurations for
3D beautification are found in Fig. 5.28. Of course, it is only allowed that a voxel
be shifted to a position that was defined as a skeleton voxel before final thinning.

5.6.4.3 Pruning

Pruning the curve skeleton is usually a good idea, to remove non-significant branches and to make the skeleton more rotation independent. The degree of pruning
that should be performed is very much dependent on the application. We present a
very general pruning method, general in the sense that it is easily adapted to various
pruning needs (Svensson and Sanniti di Baja, 2003).

We still use the original voxel classification of curve, edge, inner, and junction voxels from the surface skeleton S_s. In addition, we classify the voxels in the
thin curve skeleton as end points, internal points, and branch points, and will still
consider curve and junction voxels as classified on S_s as significant voxels. Only
peripheral branches are removed, i.e., branches that are delimited by an end point
and a branch point. (Internal branches have branch points at both ends.) Differently
from the 2D case, we either remove the branch completely, or not at all.

The pruning method requires setting three thresholds. The first is the *maximum
length*, Λ, of branches that are removed without checking their significance level.
This "brute force" pruning can be used as a first pruning step, to remove furry noise
branches, but Λ should be small. Often $\Lambda = 1$ is a good choice. Then, only end
points directly linked to branch points are removed.

The second threshold is the *branch significance*, β. We will use as branch significance the ratio between number of significant voxels (curve and junction voxels),
#SV, and total number of voxels, *#TV*, in the branch. *#SV* and *#TV* are initialized to 0 whenever a branch is being traced from its end point. Another possibility,
especially useful for thin elongated shapes is the ratio between the number of CMBs
and total number of voxels in a skeleton branch.

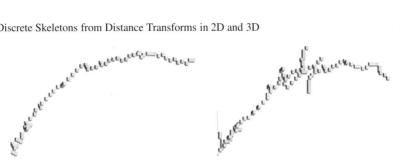

Fig. 5.29 The $\langle 1,\infty,\infty \rangle$ (*left*) and $\langle 1,1,1 \rangle$ (*right*) pruned and beautified curve skeletons from Fig. 5.27

Many branches are attached to a cluster of branch points, rather than a single one. When the branch is removed, the cluster is thinned by removing voxels that have become superfluous for topology preservation, and its voxels are re-classified.

The third threshold is the *number of iterations*, v of branch pruning. When the voxels remaining in a cluster of branch points are reclassified, some may have become end points. This means that an initially internal branch may have become peripheral and thus can be pruned. A second pruning iteration can possibly remove this branch, and so on. To handle iterative branch removal as well as possible, we store #SV and #TV of the removed branch in the remaining branch point. In subsequent iterations, these values are added to the newly traced branch. For large objects, $v > 1$ can be a good choice.

A further choice has to be made: should only *new* end points be considered in subsequent iterations, to avoid unnecessary shortening of what becomes the main branches, or should *all* end points be considered in all iterations? We describe the first choice in the algorithm.

The curve pruning algorithm can be viewed in detail in Svensson and Sanniti di Baja (2003). Again, when implementing the curve pruning algorithm, lists of end points, branch points, etc. should be used instead of complete image scans.

The $\langle 1,\infty,\infty \rangle$ and $\langle 1,1,1 \rangle$ pruned and beautified curve skeletons of the finger are found in Fig. 5.29. We have used $\Lambda = 5$, $\beta = 0.80$, and $v = 2$.

This pruning method is useful for all thin curve structures in 3D, independently of how they were created, provided that significant voxels can be identified in some way. There are, for example, a number of curve skeletonization algorithms that thin the 3D shape directly, rather than using the surface skeleton as an intermediate structure. These can equally well be pruned in this way.

Finally, the $\langle 1,\infty,\infty \rangle$ and $\langle 1,1,1 \rangle$ skeletons of the whole hand, thinned and beautified, but not pruned, are found in Fig. 5.30.

5.7 Some Applications and Extensions

The (distance valued) skeleton can be used for many different shape analysis tasks (Borgefors, 1994). In document analysis, DT based 2D skeletons serve as a useful tool in the OCR process (Rice et al., 1999). In 3D, distance valued curve skeletons

Fig. 5.30 The $\langle 1, \infty, \infty \rangle$ (*left*) and $\langle 1, 1, 1 \rangle$ (*right*) curve skeletons of the hand. No pruning has been performed

have been shown useful when extracting centerlines in elongated structures (Zhou et al., 1998). One particular example is in visualization of blood vessels in Magnetic Resonance Angiography (MRA) images (Nystrm and Smedby, 2001). A totally different application field is when measuring features in the fibre network of paper (Svensson and Aronsson, 2003).

Shape decomposition can be guided by skeleton decomposition. As described in Chapter 10, the resulting graphs can provide a basis for recognition of shapes. A first, immediate, decomposition is achieved by decomposing the (curve) skeleton, in correspondence with the branch points, into the constituting branches. Each individual skeleton branch corresponds to a significant figure, i.e., object part, with the skeleton describing the way the object curves and the radial distance describing bulging and compression. Figures may be further subdivided into elementary parts that are satisfactorily interpolated from the extremes of the part. See Sanniti di Baja and Thiel (1994) for skeleton decomposition in 2D. For the 3D case, a somewhat similar method, though not explicitly based on skeleton decomposition, can be found in Svensson and Sanniti di Baja (2002).

In the previous sections, we have introduced pruning as a way to simplify skeleton structure. Pruning could also be used to hierarchically rank skeleton branches according to their significance, a task that is also necessary for the recognition approaches described in Chapter 10. Branches with small significance disappear using a relatively small pruning threshold, while more and more significant branches remain in the skeleton even when pruning with larger and larger thresholds. Ranking skeleton branches can be used to achieve a multi-resolution representation of the shape. An alternative way to obtain a multi-resolution representation is to use a multi-scale structure, e.g., a pyramid. In Borgefors et al. (2001), a procedure was presented to decompose the discrete skeleton, computed in a binary AND-pyramid. The skeleton at each level was decomposed into meaningful parts in a consistent way across all resolutions. Each part corresponded to an elongated region of the shape and was marked with its "importance" for characterizing the shape. The importance of a part of the pattern was explicitly determined by the resolution level at which it became visible, and implicitly determined by the thickness of the region. The obtained skeleton decomposition seems to correspond reasonably well to human intuition. Topological and shape features of the parts were reflected as well as possible at lower resolution levels, within the limits of decreasing resolution; the parts at each level were linked to corresponding parts in higher and lower resolutions.

In this chapter, we have only described skeletonization in binary images. However, for many applications, especially in the biomedical field, grey-level skeletonization is preferable. While the literature on discrete skeletonization is extensive for the binary case in both 2D and 3D, a considerably smaller number of papers are available on grey-level skeletonization, especially for voxel images. This is certainly, at least partially, due to the fact that a large number of difficulties, not present in the binary case, are discovered when working with grey-level images. It also depends on the fact that it is not easy to find a common interpretation of what the grey-level skeleton should be, in particular concerning its topological properties. Decisions about topology seems to be dependent on the specific application and thus the population from which the class of objects come. Different algorithms that do not start from an understanding of the population produce different results, each with a complex structure, when applied to the same image. Chapters 8 and 9 discuss a way to determine a single branching topology for the class and fit medial models with that topology to objects in images of such objects. To conclude this chapter, we will show some results obtained by our 3D grey-level skeletonization methods (Svensson et al., 2002a) on MRA images.

To compute the grey-level surface skeleton, distance information and grey-level information were combined. In particular, a regularity condition defined in the distance transform was used to avoid excessive shortening of peripheral parts of the skeleton. The skeleton structure was then simplified by removal of some peripheral surfaces, to obtain the desired results.

In Fig. 5.31, a fuzzy segmented blood vessel tree (left) is shown together with the resulting grey-level surface skeleton (middle). For comparison, the skeleton

Fig. 5.31 *Left*: fuzzy segmented blood vessels from MR angiography. [Courtesy of Dr. Saha, MIPG, Dept. of Radiology, Univ. of Pennsylvania, Philadelphia.] *Middle*: grey-level surface skeleton of the fuzzy segmented vessels. *Right*: surface skeleton of the hard segmented vessels

computed when the blood vessel is hard segmented as a binary image is also shown (right). Indeed, the grey-level skeleton is much less "busy." This is probably due to the fact that the grey-level skeleton is mainly located in regions characterized by locally higher intensity, while all voxels have the same significance (and hence contribute equally to the skeleton) in a binary image.

We conclude by mentioning some weaknesses of distance function based discrete skeletons. The most serious weakness affects 3D skeletonization and is associated with the lack of invariance under object rotation. This is due to the fact that, so far, only the non-invariant metrics D^6 and D^{26} have been used. This weakness is not present in the 2D case, where algorithms using invariant metrics—including the Euclidean one—are available. Another problem that also mostly (but not uniquely) affects 3D skeletonization is that the methods require isotropic data, i.e., square pixels in the 2D case and cubic voxels in the 3D case, whereas many image acquisition devices produce elongated elements. Therefore, in these cases, interpolation is necessary prior to skeletonization.

A third weakness regards pruning. Pruning is a necessary part of skeletonization both to reduce sensitivity to noise and to eliminate less significant parts of the skeleton. Our pruning methods are based on information preserving criteria and involve thresholding a measure of significance. It would be interesting to consider improvements in measures of significance discussed in this chapter by making use of the integral medial measures described in Chapter 3. Even with the improvement suggested, some parameters and thresholds have to be fixed. The result is a need for fine tuning for each particular application.

Acknowledgements We thank Dr. Stina Svensson for her ongoing collaboration and for developing several of the algorithms and implementations described in this chapter.

Chapter 6
Voronoi Skeletons

Gábor Székely

Abstract In this chapter we discuss a medial axis computation technique for a discrete binary object based on the Voronoi diagram of a point sample obtained from its boundary. The method is developed for both 2D and 3D objects. A number of issues must be dealt with including the handling of medial axis topology, adequately sampling the boundary, and the pruning of branches that represent insignificant boundary details. Generally, the methods of this chapter successfully deals with these issues in 2D but face some difficulties in 3D. These issues in 3D are dealt with further in Chapter 7.

6.1 The Voronoi Skeleton and Its Extraction in 2D

In Chapter 5 we discussed the computation of voxelized skeletons of binary images using operations based on the digital Euclidean distance function. We now discuss the computation of piecewise flat, continuous skeletons from point clouds sampled from a boundary, using the Voronoi diagram, i.e., the boundary between regions whose points are closer to one boundary point than any other. Computing the Voronoi diagram (Klein, 1987a,b; Ogniewicz, 1993) uses well developed discrete methods from computational geometry and thus has the advantage of being geometrically correct and algorithmically efficient. The skeleton is derived from the Voronoi diagram using a pruning method.

6.1.1 Basics

Under the assumption that the binary object is discretized using an isotropic Cartesian grid, the image plane is tessellated into square pixels. Whereas it is common to

G. Székely
Computer Vision Laboratory, ETH Zurich, Switzerland,
e-mail: szekely@vision.ee.ethz.ch

K. Siddiqi and S. Pizer (eds.) *Medial Representations – Mathematics, Algorithms and Applications*.
© Springer Science + Business Media B.V. 2008

vertices with degree four and higher will occur. Such situations are, however, unavoidable for regular raster sampling.

The above theoretical problems can be mitigated by the use of appropriate modifications to the Voronoi generation algorithm and the use of pruning procedures to eliminate skeletal branches due to boundary noise. These problems could also be avoided by determining the object boundary with with sub-pixel accuracy. As a result, we have a method for approximating the skeleton which is geometrically correct asymptotically. As we shall show later in this chapter, using the neighborhood relations defined on the original object boundary, one can also define an appropriate notion of topology preservation and ensure that it is respected.

6.1.2 The Boundary Sampling Problem

Because uniformly dense sampling at the resolution defined by the usual image grid leads to an enormous number of generating points, many efforts have been spent on sub-sampling the boundary. Sampling homogeneously from the point of view of *information content* as to the reconstructed boundary leads to non-uniform sampling geometrically, with more samples taken in highly curved regions of the boundary (Asada and Brady, 1986; Weiss, 1986, 1990). These results are in good correspondence with perceptual evidence resulting from psychophysical investigations (Attneave, 1954).

Some of the theory of Voronoi-based skeletonization requires that the boundary be sampled in a fashion that is uniformly fine. An unprincipled use of non-uniform sampling can certainly be problematic, as illustrated in Fig. 6.4. In this example several sample points have been taken in the vicinity of the corners of the triangle (the high curvature areas), but no samples have been taken along the sides. The resulting Voronoi diagram shows considerable deviation from the expected skeleton of the continuous object. Thus when nonuniformly sampling before Voronoi skeletonization, rather than minimizing the error of the reconstructed boundary we wish to minimize the deviation of (a subset of) the Voronoi Diagram of the discrete points from the skeleton.

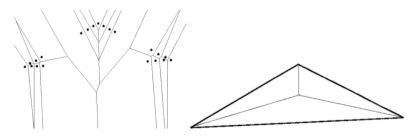

Fig. 6.4 The skeleton of a polygonal object (*a triangle*) on the right and the Voronoi diagram of a set of boundary points generated by an anisotropic sampling along its boundary (*left*)

Owing to the difficulties mentioned above, in this chapter we rely exlusively on maximally dense uniform sampling. This represents somewhat of a brute-force solution. Section 3.3 of Chapter 7 discusses minimal requirements for sampling the boundary non-uniformly.

6.1.3 Generation of the Voronoi Diagram

In computational geometry, Delaunay triangulation is extensively used for handling proximity problems on discrete point sets. Indeed, because the Delaunay triangulation is the dual of the Voronoi diagram (see Fig. 6.5), many of the methods for computing the Voronoi diagram proceed by first computing the Delaunay triangulation and then computing the dual of the result. The Delaunay method triangulates the convex hull of a discrete point set in the plane such that the circumcircle of each triangle does not contain any other generating points.

The duality of the two concepts can be seen by mapping the constituents of the Delaunay triangulation onto the elements of the Voronoi diagram. The (0-dimensional) Delaunay points (the generating points) can be mapped to the (2-dimensional) Voronoi polygons (their proximity region). The Delaunay edges (1D) connect exactly those points which share a Voronoi edge (1D) between their proximity regions defining the mapping. The Voronoi vertices (0D) are actually the centers of the circumcircle of the triangle defined by the equidistant generating points, i.e., they can be mapped onto a Delaunay triangle (2D). In the degenerate case of a Voronoi vertex where more than three Voronoi edges are joined, a polygon with the corresponding number of co-circular vertices will be generated by the Delaunay "triangulation." This can later be triangulated in an arbitrary way. In order to keep the formulation simple, the expression *Delaunay triangle* will always be used without explicitly mentioning the degenerate case of a Delaunay polygon.

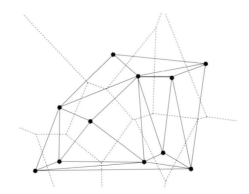

Fig. 6.5 Delaunay triangulation of the previously presented point set. The generating points are shown as black dots, the Delaunay triangulation is given by the solid lines while the Voronoi edges are shown as dotted lines

The dual relationship between the Voronoi Diagram and the Delaunay trinagulation allows us to compute the Voronoi Diagram using algorithms for Delaunay triangulation and also to edit the Voronoi Diagram to prune away unwanted branches and merge others using features directly available in the Delaunay triangulation. This strategy is especially apparent in the methods of Chapter 7.

When dealing with large point sets care has to be taken to avoid numerical instabilities that can arise due to degenerate cases. As a consequence, robust and efficient algorithms for 2D and 3D Delaunay triangulation have become available only relatively recently. The state of the art algorithms for which software is available are discussed in Section 2 of Chapter 7.

6.1.4 From Voronoi Diagrams to Skeletons

The Voronoi diagram of the boundary sample points contains branches approximating the skeleton, but it is conceptually different from the skeleton itself. In its construction only the geometric positions of the sample points has been utilized. In order to handle topological issues in skeleton generation, a topology has to be enforced on this discrete point set.

Proximity relations on the original object boundary can be used to define a neighborhood topology between these points. Informally speaking, the connecting lines between boundary neighbors define a continuous polygonal approximation of the original object that will cut the Voronoi Diagram into an internal and external part corresponding to the internal (endo-) and external (exo-) skeleton by cutting the Voronoi edges separating neighboring points on the contour. The topology of these partial Voronoi diagrams can now be investigated, and homotopy can be established between them and the original object.

The problem can be formulated and solved in an exact mathematical way through the definition of r-regular shapes in mathematical morphology (Brandt and Algazi, 1992). An open set is called to be r-regular if it is morphologically opened and closed with respect to a disk of radius r. It can be shown that if the boundary of an r-regular object is equidistantly sampled with a distance between neighboring points $d_E(\mathbf{x_i}, \mathbf{x_{i+1}}) < 2r$ (a requirement for maximally dense homogeneous sampling of the contour) the boundary of the regular shape will cut the Voronoi Diagram of the point set $\{x_i, i = 1, \ldots, n\}$ into a subpart internal and external to the object region. These are homotopic to the original object (the connectedness of the exoskeleton is only provided by a common hypothetical vertex of infinite branches at infinity); an upper bound on the regeneration error can also be given. For the proofs it is essential that the original continuous object is an r-regular shape, where r must correspond to the sampling density of the image raster. This condition may not hold for many applications, limiting the applicability of the theoretical results. Also see the related work in Section 2.1 of Chapter 7.

Fig. 6.6 The image of a key (*left*), the internal Voronoi Diagram of its sampled boundary and a rough skeleton extracted from the Voronoi Diagram

6.1.5 Topological Organization of the 2D Skeleton

We will now investigate the internal part of the Voronoi diagram, corresponding to the endoskeleton of the object. The Voronoi Diagram and the skeleton itself is a planar graph in 2D, making their topological structure very simple, as illustrated by Fig. 6.6 for the example of the outline of a key. The binary object is shown on the left side, the Voronoi Diagram in the middle, and a rough skeleton containing only axes corresponding to the most dominant symmetries of the object on the right.

Closer investigation of the skeleton presented on the right reveals that this actually consists of two different parts:

- One part of the skeleton basically represents the symmetry between the contour of the hole in the key and its external outline. This part of the skeleton is shown in bold on the figure. This skeletal part is a cycle, and no part of it can be deleted without changing the skeletal topology. This corresponds to the fact that the boundary of the key can only be represented by two unconnected contours. We shall call the part of the skeleton that cannot be further reduced without topological changes the *topological skeleton*. An object with the topology of a disk (i.e., consisting of one single component without a hole) has no uniquely defined topological skeleton, which in this case is just one single Voronoi vertex (i.e., a point). From a topological point of view any of the vertices can be regarded as the topological skeleton of such an object.
- The skeletal branch corresponding to the key-bit represents symmetries originating from one single object contour and is a connected graph without any cycles, i.e., a tree. The tree-like organization can be better observed on a more detailed skeletal representation containing branches corresponding to finer details of the outline, as shown in Fig. 6.7. The appearance of additional skeletal branches can

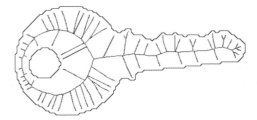

Fig. 6.7 A more detailed skeletal representation of the key contains branches corresponding to final details of the boundary

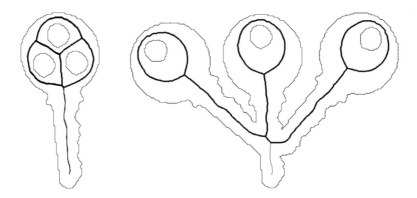

Fig. 6.8 Skeletons of topologically more complex objects. The topological skeleton is shown in bold

also be observed. All of the trees are rooted on the topological skeleton, enforcing a hierarchy between their branches. These skeletal branches can be deleted in the direction from the leaves to the root, without the introduction of topological changes, when the branches are considered to represent insignificant object features.

When an object has several holes the topological skeleton has a somewhat more complex structure, as illustrated by the two examples in Fig. 6.8.

The duality property guarantees that this simple topological organization is inherited by the Delaunay triangulation. Leaves of the skeletal tree correspond to Delaunay triangles that have two sides on the object boundary (the duals of the leave branches) and one cutting through the object (the dual of the Voronoi edge attaching the leaves to the tree). The Delaunay triangle on the other side of this edge becomes the direct descendant of this boundary triangle. By recursion, a partial ordering can be defined over the complete Delaunay triangulation corresponding to the partial order of the dual Voronoi edges imposed by the rooted tree structure of the topologically deletable branches.

6.1.6 The Salience of 2D Skeletal Branches

As discussed in Chapter 1, the issue of boundary noise is a challenge that any successful skeletonization technique must face. As a consequence of the symmetry-curvature duality theorem (Leyton, 1987) a small perturbation on the object boundary can result in a skeletal branch that represents it, as illustrated by the example of an ellipse with a bump in Fig. 6.9. Such perturbations can also be artifacts of discretization on an image raster, which in turn can depend on orientation and scaling. Thus appropriate measures of salience must be incorporated in order to distinguish branches due to noise from those which represent important shape features, such as the major axis of the ellipse. The algorithms of Chapter 4 based on shocks of a grass-fire flow handle this issue by using salience measures that are related to the object angle θ. The methods of Chapter 5 based on digital distance transforms incorporate a notion of elongation (2D) or boundary curve length (3D) as significance measures by which to prune unwanted skeletal branches.

For methods based on the Voronoi diagram of a set of boundary point samples, an enormous number of Voronoi branches are generated as the sampling density increases, as illustrated by the example of the outline of a key in Fig. 6.10. In this example the branches corresponding to salient medial loci are entirely obscured.

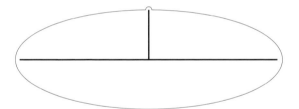

Fig. 6.9 Skeleton of a elliptical contour with a small protrusion

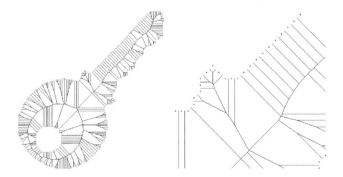

Fig. 6.10 The internal Voronoi diagram of the densely sampled contour of a key (*left*) with a portion enlarged (*right*)

Voronoi-based methods have used two fundamentally different strategies by which to associate salient Voronoi branches with the skeleton:

- *Global salience measures* express the significance of a skeletal branch by measuring the importance of a subpart for the overall appearance of the object. We will show that due to the simple topological structure of the Delaunay triangulation it is possible to associate a subpart of the object with a single Delaunay edge. Duality can then carry this association over to Voronoi edges, i.e., skeletal branches.
- *Local salience measures* estimate the importance of a skeletal point by calculations based on locally defined geometrical measurements on the Voronoi and/or Delaunay constituents.

6.1.6.1 Local Significance Measures

Blum (1973) proposed that skeleton branch importance can be defined in an intuitively appealing way using the grassfire analogy where the velocity of skeleton branch formation measures the "smoothness" of fronts as they collapse at symmetry points. Figure 6.11 illustrates this concept. The original object boundary, shown as a bold line, generates a continuous fire-front. The front positions after some discrete time intervals are shown as dotted lines. The skeleton is developing in this case by a finite speed illustrated by the vector **v**. It can be easily shown that the speed of skeleton formation is determined by the object angle θ at which the fire-fronts meet ($\|\mathbf{v}\| = 1/\sin 2\theta$). As the opposite boundary segments become more and more parallel, 2θ will converge to π, leading to infinitely fast skeleton formation in the limiting case. Thus, measures on velocity in the above sense are essentially analogous to those based on the object angle discussed in Chapter 4.

The significance of skeletal branches has also been studied by the methods of the Tichonov regularization theory (Tikhonov and Arsenin, 1977). Skeletonization can be regarded as a mapping and can be required to be a semi-continuous operation (Yu,

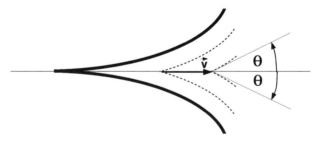

Fig. 6.11 Speed of skeleton formation by fire front quenching. The original object outline is denoted by bold lines, different discrete stages of the fire-front development are shown dotted. The skeleton formation speed (**v**) depends on the angle of the meeting fire-fronts which is twice the object angle θ

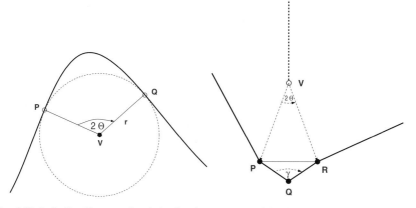

Fig. 6.12 *Left*: Significance of a skeletal point V measured by the arc length 2θ between the boundary points P and Q it is generated by. *Right*: Skeletal branch significance measured by the pointedness of the Delaunay triangles. P, Q and R are three consecutive sample points on the object boundary, forming a Delaunay triangle. The object boundary is shown as a bold line

1989; Yu et al., 1992). This means that the regularized skeleton of a perturbed set must be in a certain Hausdorff parallel set of the skeleton of the original object. It can be shown that regularization can be driven by a continuously changing parameter α, requiring that all points V on the α-regularized skeleton fulfill the following requirement

$$r(1 - \cos\theta)\,\varepsilon\,\alpha,$$

where r is the radius of the maximal inscribed disk and 2θ is the angle between V and the bitangent points P and Q, as illustrated by Fig. 6.12 (left).

It is obvious that the two characterizations are basically equivalent as they both build on twice the object angle θ. All practically used algorithms in the literature for the calculation of local significant measures realize different aspects of this idea and thus these are fundamentally related to the methods of Chapter 4. Several of the successful implementations are listed below.

- Talbot and Vincent (1992) generate a bisector function by directly calculating the angle between the meeting fire-fronts.
- Meyer (1989) as well as Attali and Montanvert (1994) approach the problem by the geometric characterization of the single Delaunay triangles. The importance of a Delaunay edge is measured by the angle between the other two sides of the triangle (γ) as illustrated in Fig. 6.12 (right). The smaller this angle (i.e., the more pointed the triangle) is, the greater the salience assigned to the edge. In fact,

$$\gamma = \pi - \theta,$$

showing the basic equivalence with the other measures in this class.

- As discussed in Chapter 1, skeletal branches correspond to ridges on the distance map of the objects. The behavior of the distance function f can be characterized by its directional behavior orthogonal to the ridge. This 1D function, $f_\perp(\lambda) = f(\mathbf{x} + \lambda \mathbf{e}_\perp)$, will depend on the angle at which the fire-fronts meet. In the limiting case of parallel fire-fronts $f_\perp(\lambda)$ will have a slope ± 1 with a first order discontinuity at the ridge position ($\lambda = 0$). This idea allows the calculation of a significance measure by sampling the discrete Euclidean distance map of the object orthogonal to the skeletal branch to be characterized and quantifying the deviations from the above function resulting from the optimal limiting case (Näf et al., 1996). A formal approach is to use the average outward flux measure developed in Chapters 3 and 4.
- Skeleton formation speed can be directly calculated from the distance map ridges during simulated grass-fire propagation as in Leymarie and Levine (1992).

6.1.6.2 Global Significance Measures

Edges of the 2D Delaunay triangulation connect two points on the object boundary. Consequently, they have the unique feature that they cut the object triangulation as well as the contour into two separate parts, as illustrated in Fig. 6.13. The Delaunay edge connecting the boundary sampling points P and Q is the dual of the Voronoi edge joining the Voronoi vertex R corresponding to the triangle. The edge separates a set of triangles denoted by dashed lines on the image from the rest of the object. The Voronoi duals of these triangles are exactly the Voronoi vertices distal from the vertex R, according to the tree structure of the Voronoi diagram. This separation of

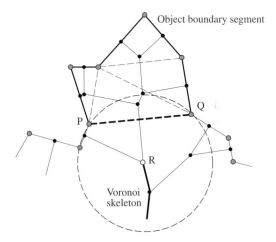

Fig. 6.13 Part-subpart decomposition of a Delaunay triangulation by a Delaunay edge connecting the boundary sampling points P and Q. This edge is the dual of the Voronoi edge joining the Voronoi vertex R corresponding to the triangle. The edge separates a set of triangles denoted by dashed line on the image from the rest of the object

triangles corresponds well to a part-subpart decomposition of the object, suggesting a way to derive an importance measure for the Delaunay edge from the significance of the subpart related to the overall object shape.

Different measures have been proposed to express the importance of a selected subpart. They are usually based on the concept of sculpting, i.e., the removal of the subpart from the object. The significance of the change in the object shape is then quantified by comparing the original outline segment of the boundary between P and Q with the new replacement, i.e., the Delaunay edge. Such a comparison can be made by, e.g., measuring the maximal distance between the original object outline and the replacing Delaunay edge (Brandt and Algazi, 1992) or by measuring the difference between the length of the boundary segment and its replacement, called the chord residual (Ogniewicz, 1993). Other similar measures have also been proposed, e.g., the use of the length of the connecting arc of the maximal inscribed disk instead of the Delaunay edge (Klein, 1987b). Katz and Pizer (2003) have produced a heuristic but apparently successful global measure for 2D based on principles of visual apertures.

Whereas local salience measures are typically used to remove skeletal branches that represent insignificant boundary details, global measures judge form changes associated with the presence or absence of the branch under consideration. These latter measures are particularly important when qualitative notions of part significance are required. An example that is developed in Chapter 10 is the interpretation of 3D skeletons as hierarchical graphs in order to facilitate 3D object model retrieval.

6.1.7 Pruning the 2D Voronoi Skeleton

The use of pruning to remove unwanted branches has been discussed in Chapters 1, 4 and 5. Pruning techniques for Voronoi-based methods depend essentially on two factors: (1) the sequence in which the Delaunay triangles (respectively, Voronoi vertices) are processed and their structural importance is decided and (2) the salience measure used to judge the significance of a single constituent.

The simple topological structure of the Voronoi diagram (respectively, Delaunay cell complex) guarantees a unique processing sequence. The partial ordering of the Voronoi vertices/edges provided by the tree-like organization of the deletable parts of the Voronoi diagram defines the necessary ordering for the establishment of a deletion sequence. Voronoi vertices that are not related by this partial ordering behave completely independently during a deletion process: the deletion of one of them does not influence the topological removability of the other one. Notable exceptions are objects with the topology of a disk, which generate a single tree without a topologically defined root. In such cases only a reasonable significance measure can guarantee the existence of a single, connected, unique, undeletable core of the Voronoi graph overtaking the role of the topological skeleton (which is not uniquely defined in this case) and cutting the single tree into unrelated

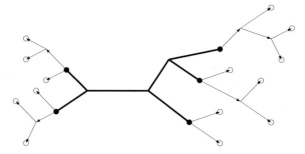

Fig. 6.14 Definition of a skeletal core, playing the role of the topological skeleton for objects that are topologically equivalent to a disk. The edges shown in bold on this hypothetical skeletal graph build the core declared to be undeletable by assigning a salience measure to them exceeding some selected threshold. The roots of the tree-components are denoted by filled circles and the imposed partial ordering from the roots to the leaves (*empty circles*) are denoted by the arrow direction on the deletable edges

rooted subparts, as illustrated by Fig. 6.14. The edges shown in bold on this hypothetical skeletal graph build the core declared to be undeletable by assigning a significance measure to them exceeding some selected threshold. The roots of the tree-components are denoted by filled circles. The imposed partial ordering from the roots to the leaves (empty circles) is denoted by the arrow direction on the deletable edges.

The dual formulation of the pruning process is provided by the sculpting of Delaunay triangles. As mentioned before, leaves of the Voronoi-trees correspond to Delaunay triangles that have two sides common with the object boundary and one cutting through the object. If such a triangle is cut away, its internal side becomes part of the new object boundary, and the other Delaunay triangle joining this edge may become deletable, as illustrated in Fig. 6.13. The recursive application of the above rule leads to an eventual pruning of the object to its topological core.

In order not to delete important object parts, at every step of the deletion process a saliency check has to be applied in addition to the topological deletability check. Any of the previously discussed importance measures can be used for this purpose. The saliency of the Delaunay triangles (respectively, Voronoi edges) are checked against a threshold, and the constituent will be kept if the threshold value is exceeded.

Closer investigation of one of the global measures, the chord residual, reveals that it can be calculated in a recursive way following the same recursion path as the sequential sculpting of the Delaunay triangles. Figure 6.15 demonstrates how the lengths of the original boundary segments can be propagated through the intermediate Delaunay edges during the recursive deletion process as summarized by the pseudo-code of the complete deletion algorithm provided in Fig. 6.16. This gives not only a computationally efficient way to calculate the residual but also proves that the salience measure grows monotonically with the partial ordering ($\prec$) of the triangles. In other words

$$T_i \prec T_j \Rightarrow \Delta R(T_i) < \Delta R(T_j).$$

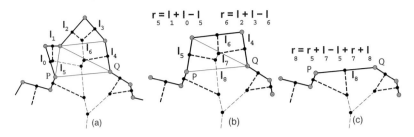

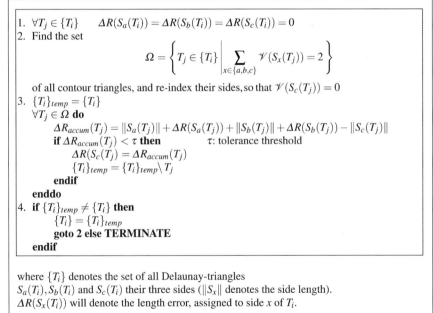

Fig. 6.15 Recursive calculation of the chord residual between points P and Q. Boundary points are shown as patterned circles. Thin solid lines denote interior line segments of the Delaunay triangulation. The sides of Delaunay triangles that are aligned with the boundary appear as bold lines. Dashed lines indicate Voronoi edges. Voronoi edges that separate adjacent boundary points are depicted as bold dashed lines. Voronoi vertices appear as black dots. (**a**) Depicts the original boundary, with the l_i representing the lengths of sides of Delaunay triangles. Triangles (l_0, l_1, l_5) and (l_2, l_3, l_6) fulfill the removability requirements (two sides – l_0, l_1 resp. l_2, l_3 – on the current object boundary), so they can be pruned during the first step. (**b**) Shows the resulting simplified object. The associated (bold) Voronoi edges have also been removed. Symbols r_5 and r_6 denote the residual values ΔR calculated for the new boundary segments l_5 and l_6. Now we can successively remove triangles (l_6, l_4, l_7) and (l_5, l_7, l_8). This final result together with the (recursively) computed residual r_8 is shown in (**c**). At this stage the boundary path between P and Q has been replaced by only one Delaunay edge (l_8). Voronoi edges generated by adjacent boundary points are removed in any case

---

1. $\forall T_j \in \{T_i\} \qquad \Delta R(S_a(T_i)) = \Delta R(S_b(T_i)) = \Delta R(S_c(T_i)) = 0$
2. Find the set

$$\Omega = \left\{ T_j \in \{T_i\} \;\middle|\; \sum_{x \in \{a,b,c\}} \mathcal{V}(S_x(T_j)) = 2 \right\}$$

   of all contour triangles, and re-index their sides, so that $\mathcal{V}(S_c(T_j)) = 0$
3. $\{T_i\}_{temp} = \{T_i\}$
   $\forall T_j \in \Omega$ **do**
   $\qquad \Delta R_{accum}(T_j) = \|S_a(T_j)\| + \Delta R(S_a(T_j)) + \|S_b(T_j)\| + \Delta R(S_b(T_j)) - \|S_c(T_j)\|$
   $\qquad$ **if** $\Delta R_{accum}(T_j) < \tau$ **then** $\qquad \tau$: tolerance threshold
   $\qquad\qquad \Delta R(S_c(T_j)) = \Delta R_{accum}(T_j)$
   $\qquad\qquad \{T_i\}_{temp} = \{T_i\}_{temp} \backslash T_j$
   $\qquad$ **endif**
   enddo
4. **if** $\{T_i\}_{temp} \neq \{T_i\}$ **then**
   $\qquad \{T_i\} = \{T_i\}_{temp}$
   $\qquad$ **goto** 2 **else TERMINATE**
   endif

---

where $\{T_i\}$ denotes the set of all Delaunay-triangles
$S_a(T_i), S_b(T_i)$ and $S_c(T_i)$ their three sides ($\|S_x\|$ denotes the side length).
$\Delta R(S_x(T_i))$ will denote the length error, assigned to side x of T_i.

$$\mathcal{V}(S_x(T_i)) = \begin{cases} 1 \text{ if the side } x \text{ of the triangle } T_i \text{ is currently on the border of the object} \\ 0 \text{ otherwise} \end{cases}$$

Fig. 6.16 Regularization algorithm for 2D Delaunay triangulations

It can be shown that all the above mentioned global salience measures share this monotonicity property of the chord residual. This provides the advantage that skeletal parts can be determined by *simple thresholding* since the monotonicity property will guarantee that topology is preserved. Unfortunately, none of the local salience measures has this property since they depend essentially on the object angle θ which varies according to the degree to which the corresponding boundary segments are parallel. Therefore, such measures have to be used in conjunction with a method for checking that topology is preserved. However, the requirement that topology be preserved can in fact be a limitation for certain applications. For example, the presence of even a short skeletal branch with a high object angle can result in longer branches that are connected to it being preserved, even when they do not have high salience.

6.1.8 A Hierarchy of Skeleton Branches

We will illustrate the skeletonization process on a single 2D coronal slice of the outline of a brain. While this example is certainly somewhat artificial, it addresses the basic aspects of brain shape characterization. It also serves as a useful test case to illustrate skeletonization in 3D in Section 6.2.

Figure 6.17 (top left) shows a coronal slice from the surface of a human brain segmented from high resolution MRI data. In this example we wish to analyze the sulcal foldings, represented by the exoskeleton. In order to avoid technical difficulties caused by infinite skeletal branches, we describe the sulcal structure by the endoskeleton of a negative mold of the brain corresponding to the image background, i.e., the black region. The Voronoi diagram of a dense sample of boundary points is shown in Fig. 6.17 (top right).

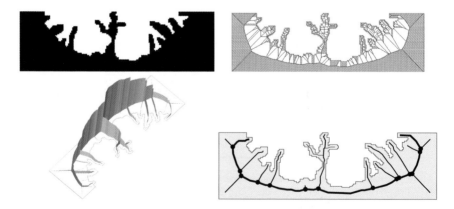

Fig. 6.17 *Top left*: A coronal slice through a brain surface segmented from high resolution MRI data. *Top right*: The Voronoi diagram generated by a dense sampling of boundary points. *Bottom left*: The value of the chord residual for the above example, shown as a height map for each of the skeletal branches. *Bottom right*: Skeleton of the brain slice after thresholding to remove branches caused by discretization effects

The value of the chord residual can then be calculated for all of the branches of the Voronoi diagram. The resulting significance measures are visualized in Fig. 6.17 (bottom left). The residual value is shown as the height of a sheet positioned at the skeletal branches. Theoretical considerations can be used to generate a reasonable threshold value for the residual in order to get rid of skeletal branches caused by discretization effects. The skeleton, shown in Fig. 6.17 (bottom right), can be extracted through simple thresholding of the Voronoi branches.

Skeletal branch significance provides a basis to build up a scale-space of a shape based on the relation of the symmetry axes (Pizer et al., 1987; Ogniewicz, 1993). Monotonic measures have been used, e.g., to generate different hierarchy levels of the branches (Ogniewicz and Kübler, 1995). The chord residual defines a Voronoi vertex with the highest residual, clearly visible in Fig. 6.17 (bottom left). It is reasonable to select this vertex as the topological skeleton of the object. Starting from this point, the Voronoi edges can be traced consecutively along the skeleton in both directions. At every branch point the path with the larger residual value can be followed. In this fashion a single non-branching skeletal line can be extracted, which is shown as a bold line in Fig. 6.17 (bottom right). This skeletal segment, which we call the first-order skeleton, represents the most salient aspects of the object according to the chord residual function. In degenerate cases where nearly equivalent skeletal branches join, the first-order skeleton may be branched as well.

In a subsequent step the branches neglected in this first pass are processed and followed in a similar fashion, resulting in a set of second-order skeletal branches. Recursive processing generates a complete hierarchy between skeletal branches, as illustrated in Fig. 6.17 (bottom right). The depth in the hierarchy is coded by the width of the skeletal lines. The first-order branch points (between the first-order skeleton and its second-order branches) are denoted by filled circles.

The skeletal hierarchy is a very powerful tool for addressing two different aspects of an object's shape which are essential to form analysis. Lower order skeletal branches represent global shape features and can be selectively used for categorical object recognition, i.e., to identify or match representatives of a specific class of shapes. In fact, this is the strategy adopted in Chapter 10 where the 3D skeleton is interpreted as a directed graph (but using a different salience measure by which to determine a part-subpart hierarchy), which then is used for 3D object retrieval. On the other hand, higher order branches describe the fine, local variations of individual objects, which can be used for their unique characterization. Thus the hierarchical skeleton can serve as a generic, multi-scale shape descriptor which unifies coarse and fine features.

6.2 The Voronoi Skeleton in 3D

The success of Voronoi skeleton generation in 2D has led to several attempts to implement the same ideas for the 3D case; see (Brandt and Algazi, 1992; Reddy,

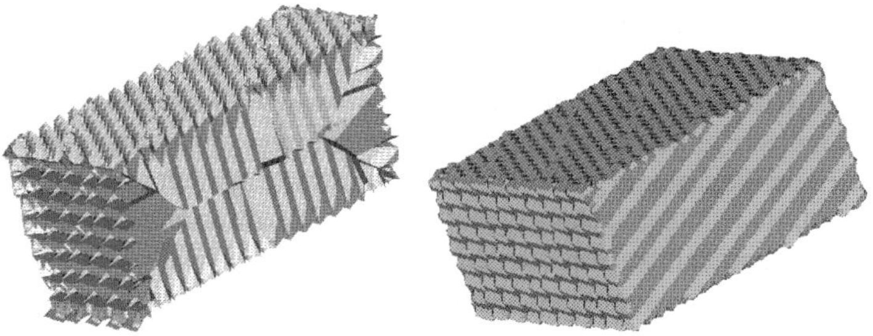

Fig. 6.19 *Left*: The internal Voronoi diagram of dense sample points on a surface of a box in an oblique position. The constituents of one of the long faces are removed to show the deeper structure. *Right*: The dual of the Voronoi diagram, i.e., the corresponding Delaunay tetrahedralization

topologically correct internal and external part, as in 2D. Figure 6.19 (left) shows the internal part of the Voronoi diagram of a more densely sampled box in an oblique position. The Voronoi edges separating neighboring boundary points are not shown, and the constituents of one of the faces are cut away to reveal the deeper structure of the Voronoi diagram.

The dual of the Voronoi diagram, the Delaunay tetrahedralization, is shown in Fig. 6.19 (right). The object is now built up from Delaunay tetrahedra. In analogy to the 2D case, co-sphericity leads to degenerate elementary cells that are actually Delaunay polyhedra. Handling such polyhedralization is more demanding than tetrahedralization, sometimes causing technical complications. Nonetheless, this is never a theoretical problem. Therefore we will speak about Delaunay tetrahedralization and usually neglect the special aspects of polyhedral representation.

The basic difficulty in 3D skeletonization using Voronoi-based methods is the topological structure of the Voronoi diagram (respectively, Delaunay tetrahedralization). In the 2D case cycles of the Voronoi graph (i.e., the graph consisting of the Voronoi edges and vertices) correspond to medial axes separating non-connected contour parts. As parts of the topological skeleton none of them can be removed from the Voronoi graph without a change in topology. However, this is not the case in 3D. There are usually many cycles representing insignificant features in the topologically deletable part of the Voronoi graph. Therefore, there is no longer a way to establish a well defined partial ordering as was possible in 2D. This lack of partial ordering causes a dependency between Voronoi edges because the deletion of one constituent can change the topological relationships between the others.

This unpleasant property is inherited by the Delaunay tetrahedralization. There are usually many Delaunay cell candidates for deletion in a sculpting process. In contrast to the 2D case they cannot be deleted independently because their topological environments interfere with each another. The situation is somewhat similar to the deletion problem in thinning procedures. Thus a deletion sequence has to be selected more or less arbitrarily.

The topological deletability of Delaunay cells can be decided locally in 3D as well. Simple rules have been proposed for Delaunay tetrahedra (Attali and Montanvert, 1994) and these are similar to the ones in 2D. Exact formulation of such rules for polyhedra in general is more complicated, but sufficient conditions for topological deletability of single Delaunay polyhedra have been proved using the theory of cell complexes (Kovalevsky, 1987, 1989). These are based on the decomposition of the surface of a Delaunay polyhedron into components containing only outside elements (Delaunay constituents that are on the surface of the object) and only inside ones (cutting through the inside of the object). A polyhedron can be deleted without topological change, if there are exactly one inside and one outside component on the boundary separated by a non-intersecting closed boundary line (Näf et al., 1996; Näf, 1996). These rules are the Voronoi counterparts of the rules for homotopy preserving thinning on a digital lattice discussed in Chapters 4 and 5.

6.2.3 The Salience of 3D Skeletal Branches

The basic constituents of a 2D Voronoi skeleton are 1D skeletal curves resulting from the association of Voronoi edges. In 3D the Voronoi skeleton consists of skeletal curves (built from Voronoi edges) and skeletal sheets (resulting from Voronoi faces). As discussed in Chapter 1, the type of local object symmetry can be detected by an eigenvalue analysis of the the Hessian matrix of the Euclidean distance function. Thus it is possible to distinguish the cases of skeletal curves (which arise from nearly tubular structures) and skeletal sheets and to use appropriate salience measures for each.

In 2D, salience measures were defined for Voronoi edges. In 3D we must come up with salience measures for Voronoi faces and Voronoi edges in a consistent way. In the following we will investigate such measures for Voronoi edges (or for their dual, the Delaunay faces) and for Voronoi faces. The extension of these measures to other type of constituents can be made heuristically. The importance of a Voronoi face can also be defined, for example, as a function of the importance of its bounding edges.

6.2.3.1 Global Salience Measures

The simple topological structure of the Delaunay triangulation in 2D allowed a natural association of an object part with a Delaunay edge since each internal Delaunay edge always cuts the object into two disjoint components. Thus in 2D we were able to define global salience measures based on the contribution of a subpart to the overall shape of the object. Unfortunately this topological cell organization is missing in the 3D Delaunay tetrahedralization. In particular, faces of a Delaunay tetrahedron cut the object into two parts only in degenerate cases. Therefore alternative strategies have to be employed to extend such measures to 3D.

The most appealing possibility would be to use some well defined properties of the 2D decomposition as a basis for generalization. In 2D, topology preservation provides enough constraints to lead to a unique solution. One could try to generalize this idea to 3D by looking for a collection of Delaunay tetrahedra making the cell attached to a Delaunay face topologically deletable. Unfortunately, in contrast to the situation in 2D, there is no unique way to select such a set of Delaunay cells.

Another alternative is to generalize the chord residual function (Székely et al., 1992) to 3D. One can regard this measure as the difference between the minimal Euclidean distance of two boundary points in the 2D image plane (the length of the Delaunay edge) and the minimal distance along the object boundary (the length of the smaller boundary segment). In 3D this would imply a comparison between the Euclidean distance in the 3D volume with the geodesic distance on the object surface. This generalization is not only computationally expensive (the estimation of geodesic distance for all surface point pairs is needed) but also compromises the monotonicity property of the 2D measure.

The definition of monotonicity in 2D relied on the partial ordering of the Voronoi edges. Since in 3D no natural partial ordering exists, the direct extension of this definition is questionable. It may be possible to define monotonicity indirectly by its most important property in 2D. That is, a 3D salience measure would be regarded as monotonic if simple thresholding guaranteed the preservation of skeletal topology.

Due to the difficulties caused by the topological cell structure, some decomposition strategies use geometric rather than topological criteria. For example, the use of "visibility criteria" based on a viewcone defined by the local geometry of the Voronoi or Delaunay cells, has been proposed by Attali (1995) and Näf (1996). The selection of minimal closed curves on the object surface between generating points has also been studied (Näf, 1996). Unfortunately, neither of these criteria seem to deliver reasonable decomposition strategies.

6.2.3.2 Local Salience Measures

Owing to the lack of reasonable global measures, local ones play a central role in 3D skeleton generation. In fact, several of the local salience measures have been generalized to 3D. The pointedness measure can be calculated from the angle of the two opposite sides of the tetrahedron to be removed, as illustrated by Fig. 6.20 (Attali and Montanvert, 1994; Attali, 1995). The analysis of the ridges of the Euclidean distance function, based on comparison with the optimal 1D ridge function, can be done similarly as in 2D (Näf, 1996). Finally, the speed of skeleton formation can be estimated from the local analysis of changes of the radius function along the skeleton (Brandt and Algazi, 1992); as already mentioned, this measure is fundamentally related to the object angle based measures discussed in Chapter 4.

Due to the lack of monotonicity, these measures cause similar problems during skeletal pruning as in 2D. Locally salient but globally insignificant Delaunay cells can block the pruning process, and the measures are unable to detect connection branches of the skeleton. Still, due to the absence of alternatives, local measures are the preferred choice for simplifying Voronoi skeletons in 3D.

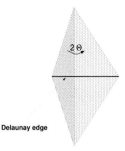

Delaunay edge

Fig. 6.20 The pointedness measure in 3D based on the angle of the two Delaunay faces opposite to the edge to be characterized

6.2.4 Pruning the 3D Voronoi Skeleton

6.2.4.1 Sequential Pruning in 3D

The strategy of simplifying the Voronoi skeleton by sequential pruning is identical to the 2D procedure. We need to select a processing sequence for the Delaunay cells and to check the deletability of the cells according to this predefined sequence. Topologically deletable tetrahedra will be cut away if their importance falls below a predefined threshold according to a selected significance measure.

Due to the lack of topological order discussed above, the critical component of the pruning procedure is the definition of a reasonable processing sequence. Different strategies can be followed to select among the numerous solutions which respect the existing topological dependencies. In analogy to the 2D case the uniqueness of the topological skeleton is lost for objects with spherical topology, where any Voronoi vertex can take on this role. As illustrated in Fig. 6.21, it is also lost for objects having handles, where any closed line on the medial surface can represent the object's topology.

Once the processing sequence has been established and the Delaunay tetrahedra are ordered, we can speak more easily about monotonic importance measures, by requiring that the relevance of the Delaunay cells grows continuously during the sculpting process. Theoretically one could imagine the construction of a cumulative 3D measure similar to the 2D procedure summarized in Fig. 6.15. The basic problem that arises in the generalization is that the deletion of a Delaunay cell may (and usually does) lead to the birth of more than one new boundary face. This is an unpleasant difference from the 2D situation, where a single new boundary face replaces the old ones. It is still possible to propagate the measure to all of the new boundary faces, but care has to be taken to not count the contributions multiple times. Two possible strategies can be used to avoid multiple contributions:

- The measure to be propagated is distributed between the uncovered new faces. This solution has the problem that the sharing of the accumulated measure between the new faces has to be decided arbitrarily.

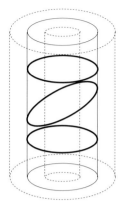

Fig. 6.21 The skeleton of an infinite cylinder with a bore hole. The object boundary is sketched by dotted lines, the skeletal sheet is shown by thin solid line. The bold solid closed curves on the medial surface are examples of skeletal curves representing the object's topology. Any of them can serve as the topological skeleton of the object

- It is possible to propagate the contribution fully to all new faces while maintaining a bookkeeping in order to avoid duplication during later stages of the propagation. Because this solution can be implemented without arbitrary selections, it is more attractive even though it is technically somewhat more demanding.

Although such a measure can be realized in practice, this process still cannot solve the basic problem of 3D skeleton simplification as it is dependent on the deletion sequence, which was selected arbitrarily.

Different strategies have been proposed in the literature to provide a controlled and possibly reproducible selection of a processing sequence. For example, similar to parallel 3D thinning, it seems reasonable to simulate a peeling of the object by defining deeper and deeper topologically deletable layers of the Delaunay cells (Näf, 1996; Attali, 1995). Different geometric strategies have been investigated to define a perceptually appealing compromise between such layer definition and depth-first processing strategies trying to minimize the number of deleted tetrahedra for reaching a selected Delaunay face (Näf, 1996). A second approach is to use a directed graph between the Delaunay tetrahedra to define a sufficient partial ordering between them. This graph must be built based on the existing topological dependencies and must be extended to a sufficiently complete ordering to perform the sequential deletion. The definition of these additional dependencies can rely on a peeling-like strategy with additional heuristics (Attali, 1995).

6.2.4.2 Skeleton Simplification by Incremental Growing

Since the structure of the 3D skeleton does not support the approach of processing inwards from the leaves used in 2D, the strategy of dealing with the the most prominent branches before the more complex collection of the typically less important

parts seems preferable in 3D. Such an outward growing strategy is strongly rem-
iniscent of the process of the creation of the hierarchical skeleton in 2D. It also
bears a similarity to the topological reconstruction process used in Malandain and
Fernandez-Vidal (1998).

This outward processing (which is not a pruning, but a successive creation of
the skeleton) can be summarized as follows. We begin by selecting the topological
skeleton. We have seen that in general it is not unique and that therefore additional
constraints have to be used to select it. One possibility is to prefer the selection of the
components for the construction of a topological skeleton which lie deepest within
the object. This skeletal seed is extended by following the most prominent branches
until the object boundary is reached. One can then search for unrepresented but
salient object features by looking for prominent Voronoi constituents that are not
yet included in the skeleton, or by trying to reconstruct the object from the actual
skeleton and comparing it with the original. This growing process can be applied
iteratively until satisfactory object representation is achieved.

The implementation of such growing strategies are strongly impeded by two
basic difficulties. First, the local measures are usually not sensitive enough to reliably determine which path should be followed at branching. Second, the selection
of a reasonable topological skeleton is only easy for objects with simple topology.
For more complex cases the enforcement of topology preservation is much harder
than for inward processing.

6.2.5 Interactive Generation of Skeletal Hierarchy in 3D

We have seen that in the 2D case a skeletal hierarchy provides a coarse to fine
organization of skeletal branches that in turn can be useful in a number of applications. Owing to various issues discussed above, including the lack of a unique
partial ordering of skeletal constituents and the lack of salience measures that have
a monotonicity property, the automatic generation of a unique hierarchy of skeletal
branches in 3D is a challenge. It is, however, possible to enforce a hierarchical
representation by settling on a particular combination of global and local salience
measures while enforcing a notion of topology preservation. This is the strategy
adopted in the 3D object retrieval application discussed in Chapter 10.

An alternative is to allow some degree of user interaction. In fact, experiments on
large and complex datasets have been shown that the development of computer supported interactive navigation tools for the definition of skeletal hierarchy is feasible
and practicable. Such procedures are only possible if the single Voronoi faces are
aggregated to larger global skeletal sheets. Corresponding aggregation algorithms
have been proposed and implemented even in a recursive manner, grouping the small
Voronoi faces into a computationally and manually manageable set of larger skeletal sheets (Näf et al., 1996; Näf, 1996). A skeletal hierarchy can then be defined by
manual guidance using the developed interactive tool-box.

6.3 Application Examples

We now provide a number of examples of the use of Voronoi skeletons for 3D shape analysis and characterization. These computational results are feasible, despite the challenges discussed above, by careful implementation and by the appropriate choice of salience measures for each application.

6.3.1 Skeletons of Artificial 3D Objects

Figure 6.22 demonstrates the power of existing 3D skeletonization techniques on a few artificially created objects (Näf et al., 1996; Näf, 1996): a box, a cylinder, and a more complex object resulting from the union of a box and an intersecting cylinder. These examples illustrate how the dimension of the 3D Voronoi skeleton changes to adapt itself to the type of symmetry being represented (tube or slab). These results compare very favorably to the results in Chapter 4, Figs. 4.4 and 4.6.

6.3.2 Bone Thickness Characterization Using Skeletonization

The optimal fixation of the cup in total hip joint replacement procedures requires general knowledge about the thickness of the surrounding bone structures. Such knowledge can be extracted from the analysis of the anatomical variation in a selected training population. Quantitative evaluation, however, is not possible without adequate representation of the hip bone. Skeletal description provides a natural basis for this analysis and has proved to be useful for the characterization and visualization of the thickness of the hip bone around the acetabulum.

The upper row of Fig. 6.23 shows a 3D rendered fusible stereo pair of a hip bone to be analyzed. The dataset contained 49,733 boundary points, which produced 65,073 Delaunay polyhedra and 102,447 elementary Voronoi faces. After a first regularization step using the ridge strength as significance measure, 29,928 elementary Voronoi faces were left. The remaining faces can be sewn together at edges where only two of them meet. This aggregation of individual faces into groups representing figures produced 1,218 face groups. An analysis of these groups shows that many of them represent insignificant skeletal parts. Hence a second regularization step acting on these face groups and using the overall area of the group as significance measure was carried out. All groups with an overall area of less than 10 pixels have been removed provided that homotopy equivalence to the object could be guaranteed, leaving a total of 28,196 faces. Obviously this produced an additional number of edges joining only two elementary Voronoi faces. Hence the aggregation procedure has been repeated which led to 179 face groups. The lower row of Fig. 6.23 presents a fusible stereo pair displaying the resulting skeleton.

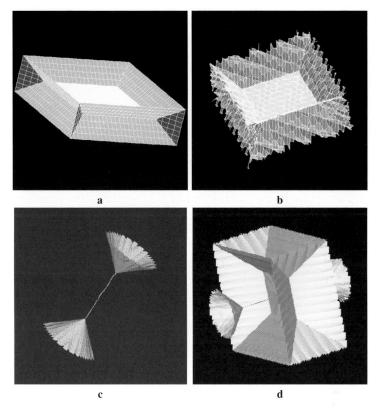

Fig. 6.22 Simplified Voronoi skeletons of some artificial objects. The upper row shows skeletons of a box, once aligned along the coordinate axes (**a**) and once in an oblique position (**b**). (**c**) Shows the skeleton of a cylinder, while (**d**) demonstrates the generated medial axis of a more complex object resulting from the union of a box and an intersecting cylinder

For the visualization of the skeleton the use of elementary Voronoi faces is acceptable. However, for the representation and analysis of the 3D shape of the organ, the face groups that represent global medial surfaces are used.

Figure 6.24 shows the single-figure acetabulum medial surface that has been produced by the aggregation procedure. The planning of surgical procedures require the identification of thick bone areas capable of supporting the hip joint replacement prosthesis. In order to visualize and identify such areas, the medial surface of the acetabulum has been colored by the local bone thickness. The light areas, marked by arrows, denote optimal regions for prosthesis support.

6.3.3 Analysis of the Cortical Structure of the Brain

Neuroanatomical and histological findings from post-mortem brains and in vivo findings from MRI studies suggest the presence of morphological temporal lobe

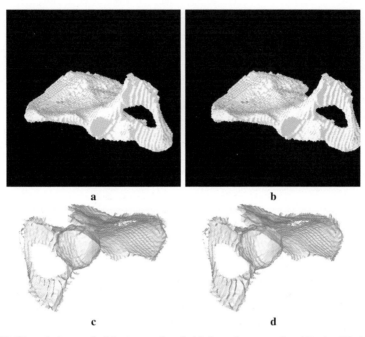

Fig. 6.23 3D renderings as fusible stereo pairs of a hip bone (*upper row*) and its simplified Voronoi skeleton (*lower row*)

Fig. 6.24 The medial surface representing the acetabulum of the hip bone colored by the local bone thickness (*from black to white*). The arrows indicate optimal areas for prosthesis support

abnormalities in schizophrenia. To determine whether or not sulco-gyral pattern abnormalities in the temporal lobe could be detected in vivo, computerized surface rendering techniques for MR data have been developed. These facilitate qualitative and quantitative analysis of the three-dimensional structure of the temporal and frontal cortex. 3D renderings of the brain surface have been used to determine characteristics of the sulco-gyral patterns correlating with clinical findings (Kikinis et al., 1994).

One of the serious limitations of this analysis is that the structural description of the brain surface has been derived from a single 2D view of the rendered data, ignoring the 3D structure of the cortex. The usage of true 3D shape features for the description of the cortical structures is essential for more precise and reliable statistical analysis of the data.

Skeletal representation offers a promising way to generate more precise descriptors of the sulco-gyral foldings. In one study, the white brain matter extracted from an MRI acquisition was processed by 3D Voronoi skeletonization software. The dataset, shown in the upper row of Fig. 6.25, contained 205,848 boundary points, which produced 300,563 Delaunay polyhedra and 488,504 elementary Voronoi faces. After regularization 87,205 elementary Voronoi faces were left. The original data and the resulting elementary Voronoi faces of the skeleton are shown as a fusible stereo pair in the lower row of Fig. 6.25. In this case the Voronoi faces are

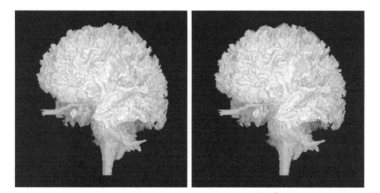

a   b

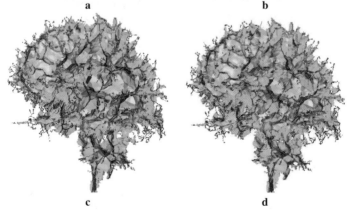

c   d

Fig. 6.25 Original object (*upper row*) and 3D Voronoi Skeleton (*lower row*) of a human brain's white matter, shown as fusible stereo pairs. The skeletal faces are colored according to the importance measure (ridge-strength of the distance map) used in the regularization process

colored according to the salience measure used in the regularization process, which is based on ridge-strength of the distance map.

The resulting skeletal structure is still very complex and cannot be used directly for the generation of 3D shape features. The extraction of a few dominant sheets is carried out using a procedure similar to that peformed for the hip bone skeleton. Following regularization and aggregation of face groups the number of face groups is reduced to 10,684, and these groups are in turn aggregated a second time, representing the skeleton of the temporal lobe. A further manual aggregation is done with the help of tools supporting the visualization of the face groups according to their size and neighborhood relations and the generation of a hierarchical ordering between them. This procedure results in a compact skeletal description consisting only of a few most prominent skeletal branches. The shape of the cortical surface, particularly its gyration pattern, is then coded in the branching of the individual skeletal sheets. This allows for the extraction and quantitative description of the branching lines between neighboring skeletal sheets, leading to truly 3D descriptors of the sulco-gyral structure of the temporal lobes.

A similar application of the 3D skeleton to the analysis of brain sulci is discussed in Chapter 11, and Chapter 9 develops the use of the 3D Voronoi skeleton for statistical shape analysis of a data set of hippocampi segmented from MRI data.

6.4 Discussion

This chapter has reviewed the use of Voronoi diagrams for skeletonization in 2D and 3D. This construction proves to be a powerful tool for determining medial loci when the bounding surface of an object is represented by a set of discrete point samples.

The technique has several strengths, which include the following: (1) The construction of skeletons in 2D and 3D essentially follow the same steps of boundary point sampling, Voronoi diagram construction and skeletal branch detection while ensuring homotopy type. (2) Efficient and robust methods for computing Voronoi diagrams are now available for both 2D and 3D data sets. (3) The methods permit the use of both local and global salience measures to determine which skeletal constituents to retain and which to delete. (4) Hierarchical representations based on an interpretation of the Voronoi skeleton are possible, supporting applications to object retrieval and coarse to fine shape analysis.

Voronoi skeletonization also suffers from some limitations, which include the following: (1) The method is sensitive to the density of boundary samples and can suffer from artifacts due to boundary perturbation and/or discretization, thus requiring the use of salience measures. (2) The order of processing for pruning in 3D is heuristic due to the inherent lack of a partial ordering of Voronoi faces. Furthermore, global salience measures that respect a monotonicity property are not available. (3) The aggregation of Voronoi faces in 3D to form larger skeletal sheets is based on heuristics. (4) The association between the resulting Voronoi skeletons and the generic types of medial loci reviewed in Chapters 1 and 2 remains a largely open problem, particularly in 3D.

The following chapter develops new methods for selecting the boundary points from which the Voronoi diagram is computed and for editing the resulting set of Voronoi edges to produce better approximations to medial loci. These methods, appplicable in both 2D and 3D, are based on theoretical results on the quality of such approximations that are reviewed in that chapter.

Chapter 7
Voronoi Methods for 3D Medial Axis Approximation

Nina Amenta and Sunghee Choi

Abstract As shown in the previous chapter, the medial axis of an object can be approximated using a subset of the Voronoi diagram of a sample of points taken from the object surface. In this chapter we focus on the 3D case and review some theoretical results on the approximation of medial axes by Voronoi diagrams. We present algorithms for selecting the boundary point samples used to produce the Voronoi diagram and for selecting parts of the Voronoi diagram as a step in constructing more accurate medial axis representations. We also state some open questions that concern the quality of these algorithms and their dependence on the sampling density.

7.1 Introduction

In Chapter 6 the use of the Voronoi diagram as a tool for detecting medial loci of discrete objects starting with point samples from their boundaries was discussed. Now that there are good programs available for computing 3D Voronoi diagrams, this method is both simple and practical. However, we must define ways to determine the right boundary sampling for 3D and to edit the result so as to produce a good approximation of the 3D medial axis of a surface F using a subset of the Voronoi diagram of a set of surface points. Theoretically, we provide a variety of results that show that given an appropriate sample S from the surface F, some subset of the Voronoi diagram $\mathcal{V}(S)$ forms a good approximation to a stable subset of the medial axis $\mathcal{M}(F)$ of F. Practically, we describe how to edit in the Delaunay domain so as to produce the desired approximation.

N. Amenta
Department of Computer Science, University of California at Davis, USA,
e-mail: amenta@cs.ucdavis.edu

S. Choi
Computer Science Division, EECS, Korea Advanced Institute of Science and Technology, Korea,
e-mail: sunghee@cs.kaist.ac.kr

K. Siddiqi and S. Pizer (eds.) *Medial Representations – Mathematics, Algorithms and Applications*.
© Springer Science + Business Media B.V. 2008

As discussed in the previous chapter, methods based on Voronoi diagrams are particularly appropriate in the situation in which the input consists only of the sample S, for instance when the input is a set of points in 3D produced by a set of aligned laser range scans. In this case the medial axis approximation provided by the Voronoi diagram of the points is as good as one is likely to get. Most of the available algorithms simultaneously produce both an approximate medial axis M and a dual approximate surface representation[1] T.

Voronoi-based methods are also used when F is given as an implicit, parametric, or polygonal surface, for instance in CAD applications. In this case we first choose a set S of point samples and then use its Voronoi diagram to approximate the medial axis. In this situation we can control the choice of S and hence the quality of the approximation.

One of the attractive features of Voronoi diagram-based approximations to the medial axis is that there are some theoretical bounds on the approximation quality. Certainly when F is given explicitly, we can show that the medial axis approximation M is topologically correct. Specifically, if we assume for simplicity that $F \subset Q$ where Q is a bounded region of $\mathbb{R}^3$, we say that M is topologically correct if it is homotopy equivalent to $Q - F$. Even when F is not given explicitly but as a dense sample of points, we can usually prove that M is topologically correct under the assumption that S is indeed sufficiently dense. Results related to those presented in this chapter are also discussed in the recent survey article of Attali et al. (2004).

7.2 Approximating the Medial Axis

It has long been observed (Blum, 1967) that if S is a dense set of point samples from a curve in the plane, part of the Voronoi diagram of S forms an approximation M of the medial axis of the curve, as in Fig. 7.1. In the plane, it is easy to see which part of the Voronoi diagram to choose: the Voronoi edges that do not cross the curve are exactly those that belong to M. Recall from Chapter 6 that in 2D a Delaunay edge between points p and q of S is dual to a Voronoi edge separating the Voronoi cells of p and q (see any textbook on computational geometry, for instance deBerg et al., 1997). The Delaunay edges can be separated into two subsets, those dual to edges in M, and the remainder, T, which forms a piecewise-linear approximation to the original input curve. So the Delaunay triangulation $\mathscr{D}\mathscr{T}(S)$ contains a curve approximation T, and the Voronoi diagram $\mathscr{V}(S)$ contains a medial axis approximation M, with the Delaunay edges partitioned into $T + \text{dual}(M)$, and the Voronoi edges partitioned into $\text{dual}(T) + M$. Notice that in 2D, all the Voronoi vertices are adjacent to edges in M. However, as discussed in the previous chapter, extending this insight to 3D is not completely straightforward. Nonetheless, this basic approach has led to practical methods for computing an approximation to the medial axis.

Approximating the medial axis is important since computing the exact medial axis of a given surface directly is difficult, because of the high algebraic degree of

---

[1] In the rest of the book this set is notated **b**.

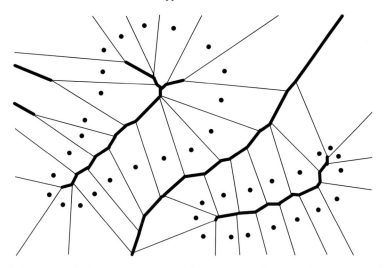

Fig. 7.1 Given a set S of points densely sampled from a curve, the Voronoi edges can be divided into two sets: those that cross the curve (light) and the remaining set M that approximate the medial axis (heavy)

the equations that must be solved exactly, even for polyhedral inputs. In 2D there is a robust program due to Held (2001) for the Voronoi diagram of line segments and circular arcs. Culver et al. (1999) give an exact medial axis computation for 3D polyhedra. It uses a tracing algorithm, which traverses the edges of the medial axis, discovering the vertices in the process; earlier tracing algorithms include those of Hoffmann (1990), Milenkovic (1993) and Sherbrooke et al. (1996).

Approximating the medial axis using the Voronoi diagram avoids these difficult algebraic computations, but requires robust programs to compute the Voronoi diagram of many points, which have only recently become available. A major problem was that a naively written Delaunay triangulation program will often crash because of numerical instabilities. In 1995, Jonathan Shewchuk introduced the program *Triangle*, which is reliably robust and quite efficient; it can compute the two-dimensional Delaunay triangulation of millions of points per minute. In three dimensions, the CGAL computational geometry library (CGAL, 2004) includes a robust three-dimensional Delaunay triangulation program which can reliably compute the Delaunay triangulation of a million points in about five minutes. In higher dimensions, Ken Clarkson's program *Hull* (Clarkson, 1992) can be used.

7.2.1 A Few 2D Results

The idea of using the Voronoi diagram to approximate the medial axis in 2D has been developed in detail in Chapter 6. Prominent work in this direction was that of Brandt and Algazi (1992) and this idea formed the basis of a widely used software package by Ogniewicz (1994). Even in 2D, slight noise in the positions of the point samples in S, due for instance to discretization, can introduce long spurious branches or "hairs" in the medial axis approximation. Ogniewicz implemented some criteria for removing these spurious branches; somewhat simpler criteria were proposed by Attali and Montanvert (1996). We shall discuss analogous methods for the 3D case in Section 7.4.2.

For the case in which the 2D input is given only as a point set, several algorithms were developed that output a topologically correct curve reconstruction and medial axis approximation, given some assumption about the quality of the input point sample. Attali (1997) gave a simple algorithm and proved that the output piecewise-linear curve approximation correctly connects adjacent points on the input curve. This implies that the corresponding medial axis approximation is also topologically correct. Attali's proof applies if the input S is a point set sampled from a smooth curve, such that the distance from any point on the curve to the nearest sample is at most a small constant δ; we shall call this model *uniform sampling*. The upper bound δ depends on the geometry of the curve, in a manner roughly similar to that described in Section 7.3. This result of Attali generalizes the earlier work of Brandt and Algazi described in Section 1.4 of Chapter 6 by removing the requirement that the r-regular input curve F be given; only a dense enough set of samples from F is required. Amenta et al. (1998a) introduced a more relaxed sampling requirement, known as ε-*sampling*; again, see Section 7.3. They proposed a different algorithm for selecting Delaunay edges as a curve approximation—the *crust*—which they showed gives correct results, again implying that the corresponding medial axis approximation—the *anti-crust*—is topologically correct. A simpler crust and anti-crust algorithm proposed by Gold and Snoeyink (2001) turns out, after some consideration, to be the same as Attali's. Their proof improves on hers, however, since it shows that the algorithm works under the relaxed ε-sampling assumption. The best result for curve reconstruction along these lines is the algorithm of Dey and Kumar (1999), which can also be used to give a dual medial axis approximation M.

7.2.2 Slivers

The main difficulty in transferring the key idea of Fig. 7.1 to a 3D algorithm is that for a curved surface in 3D, there are usually parts of the Voronoi diagram that cannot reasonably be assigned either to the medial axis approximation or to the dual surface approximation. To see this, assume we have a smooth curved surface F, sampled by a finite point set S. Consider a small ball B, intersecting F in a topological disk that does not contain any point of S. Generally the intersection of the boundary of B with F is non-planar, so we can add four non-cocircular sample points to S along this boundary. The center x of B is now a Voronoi vertex of S. By choosing a very small ball B, we can cause any arbitrarily dense set S to have a Voronoi vertex that is arbitrarily far from the medial axis.

The four new samples are nearly coplanar, so the Delaunay tetrahedron on the four samples, dual to x, is very flat. Such tetrahedra are known as *slivers*. In 3D these slivers result in the Voronoi diagram containing Voronoi vertices far from the medial axis. These slivers thus need to be removed before the dual Voronoi diagram is computed.

The Delaunay tetrahedralization of a sample of points drawn from the boundary of a smooth surface in 3D usually contains many slivers. In 2D, in contrast, every Voronoi vertex belongs to the medial axis approximation M, as in Fig. 7.1. Thus, as also developed in Chapter 6, Voronoi diagram-based medial axis computation is far simpler in 2D than in 3D.

7.3 Sampling and Approximation

In this section we outline models that have been used to define point samples that can be used for medial axis approximation. The choice of sampling model is naturally related to the subset of the medial axis that will be approximated by the the the Voronoi diagram. We will describe a number of purely structural results, which show that under particular sampling models, particular subsets of the Voronoi diagram approximate particular stable subsets of the medial axis.

7.3.1 Stable Subsets of the Medial Axis

The well-known phenomenon that some parts of the medial axis are unstable with respect to small perturbations of the surface means that some parts of the medial axis are more difficult to approximate by surface sampling than others. A point sample might miss small surface features that induce large features of the medial axis, and the more unstable a medial axis feature is the denser the sampling required to capture it.

This relationship is illustrated in Fig. 7.2. To characterize the stability of a point m on the medial axis, we introduce some notation. We use $\rho(m)$ to indicate the radius of the medial ball $B_{m,\rho(m)}$ centered at m, and $\gamma(m)$ to indicate the largest angle formed by the vectors to the contact points of $B_{m,\rho(m)}$ with the surface F, that is,

$$\gamma(m) = \max_{p_1,p_2 \in B_{m,\rho(m)} \cap F} \angle p_1 m p_2$$

This parameter $\gamma(m)$ represents the object angle saliency criterion discussed in Chapter 4. Finally, we use $d(m)$ to indicate the maximum distance between two contact points belonging to $B_{m,\rho(m)} \cap F$. At a regular point of the medial axis, this intersection consists of exactly two points p_1, p_2 and $d(m) = d(p_1, p_2)$. These three parameters $\rho(m), \gamma(m)$ and $d(m)$ are not independent; at a regular point of the medial axis any one of them can be computed from the other two. The angle parameter γ is scale invariant while the other two are not.

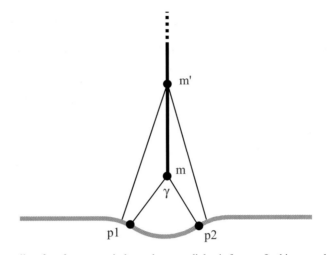

Fig. 7.2 A small surface feature can induce a large medial axis feature. In this example, perturbing the surface (bold curve) by removing the depression would eliminate the long medial axis "hair" (the heavy vertical line) ending at the medial axis point m. If the surface is sampled sparsely, so that no samples fall on the depression, then there will be no features of the Voronoi diagram of S corresponding to the "hair." To argue that m will be near some point of the Voronoi diagram of a set of point samples from the surface, we at least need the distance between samples to be less than distance between the contact points $d(p_1, p_2) = d(m)$. This idea can be formalized either directly in terms of the distance $d(m)$, which makes sense for uniform sampling, or in terms of the ratio $d(m)/\rho(m)$ or equivalently the angle $\gamma(m)$, which makes sense in a scale-invariant sampling model

If we sample the curve in Fig. 7.2 in such a way that the distance between samples is larger than $d(m)$, the bump on the surface might not be sampled, in which case the "hair" on the medial axis will not be reflected in any way in the Voronoi approximation.

Thus the results we can achieve for describing the quality of a Voronoi approximation to the medial axis take the following form. For a fixed sampling density, some selected subset M of the Voronoi diagram approximates some stable subset A of the medial axis, where by "approximates" we mean that the Hausdorff distance between A and M is upper bounded by some function of the sampling density. Recall that the Hausdorff distance between two sets A, M is defined as follows. The distance from a point x to A, $d(x, A) = \min_{y \in A} d(x, y)$. The Hausdorff distance between A, M is

$$d(A, M) = \max \left[\max_{x \in A} d(x, M), \max_{y \in M} d(y, A) \right]$$

In all of the following theorems, the stable subset A of the medial axis is defined using the parameters defining the sampling density, and converges to the entire medial axis as the sampling density increases. The Hausdorff distance between M and A also decreases with the sampling density, so that in the limit both M and A coincide with the entire medial axis. Both the upper bound on the Hausdorff distance

$d(M,A)$ and the rate at which A converges to the entire medial axis are interesting functions.

A recent example of a non-Voronoi medial axis approximation along these lines appears in Foskey et al. (2003), which efficiently computes an approximation to a subset A of the medial axis for which $d(m)$ and $\gamma(m)$ are both bounded by constants. In their work the input is a polyhedral surface F in $\mathbb{R}^3$ and the output medial axis approximation is extracted from an estimate of the distance function of F, computed on a voxel grid using hardware acceleration.

The rest of this section reviews the theoretical results of this form for Voronoi-based approximations to the medial axis. This area can be confusing since the proofs tend to be difficult, there have been a number of errors, and terminology, sampling model and assumptions differ from paper to paper, so comparing different results requires some restatement.

7.3.2 λ-Medial Axis and Uniform Sampling

The observation in Fig. 7.2 is related to a recent characterization of the stability of medial axis features due to Chazal and Lieutier (2005a). Given a medial ball $B_{m,\rho}$, they define $\beta(m)$ to be the radius of the smallest enclosing ball of the set $F \cap B_{m,\rho}$. This is a refinement of the parameter $d(m)$; at a normal point of the medial axis $\beta(m) = d(m)/2$, while for a medial ball that touches the surface in more than two points $\beta(m)$ might be larger than $d(m)/2$. They define the λ-medial axis to be the parts of the medial axis for which $\beta(m) \geq \lambda$.

Their main theorem is that if the Hausdorff distance $d(F,G)$ between F and some other set G is at most a constant $\delta = O(\lambda^4)$, then the Hausdorff distance between the λ-medial axis of F and the λ-medial axis of G is $O(\sqrt{\delta})$. The proof is given in terms of a fixed medial axis point m, treating $\rho(m)$ and $\gamma(m)$ as constants, so it is not really clear how the quality of the approximation depends on these parameters.

They observe that this result has an immediate application to Voronoi approximation of the medial axis. Consider a *uniform* sample, in which we require every point of the surface $x \in F$ to have a sample point $s \in S$ within distance δ. Then the Hausdorff distance $d(S,F) < \delta$, which implies the following.

Theorem 7.3.1. (Chazal and Lieutier). *The set M consisting of the closure of the points v of the Voronoi diagram (including points in edges and 2D faces) for which the distance between two of the nearest samples to v is at least δ is an approximation of the λ-medial axis of F, with $\lambda = O(\delta^{1/4})$, and where the Hausdorff distance between M and the λ-medial axis is $O(\delta^{1/8})$.*

Interestingly, the samples in S need not lie on the surface; they might be noisy, just so long as the Hausdorff distance $d(S,F)$ remains less than δ.

Here the λ-medial axis forms the stable subset A of the medial axis of F which is approximated by M. This result establishes the convergence of M to the medial axis of F as $\delta \to 0$, but does not tell us that at any specific small enough value δ_0, M is homeomorphic to the medial axis.

In Chazal and Lieutier (2005b) they address this problem. One way to understand it is to consider the λ-offset surfaces of F, which are the λ iso-surfaces of the distance function:

$$\{x \,\|\, d(x, F) = \lambda\}$$

When F is smooth and λ is small, there will be two components of the offset surface, one outside and one inside, and each component is homeomorphic to F. At larger values of λ, one or both components will no longer be homeomorphic to F. The values at which the topological changes of the offset surfaces occur are the *singular values* of the distance function. Zero is a singular value. Chazal and Lieutier (2005b) define the *weak feature size* to be the next smallest singular value. They prove that the λ-medial axis is homotopy equivalent to the medial axis when λ is less than the weak feature size.

These results raise the following questions.

Question 7.3.1. Are the bounds on λ and on the Hausdorff distance in Theorem 7.3.1 optimal as functions of δ, or can one prove a tighter bound?

Question 7.3.2. The big-O notation in this result hides dependence on ρ and γ. Clarify this relationship. Is it the best possible?

Clearly uniform sampling is not scale invariant. The problem with this is that since δ must be small enough to capture the smallest feature of F, uniform sampling usually requires a lot of unnecessary samples in other parts of the surface and is thus inefficient.

7.3.3 γ-Medial Axis and Scale-Invariant Sampling

Figure 7.2 seems to indicate that a scale-invariant approach to the analysis of Voronoi approximation of the medial axis should be based on $\gamma(m)$, rather than $d(m)$. Amenta et al. (2001a) defined the γ-*medial axis* as the set of medial points m such that $\gamma(m) \geq \gamma$, for some constant γ.

In Fig. 7.2, the point m belongs to the $\gamma(m)$-medial axis, but m' does not; the γ-medial axis is discontinuous where the medial axis itself is not. The entire hair does belong to the $\lambda(m)$-medial axis, for $\lambda < d(m)$.

For approximating the γ-medial axis, we use a scale-invariant sampling model. This gives us a formal way to require dense sampling near intricate details of the surface, but not in featureless areas where it is not needed.

We say that a point set S is an ε-*sample* from a surface F if, for every point $x \in F$, the minimum distance from x to any point in S is at most $\varepsilon f(x)$, where $f(x) = d(x, \mathcal{M}(F))$. This definition is illustrated in Fig. 7.3. For several of the algorithms discussed below, we can prove results on the quality of the medial axis

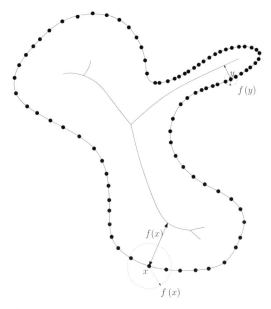

Fig. 7.3 ε-sampling figure

approximation that hold for any $\varepsilon \leq \varepsilon_0$, where ε_0 is a constant independent of any particular surface F.

For a point x at a sharp corner of the surface F, we have $f(x) = 0$, which implies that the sampling has to be infinitely dense. So the ε-sampling definition is only useful for smooth (C_1) surfaces. This is a major drawback.

Amenta and Bern (1999) and Amenta et al. (1998b) point out that when S is a dense sample from a smooth surface F, each Voronoi cell is long and skinny and roughly perpendicular to the surface (see Fig. 7.1 again for intuition). This motivates the definition of the *poles* of a sample $s \in S$, as the Voronoi vertices that lie at the opposite ends of the long skinny cell. The first pole p_1 of a sample $s \in S$ is the vertex of the Voronoi cell of s farthest from s. The second pole p_2 of s is the Voronoi vertex on the opposite side of the Voronoi cell (the vector product $\mathbf{s}, \mathbf{p_2} \cdot \mathbf{s}, \mathbf{p_1} < 0$) farthest from s.

Amenta et al. (2001a,b) proved that the set of poles converges to the medial axis as $\varepsilon \to 0$. The following restates their theorem in terms of ρ, γ and ε:

Theorem 7.3.2. (Amenta, Choi and Kolluri). *Let $B_{c,\rho}$ be a medial ball with two points $p_1, p_2 \in B_{c,\rho} \cap F$, and let $\gamma = \angle p_1 c p_2$. Let t be the nearest sample to c. Then the distance from c to the nearest pole p of t is $O(\rho \varepsilon (1/\gamma))$.*

One way to interpret this theorem is that every point of the γ-medial axis, for $\gamma = O(\varepsilon^\kappa)$, for $0 \leq \kappa \leq 1$, is within a distance $O(\rho \varepsilon^{1-\kappa})$ of some pole. This is the original statement of the theorem found in Amenta et al. (2001b). For comparison with the results in the previous section, we can instead consider a fixed medial axis point m, so that $\rho(m)$ and $\gamma(m)$ are constants; in this context the theorem tells us that the distance from m to the nearest pole decreases linearly with ε.

In addition, just as the medial axis transform can be used to reproduce the surface representation, the anti-crust can be used to exactly reproduce the piecewise-linear surface approximation T. This process is described in Section 7.4.3, below.

7.4.2 Thinning Algorithms

As discussed in Chapter 1, thinning algorithms are designed to iteratively peel away layers of an object with constraints enforced to preserve homotopy type. The incorporation of thinning procedures on a rectangular lattice has already been discussed in the medial loci detection algorithms of Chapters 4 and 5. In the context of the current Voronoi-based method, a thinning algorithm must remove a face from the anti-crust M at each step, maintaining the topological correctness while seeking to improve the geometric approximation of the medial axis.

7.4.2.1 Attali and Montanvert's Method

The method of Attali and Montanvert (1997) is one example. This algorithm focuses on the part of the medial axis interior to an object bounded by a closed surface F. They initialize M to be the interior part of the anti-crust, a two-dimensional set which is a discrete analog of the medial axis. M is associated with a dual interior solid, initialized as the union of all tetrahedra dual to vertices of M. Then they define conditions under which tetrahedra can be removed from the dual solid, or equivalently, vertices and the faces they contain can be removed from M, while preserving the topology.

A dangling edge e of M is an edge contained in no two-face in M, with one vertex contained in other faces of M and the other vertex v adjacent only to e. This vertex v is dual to a tetrahedron with three two-faces on the boundary of the dual solid, which can clearly be removed without changing the homotopy type of the solid, removing the end vertex and the dangling edge from M. Such tetrahedra are always removed as soon as they are created.

They also define a *salient tetrahedron* as one with two two-faces on the boundary of the dual solid. Salient tetrahedra can also be removed without changing the homotopy type of the solid. The Voronoi vertex v dual to a salient tetrahedron τ is adjacent to two border edges of M, both contained in the same two-face f (dual to the edge of τ interior to the dual solid). Removing the salient tetrahedron corresponds to removing vertex v and f from M. Removing f from M might create dangling edges, which will also be removed immediately.

According to the scheme of Attali and Montanvert, salient tetrahedra may be, but are not always, removed. They propose removing a salient tetrahedron if the dihedral angle between the two facets on the boundary of the dual solid is less than a given threshold. Clearly this removes sliver salient tetrahedra, but it could also remove others as well.

Question 7.4.1. Give a thinning algorithm that begins with the anti-crust and produces a set M of Voronoi faces homotopy equivalent to the solid bounded by F, and show that the Hausdorff distance between M and some stable subset of the medial axis is upper-bounded by some function of the sampling density, ρ and γ.

It is still not clear whether or not the thinning criteria proposed by Attali and Montavert could be used in such a provably correct algorithm.

7.4.2.2 Filtering for the γ-Medial Axis

Dey and Zhao (2003) propose another method to use a subset of Voronoi facets as an approximate medial axis. Their algorithm gives something like an approximation to the γ-medial axis and hence does not necessarily preserve the topology of $\mathcal{M}(F)$. They eliminate Voronoi facets from the anti-crust by filtering the dual Delaunay edges, using two criteria, each of which is a way of estimating something like γ at the points of the Voronoi approximation M.

They use the fact that the vector v_p from a sample p to its pole approximates the surface normal at p. The *umbrella* for a sample p is the topological disc made by the Delaunay triangles incident to p that are dual to the Voronoi edges intersected by the approximate tangent plane at p with normal v_p. If the Delaunay edge pq makes a small angle with the surface normals of the umbrella triangles of p, then we say it passes the *angle condition*. The *ratio condition* is that the ratio of Delaunay edge to the radius of the circumcircle of its umbrella triangles is big. Note that we would expect both conditions to be true when γ is large. 2D faces of the Voronoi diagram are selected as part of the medial axis approximation M if they satisfy either the angle condition or the ratio condition. Since the two conditions are scale and density independent, this algorithm is appropriate to use with ε-sampling. The examples in (Dey and Zhao, 2003) show excellent results.

7.4.3 Power Shape

The difficulties in selecting a medial axis approximation from the 3D Voronoi diagram arise because slivers make the choice ambiguous. We now consider an algorithm that handles the problem in a different way; it constructs a second Voronoi diagram which does not have slivers, and from which the choice of M is unambiguous. This *power shape* is the dual structure of another surface representation called the power crust. This construction uses Voronoi diagrams in a different way than the other constructions in this chapter. It is based on a kind of Voronoi diagram called a *power diagram*, which takes balls as input instead of points, and its dual triangulation, analogous to the Delaunay triangulation, known as a *regular triangulation*. The *power distance* between a point $x \in \mathbb{R}^3$ and a ball $B_{v,\rho}$ with center v and radius ρ is

$$d_{pow}(x, B_{v,\rho}) = d^2(v, x) - \rho^2.$$

The power diagram is the subdivision of the spatial domain Q into cells, each cell consisting of the points $x \in Q$ closest, in power distance, to a particular input ball $B_{v,\rho}$. Like the Voronoi diagram, the cells of the power diagram are convex polyhedra. In the dual regular triangulation, every four points whose Voronoi cells meet in a vertex of the power diagram form a tetrahedron.

An application of a 2D version of the power crust algorithm is shown in Fig. 7.4. Recall that each pole v in the usual Voronoi diagram $\mathcal{V}(s)$ is the center of a Voronoi ball $B_{v,\rho}$. We compute the power diagram of the set of these Voronoi balls of the poles. We label the cells of the power diagram according to whether the cell belongs to an inside or outside pole, using a traversal of the power diagram structure and some rules on propagating inside/outside labels. We then connect the inner poles according to the connectivity of their power diagram cells, forming a simplicial complex M, a subset of the regular triangulation. We output this M, which we call the *power shape*, as the medial axis approximation. The corresponding surface approximation is the set of 2D power diagram faces separating the inner from the outer cells.

The vertices of the power shape are the poles, which converge to the medial axis as the sampling density increases (Theorem 7.3.2). Although it does not contain vertices far from the medial axis, like the anti-crust does, there is also no proof that its 1D, 2D and 3D faces converge to the medial axis. Topologically, like the medial axis, the anti-crust, and the output of thinning algorithms, the power shape is homotopy equivalent to $Q - F$. But unlike the medial axis and the other approximations, the power shape is not in general a 2D set; it usually includes solid tetrahedra. This is a drawback in some applications.

Question 7.4.2. Prove a bound on the Hausdorff distance between the power shape and the γ-medial axis.

The power crust surface representation can be computed from the power shape, given the radii of the Voronoi balls centered at the poles, by reconstructing the power diagram. A similar approach can be used to reproduce a surface representation T from the vertices of the anti-crust, assuming each vertex v is labeled with the radius of the Voronoi ball centered at v. We reconstruct the Delaunay triangulation from the structure of the anti-crust, and output the faces separating its inner and outer dual solids.

7.5 Medial Axis Algorithms for Input Surfaces

We now review some algorithms that take surface representations as input and produce medial axis representations using the Voronoi diagram.

When the surface F is given explicitly, usually as a set of algebraic surface patches, we can hope to produce a very accurate representation of the medial axis. The Voronoi diagram can be used to sketch out the general shape, and then the vertices (at least) of the medial axis could be computed more precisely.

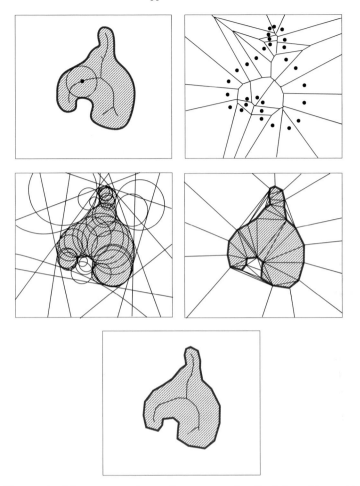

Fig. 7.4 2D example of the power crust/power shape construction. *Top left*: an object and its medial axis, with one medial ball drawn. *Top right*: first we construct the Voronoi diagram and select the poles, a subset of the Voronoi vertices which lie near the medial axis (in 2D all of the Voronoi vertices can be selected). *Center left*: each pole is the center of a Voronoi ball, which approximates a medial ball of either the object (shaded) or its complement. The straight lines are the boundaries of infinite Voronoi balls, associated with Voronoi edges that go off to infinity. *Center right*: the power diagram of this set of balls contains a region for each ball (in 2D, this is identical to the Delaunay triangulation, but not in 3D). We assign each region either to the inside or the outside of the object. *Bottom*: we use the connectivity of the corresponding regions in the power diagram to connect the poles together, forming a simplicial complex, which we output as the power shape. Only the power shape of the inner poles is shown

An algorithm along these lines producing fairly clean medial axis approximations for inputs defined by smooth implicit functions was given in Turkiyyah et al. (1997). This algorithm begins with a medial axis approximation computed as a subset of the Voronoi diagram and then improves it using local numerical optimizations. Voronoi faces dual to slivers are eliminated. Then using the given surface F, the positions of Voronoi vertices are moved onto the medial axis. Points on junction and boundary curves in the Voronoi approximation are moved onto corresponding junction and boundary curves of the medial axis. This approach is appropriate for objects that are simple enough and sufficiently well sampled that features of the Voronoi approximation do indeed correspond exactly to features of the medial axis. It would be very interesting to revisit this algorithm using some of the subsequently developed theory of sampling and convergence reviewed in the previous sections.

In the mid-nineties Sheehy et al. (1995) gave an algorithm in which the Voronoi diagram is used to guide a geometric construction of the medial axis. Point samples are distributed on the surface of the model, and the Delaunay triangulation of the point samples is computed. Samples are added to make the Delaunay triangulation dense enough such that the model surface can be represented as a union of Delaunay triangles. Each tetrahedron is then treated as a hypothesis for a feature of the medial axis. Hypothesized medial axis vertices are tested explicitly against the original set of model faces. This algorithm was implemented in Sheehy (1994) using iterative computations for the test against the model faces.

In Dey et al. (2003), the Voronoi-based approximate medial axis construction of Dey and Zhao (2003) is extended to handle typical CAD inputs with sharp corners and edges. The medial axis of an object with sharp creases includes parts of the medial axis that touch the sharp interior edges and vertices. These are constructed by adding samples along and around sharp edges; the Voronoi faces of these vertices are used to guide the construction of the approximate medial faces.

Faces of the medial axis approximation produced using a Voronoi diagram generally have noisy normals, since in general the Voronoi faces that make up the approximate medial axis separate the Voronoi cells of two sample points that do not correspond to exactly the same point on the medial axis. With modern Voronoi diagram algorithms, however, we can use sample point sets that are large enough to make the Hausdorff distance between the approximate medial axis and the true medial axis quite small.

For a smooth surface given explicitly, Schröder et al. (2003) give an algorithm that mostly eliminates this discretization noise. Instead of using faces of the Voronoi diagram, they define two approximate medial points for each sample as the intersections of the Voronoi cell with the rays from each sample with the surface normal directions. The set of approximate medial points are connected using the same connectivity as its defining samples. They show excellent results for parabolic input surfaces.

Some discretization problems remain for the boundary curves where the medial axis ends, which are still resolved raggedly. These boundary curves are interesting; they are swept out by the centers of the maximal spheres that avoid F and that are tangent to the surface at crests. Finding the crests on the surface, and from them the boundary curves of the medial axis, would be quite useful.

7.6 Discussion

It has proved to be quite difficult to get good bounds on the convergence of Voronoi approximations to the medial axis in $\mathbb{R}^3$, although the sophisticated work that has been done in this area has certainly served to make many of the subtleties apparent. One complication in the theory has to do with eliminating the parts of the Voronoi diagram dual to slivers. The early straightforward argument offered by Goldak et al. (1991) that the Voronoi diagram approximates the medial axis because in the limit, when the sampling is infinitely dense, the Voronoi diagram becomes the medial axis does not in fact apply at any finite sampling density. The issue is that slivers can always produce Voronoi vertices far from the medial axis and hence convergence is not guaranteed. Interestingly, a number of questionable claims in the more recent theory have had to do with the situation in Theorem 7.3.3, which establishes conditions under which 2D faces of the Voronoi diagram are close to the medial axis.

While the algorithm of Turkiyyah et al. (1997) is quite robust, the argument in that paper that a subset of the Voronoi diagram converges to the medial axis is questionable, in particular the argument that every point on the medial axis is within $O(\varepsilon)$ of a point on a Voronoi face, which essentially assumes that both the requirements of Theorem 7.3.3 are met. Boissonnat and Cazals (2001) offer a proposition similar to Theorem 3, but there seems to be a problem in the supporting argument (specifically, in their Lemma 15). This proposition of Boissonnat and Cazals was used in Dey and Zhao (2003); possibly it could be replaced by Theorem 7.3.3. All of the above difficulties seem to point to the importance of understanding the set of Voronoi faces described by Theorem 7.3.3.

From a practical standpoint, Voronoi methods are an obvious choice for medial axis approximation when the input is given as a set of sample points. In practice such inputs contain noise. For uniform samples noise is handled nicely (see Section 7.3.2), but there are not yet corresponding results that apply to the scale-invariant case. The non-uniform case is generally more realistic, at least for the case of laser range scanning. Also such inputs generally contain gaps where the sample set fails to cover the surface completely. A practical implementation would have to handle these somehow (perhaps by omitting corresponding parts of the medial axis approximation).

A strength of Voronoi based methods is that there are some precise results describing the convergence of various subsets of the Voronoi diagram to the medial axis as the sampling density increases. However, as discussed in this chapter there remain a number of interesting open questions in this area.

For explicitly given input surfaces, several implementations have combined Voronoi approximation, careful placement of samples, and computation using the surfaces themselves to produce good medial axis implementations. Unfortunately, as far as we know, none of these somewhat more complicated algorithms has a freely available software implementation, and they seem challenging to implement from scratch. It seems possible that our current understanding of Voronoi approximation might lead to simpler algorithms in this case.

Chapter 8
Synthesis, Deformation, and Statistics of 3D Objects via M-Reps

Stephen Pizer, Qiong Han, Sarang Joshi, P. Thomas Fletcher, Paul A. Yushkevich, and Andrew Thall

Abstract The m-rep, a representation of the interior of one or more objects, from which boundaries can be synthesized, is described in detail. An m-rep consists of sheets of medial atoms; both sampled and parametrized representations of these sheets are described. Means of forming objects made from a main sheet (*figure*) and attached protrusion or indentation subfigures are described, as are multiscale hierarchies of object complexes, objects, figures, atoms, and voxels. The object-relative coordinate system provided by m-reps is presented. To allow the estimation of probabilities on populations of m-reps, the m-rep can be understood as an element in a feature space that takes the mathematical form of a symmetric space. Doing this provides the ability to estimate probabilities by a generalization of principal component analysis to these curved spaces.

S. Pizer
Medical Image Display & Analysis Group, University of North Carolina at Chapel Hill, USA,
e-mail: pizer@cs.unc.edu

Q. Han
Medical Image Display & Analysis Group, University of North Carolina at Chapel Hill, USA,
e-mail: han@cs.unc.edu

S. Joshi
Medical Image Display & Analysis Group, University of North Carolina at Chapel Hill, USA,
e-mail: sjoshi@cs.unc.edu

P.T. Fletcher
Department of Computer Science, University of Utah, USA,
e-mail: fletcher@sci.utah.edu

P.A. Yushkevich
Department of Radiology, University of Pennsylvania, USA,
e-mail: pauly2@grasp.upenn.edu

A. Thall
Department of Computer Science, Allegheny College, USA,
e-mail: athall@allegheny.edu

K. Siddiqi and S. Pizer (eds.) *Medial Representations – Mathematics, Algorithms and Applications*.
© Springer Science + Business Media B.V. 2008

8.1 Introduction

Chapters 4–7 have taken the point of view that objects begin from a boundary representation and a medial representation is derived from the boundary. An alternative
view is that an object is a member of a population of instances of the object and
that a fixed topology of the medial locus can be derived from this population, as
well as a probability distribution on the geometry of that medial locus. Thus every
medial instance can be seen as a deformation of the mean of this probability distribution, and every boundary can be seen as synthesized from that medial instance. The
medial representation called the *m-rep*, discussed in Chapter 1, Section 2.4, enables
the formation of these fixed topology medial structures and the estimation of their
probabilities from training samples. This chapter covers the details of m-reps, the
view of an m-rep as a point on the mathematical entity called a *symmetric space*,
and the ideas of probability distributions on symmetric spaces. Moreover, m-reps
are described in two forms: grids of order 2 medial atoms called *discrete m-reps* and
splines of order 1 medial atoms called *continuous mreps*.

8.2 M-Reps, Medial Atoms, and Figures

We seek a means for representing an object, such as a liver or a car, and ensembles
of objects, such as the whole abdomen or a street scene. Following the philoso-
phy of Grenander's *Pattern Theory* (Grenander, 1996) that an object's very shape is
described by its deformations into its various instances in a population and also real-
izing that objects deform mechanically in time, we need a representation in which
the associated deformations are rich, natural, and efficiently implemented.

Let us focus on single objects first. What constitutes an object? The intuitive view
is that an object is not simply a shell but rather consists of interior material that can
be locally transformed by elongation, bending, twisting, swelling or contraction,
and displacement. We follow this useful view and choose to synthesize objects from
such a description of the object interior.

The medial representation, originally promulgated as a locus of bitangent spheres,
is well suited to providing this description because its sphere primitive is locally
maximally interior to the object. However, it fails to integrate to the interior. This
aim is achieved by a small modification (Fig. 8.1)—replacing the sphere by a hub
formed by its center and the two equal-length spokes to the points of sphere tangency. As explained in Chapter 1, we call this object-interior-component primitive
a *medial atom*, and we use the term *m-rep* to refer to a locus of medial atoms that
sweep out an object interior.

The locus of medial atoms can be a manifold with boundary; an object or object
part represented by such a manifold is called a *figure*. As illustrated in Fig. 8.2,
in 2D the manifold can be a curve, in which case we call the figure a *bar*; in a
limiting case the curve degenerates to a point, and the bar degenerates to a disk.
In 3D the manifold can be two-dimensional, in which case we call the figure a

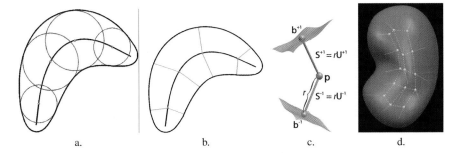

Fig. 8.1 (**a**) Sample tangent spheres representing an object in 2D. (**b**) Medial atoms representing the same object. (**c**) A medial atom in 3D. The hub is at position $\mathbf{p}$; the two spokes, of length r, are named $\mathbf{S}^{+1}$ and $\mathbf{S}^{-1}$, respectively, the unit vectors in those direction are $\mathbf{U}^{+1}$ and $\mathbf{U}^{-1}$, respectively, and the spoke ends are at positions $\mathbf{b}^{+1}$ and $\mathbf{b}^{-1}$, respectively; finally, fractional distance from the hub to the spoke end is given by τ. (**d**) A 3D kidney with a grid of samples of its medial atoms. This representation is called a *sampled m-rep* or a *discrete m-rep*. The third spoke on the atoms at the edge of the grid was discussed in Chapter 1 and is detailed in Section 8.8 of this chapter

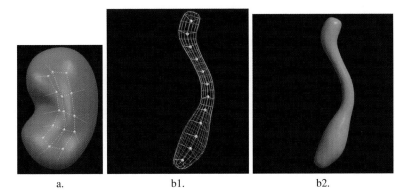

Fig. 8.2 (**a**) A slab, in which the medial locus forms a curved 2D manifold of medial atoms. Sample atoms forming a grid are shown, and the balls illustrating their hubs are samples of the medial sheet. (**b**) A tube, in which the medial locus forms a curve of medial atoms. In b1 sample atoms forming a chain are shown; the balls illustrating their hubs are samples of the medial axis. In b2 the boundary implied by the m-rep is shown

slab and its locus of hubs the *medial sheet*. In 3D the manifold can also be one-dimensional, in which case we call the locus of the hubs the *medial axis*, the atom must be interpreted as representing all spokes obtained by spoke rotation about the atom spokes' bisector, and the figure is called a *tube*. In the limiting 3D case the axis degenerates to a point, and the figure degenerates to a sphere.

A collection of figures that are attached among themselves form a multifigure object. Each figure can form a protrusion added to the collection (Fig. 8.3a), or a figure can form an indentation that is subtracted from the collection (Fig. 8.3b). Indentations can even pass entirely through the host figure, forming a hole. The form of attachment of subfigures to host figures is described in Section 8.8. Finally,

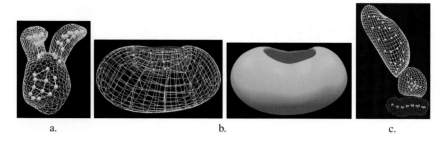

a.   b.   c.

Fig. 8.3 Multifigure objects and multi-object complexes. (**a**) A multi-figure male prostate with two seminal vesicles. (**b**) The kidney minus the renal pelvis. (**c**) The multiple objects making up a male pelvis: bladder, prostate, and rectum

the collection of figures can form multiple inter-related objects (Fig. 8.3c) that are non-intersecting but possibly abutting.

8.3 Object-Relative Coordinates

A special strength of the medial atom is that it carries a natural local coordinate system, complete with origin (its hub), coordinate frame (see Fig. 8.4), and metric (its spoke length), though strictly, medial atoms whose spokes are exactly opposed have one dimension of ambiguity in their coordinate frame; see below. Other medial atoms can be expressed in terms of a reference atom as

1. A translation of the reference atom's hub, in the metric of the reference atom (3 parameters). The origin of the atom's coordinate system is thus its hub location.
2. A magnification or demagnification of the spokes in common (1 parameter). The spoke length thus provides the distance metric of the atom's coordinate system and thereby makes local shape description of the figure invariant to local magnification.
3. Rotations of the two spokes' directions, or alternatively a 3D rotation of the frame fitted to the spokes together with a 2D rotation of the spokes towards or away from each other in their common plane (both 4 parameters). The latter view breaks down in the singular situation when the spokes are back-to-back, i.e., one is a rotation by π of the other, which occurs at all critical points of the spoke width function. Nevertheless, it exposes the *object angle*, namely half of the angle between the spokes, and the bisector of the spokes. This bisector and the vector orthogonal to it in the plane of the spokes along with the vector orthogonal to this plane forms a natural medially fitted frame. The second of these, which is in the direction of the difference of the spokes, can be shown to be normal to the medial sheet. A difficulty with this frame is that as one of spokes rotates continuously from nearly back to back with the other spoke, through back to back (object angle $\pi/2$), and beyond, the sense of the bisector changes discontinuously due to

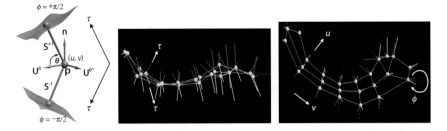

Fig. 8.4 *Left*: A medial atom's figural coordinates (u, v, ϕ, τ) and its atom frame $(\mathbf{U}^0, \mathbf{U}^{0\perp}, \mathbf{n})$, where $\mathbf{n}$ is the normal to the medial sheet. *Right and middle*: Two views of the figural coordinates for a sheet of medial atoms (shown sampled)

the fact that back-to-back spokes is a singular situation ($\nabla r = 0$, and the bisector is not unique).

By this approach any medial atom can be written as an 8-parameter (9 for end atoms; see Section 8.8) transformation of the primal medial atom, whose hub is at the origin, whose spoke length is unity, and whose spokes are both along the cardinal x-axis. These transformations, at various locations on the medial locus, provide an ability that is absent in many other object representations, namely, to provide combinations of local translations, twisting, bending, and swelling/contraction of interior material.

Writing each medial atom in a figure in terms of its immediate neighbors yields an object-relative coordinate system. In this coordinate system, whose coordinates we will write as $\mathbf{u}$, the object is seen as a spoke-length-proportional dilation from the medial sheet or axis, according to the theory of Damon (Chapter 3). That is, one of the coordinates in $\mathbf{u}$, which we call τ, is the fraction of the distance from the medial sheet or axis to the implied boundary at the spoke ends. $\tau - 1$ has the useful property that its sign distinguishes the inside from the outside of the figure. The spoke-length distance metric applied along the medial sheet provides two coordinates that can be called u and v. Applied to a tubular axis, it provides a single coordinate u.

An additional coordinate is needed to distinguish the two sides of the medial sheet and to take one around the crest from one side to the other. In the case of the tube this coordinate ϕ takes one around the tube by varying from $-\pi$ to π. In the case of the slab this coordinate has a constant value of $\pi/2$ on one side of the object and of $\pi/2$ on the other side and changes smoothly between these while passing around the crest. In the Blum formulation ϕ changes discontinuously between $-\pi$ and π at the crest. However, letting ϕ change continuously between $-\pi$ and π and be zero at the crest is more consistent with the tube representation and is friendlier to computer representation because it stabilizes the end definition provided by image data (see Section 8.8). This decision, however, leaves open the definition of the point at which the end atom is placed, truncating either u or v such that the surface begins to be parametrized with ϕ and the non-truncated of u or v. Since the object angle must begin to move quickly towards 0 as the crest is approached, one possibility is to let ϕ begin to transition when the object angle magnitude falls below some threshold, e.g., $\pi/4$ (Fig. 8.4).

8.4 Figures, Subfigures, and Multi-Object Ensembles

When we divide an object into its figures, each of which forms a slab or a tube, we must describe the connection between the figures. Most commonly we think of one figure as a subfigure of a host figure, attached to the host along a connected locus (Fig. 8.5). However, it is possible for a figure to be attached to itself, or it can be attached to one or more other figures at a disconnected locus of points—consider the handle of a mug. We now treat the means by which the subfigure is attached to its host at any one of the one or more connected loci of points of attachment.

One view is to form an attachment of a medial locus branch to the host's medial locus—this is the view that Blum and many mathematicians have taken. In that view, the patch metric on the medial locus described by Damon in Chapter 3 is typically quite small between the branch point and near the host figure boundary, as those medial atoms are responsible for little interior material. In the region of the branching the medial surface has a corner that has complex geometry, especially at the two ends of the corner.

Another view is to describe the subfigure fully by itself and to form a blend region to attach the subfigure to its host. Of course, this requires a description of the blend region and the means of smoothly attaching it both to the host figure and to the subfigure. But this view has the advantage that the subtractive (indentation) subfigure connects in a completely equivalent way as the additive (protrusion) subfigure.

In either case the part of the subfigure near the host figure (we call that part the *hinge*) needs to be understood in the coordinate system of the nearby medial atoms of the host figure. This representation allows the subfigure to be translated, rotated and scaled in the figural coordinates of the host figure (Han et al., 2004). More precisely, the subfigure atoms near the connection to the host figure translate, rotate, and magnify their coordinates relative to the nearby host atoms while maintaining

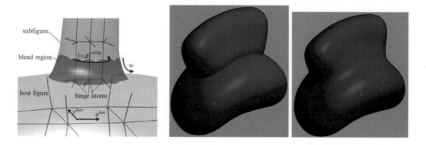

Fig. 8.5 Host figure and subfigure. *Left*: representation. *Middle*: as two separate figures. *Right*: blended

their coordinates in terms of the intrafigural neighboring atoms, and the remainder of the atoms maintain their coordinates in terms of the intrafigural neighboring atoms, all the while keeping the figural shape as close to constant as possible. The means of maintaining figural shape while deforming is a subject of the next section.

Multiple objects may be described in a similar fashion. That is, each object has its own m-rep, but in addition, to describe the relations among objects the atoms in one object that are near a second object also need to be understood in relation to the nearby atoms of the second object. This arrangement should be held mutually. With such an arrangement interpenetration among objects can be avoided by seeing whether the spoke ends of one object have a negative value of τ in the figural coordinate system of the second object.

8.5 Synthesis of Objects and Multi-Object Ensembles by Multiscale Figural Description

One point made in the first paragraph has not received adequate emphasis till now: the medial representation is being recommended as the basic means of representing objects, i.e., the representation from which other aspects of the object, such as its boundary and locations in its interior, are *synthesized*. To convey this point better, we jump off from the relation between the figures that make up an object and the object as a whole. Just as one cannot understand the leaves of a tree and the tree as a crown on a trunk at the same spatial scale (Koenderink, 1990), one needs to understand figures and the object as being at separate spatial scales. The figures can be understood as determining with finer tolerance information that is already conveyed by the object as a whole, at a larger tolerance. Moreover, the figural sheets can be allowed to have only two coordinates of parametrization precisely because small deviations from smoothness that the Blum medial axis might require in the sheet's boundary, leading to arbitrarily many subfigures, can be handled at a smaller scale level.

Continuing to a larger scale level than figures, the objects forming a multi-object ensemble determine with finer tolerance information that is already conveyed by the object ensemble as a whole, at a larger tolerance. Similarly, figural sections, each an interior region of the figure corresponding to a neighborhood of medial atoms (Fig. 8.6), determine with finer tolerance information that is already conveyed by the figure as a whole, at a larger tolerance. This process can be continued, down to the scale of the voxel, where individual voxels may be very locally displaced, rotated, or scaled to refine the tolerance given by the medial representation. In this sense, an m-rep is a representation that is medial at the large and moderate spatial scales corresponding to objects, figures, and figural interior sections but even more local at the smaller spatial scales.

Why do we represent objects in this large-scale-to-small fashion? It is because it is not useful to synthesize or deform an object in much more than O(N) time, where N is the number of primitives at the smallest scale, since an object or object

Fig. 8.6 Figural interior section corresponding to the neighborhood of a medial atom

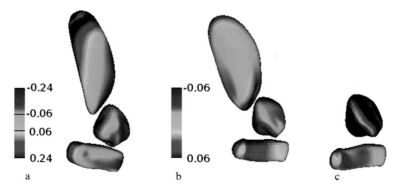

Fig. 8.7 The multiscale deformation of an object ensemble made from (top to bottom) bladder, prostate and rectum during a segmentation. The color bars next to panel a and panels b–c, respectively, show the amount and direction of the boundary displacement corresponding to each color. Positive values show displacements towards the exterior of the object, and negative values show displacements towards the interior of the object. The scale for panel a has larger values of displacement than that for the other panels. In each case the displacement is shown on the result from the next larger scale level, with scale levels processed from large to small. (**a**) Ensemble deformation, displayed on initial m-rep. (**b**) Deformation of the bladder produced at the bladder object stage, and sympathetic deformations of the prostate and rectum, displayed on ensemble stage result. (**c**) Deformation of the prostate produced at the prostate object stage, and sympathetic deformations of the rectum, displayed on the prostate and rectum in the bladder stage result

ensemble will typically have thousands of primitives at that scale. Working at the smallest scale only and yet reflecting relationships of arbitrary degrees of locality would yield an $O(N^2)$ algorithm, which is unacceptably slow. The large-scale-to-small approach is designed so that calculation of the value of a primitive at one scale need only account for that primitive and its relation to neighboring primitives separated by its own scale. The hierarchical application of this principle provides $O(N)$ algorithms for synthesis and deformation.

The synthesis of an object then occurs from large scale to small (Fig. 8.7). At the largest scale, that of the object ensemble, a base model is geometrically positioned, rotated, scaled and grossly warped, as described by just a dozen or fewer variables. To do this requires knowledge of the gross variations among the family of object ensembles being modeled, information that can be garnered from statistics on samples of the ensemble. How to represent the associated probability densities and estimate its parameters is discussed in Section 8.7.

For all of the remaining levels of scale, the next smallest scale is one at which the components of the previous scale are each refined. These successively smaller scales will be, at the least, the multi-object ensemble (if there are multiple objects being modeled), the objects, the figures of which they are made up, interior sections of those figures, and smaller subsections of those figures. More finely spaced scale levels are also possible. For example, one might choose to place larger figures in one scale level and smaller ones in the next smaller scale level.

Deformations of an object ensemble occur from large scale to small, with the smaller scale deformations being *residuals* from the larger scale ones, i.e., being applied on the result of the large scale deformations. Thus, for example, the total deformation applied to a figure making up one object consists of the effect on it of a deformation global to the object ensemble, combined with a deformation global to the object of which it is a part, combined by the residual deformation of the figure itself.

When dealing with the residual deformation of an entity at one scale level, one needs to be aware that smooth deformations by arbitrary transformations on the atoms forming an m-rep figure may produce the illegality of one Euclidean point having two different figural coordinates. This may occur by the two local conditions of the medial sheet kinking and then, in further deformation, self-intersecting, or of the boundary kinking and then, in further deformation folding. Or it may occur by non-local self-penetration. The mathematics of Damon given in Chapter 3 have given local conditions on the medial locus that will allow one to prevent the local conditions, but non-local self-intersection may require a more expensive search.

Until now we have acted as if the topology of the m-rep, i.e., what are the figures and what is a subfigure of what host figure, is a given. But where should the model topology come from in the first place? Two possibilities present themselves.

First, based on knowledge of the application area, the user can understand what the parts are. The first example is of human anatomy: the lobes of the liver come from that discipline, and the smaller lobe can be seen as attached to the larger one. A second example is of the automobile: a manufacturer understands that the bumper should be a part separately modeled from the car body.

Second, the m-rep topology can come from the statistics of instances of the object, each geometrically analyzed into medial components. Styner (Styner et al., 2003c) has suggested how the Blum medial analysis of instances from their boundaries can lead to a stable set of figures and figure/subfigure relationships even though each individual medial analysis is rather unstable. Essentially, in each instance portions of medial axis are grouped into sheets based on continuity of the medial sheet and of r values, sheets corresponding to an appropriately small fraction of the volume are deleted, and the branchings of the remainder are identified. Then sheets and branchings held in common across the cases are found. Indeed, some populations may need models of multiple branching structures to encompass the whole population, but for anatomic objects it is impressive how frequently a single branching structure will do.

8.6 M-Reps as Symmetric Spaces

M-reps are designed to be deformed. One major use of deformation is in a statistical study of a population, in which an m-rep model is deformed into each instance in the population and then a probability density is derived from the collection of deformations. Another major use of deformation is in mechanical simulations on 3-space that includes one or more objects. Therefore the mathematical relationships among deformations on m-reps is important to understand. In the following we will see that the set of deformations forms what is called a *symmetric space*. A background to the mathematics of symmetric spaces and a description of m-rep deformations as a symmetric space can be found in (Fletcher et al., 2006). Also included in that chapter is a more thorough discussion of statistical analysis of m-rep objects than that included in Section 8.7.

Let us begin with a single medial atom. Let us consider all transformations on a medial atom, i.e., all combinations of translations of the hub, magnifications or demagnifications of the common spoke length, and rotations of the spokes. We shall see that this set of all transformations on the atom forms what mathematicians call a *symmetric space* made up of a Cartesian product of *Lie groups* and *quotients of Lie groups*, and the medial atom is a single point on this space. In the following we explain these terms and ideas and generalize them to full m-reps, preparing their application to statistics of m-reps, to interpolations of the transformations between m-reps, and to interpolations in space between samples of sampled m-reps (Fig. 8.1d).

Each medial atom can be understood as a transformation of the primal medial atom, which, given a base (x, y, z) coordinate system, we will take to be the atom whose hub is at the origin, both of whose spokes point in the positive x direction, and whose spokes have unit length. The medial atom **m** then is understood as the translation of the hub from the origin to the hub position of **m**, combined with the magnification (multiplication) of the unit spoke length by the radial length of the spokes in **m**, combined with the rotations of each of the spokes to their respective latitudes and longitudes on the sphere. The first two of these transformations are algebraic group operations, with the operations of vector displacement and multiplication respectively. That is, (1) we can compose two vector displacements or two spoke length multiplications, (2) for each of these operations there is an identity transformation, and (3) there is a unique inverse to each transformation. This group property, together with the fact that the set of transformations and the composition and inverse operations are smooth, are the definition of a Lie group.

The spoke rotational transformations do not form a group, among other reasons for lack of a unique inverse. However, each spoke rotation can be thought of as rotation of a sphere that includes no rotation about the spoke axis. Mathematicians call

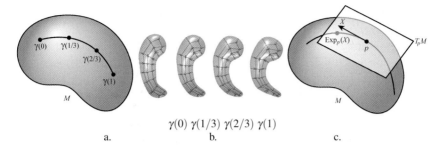

$\gamma(0)$ $\gamma(1/3)$ $\gamma(2/3)$ $\gamma(1)$

a.   b.   c.

Fig. 8.8 (**a**) A high-dimensional curved manifold of m-reps. On this manifold a point is a whole m-rep. The geodesic path shown between two such points (m-reps) gives a shortest route of transformations between the m-reps, a sequence shown for a hippocampus in part (**b**). This same diagram (with the surface being of lower dimension) can be taken to illustrate the manifold of medial atoms. In that interpretation a point is a single medial atom. (**c**) The tangent plane to a symmetric space and the Exp mapping of a vector in the linear tangent space to a geodesic path on the symmetric space

this *the quotient* $SO(3)/SO(2)$ of the Lie group of sphere rotations (3D rotations called $SO(3)$) and the Lie group of rotations about the spoke axis (2D rotations called $SO(2)$). This quotient of Lie groups has adequate smoothness due to the smoothness of the contributing Lie groups.

The result of a medial atom being in a set that is a Cartesian product of Lie groups and quotients of Lie groups is that the set of medial atoms can be understood as an 8-dimensional smooth manifold (9-dimensional for end atoms), albeit one that is curved (see Fig. 8.8). On this manifold, we can define a distance function and thus geodesic paths. For m-reps, the associated distance-squared function for an appropriately small m-rep transformation might be formed from the sum over the atoms of the sums of the squared displacements of the two spoke ends associated with hub translation, spoke length magnification/demagnification, and swings of the spokes, respectively. Alternatively, the squared displacements from these four atom components might be normalized according to the figural volume changes they produce.

A manifold is called a *symmetric space* if it satisfies the property that at each point there is distance-preserving diffeomorphism (bijective, smooth warp whose inverse is also smooth) on the manifold such that all geodesics through that point are reversed by that diffeomorphism. For the linear space of translations, the diffeomorphism is negation of the translation transferred to the origin. For the multiplicative space, the idea works by applying the linear theory to the logarithm of the magnification. For the rotations of axes on the sphere, the reversing map is rotation about that axis by π.

Cartesian products of symmetric spaces are also symmetric spaces. For example, a medial atom is a Cartesian product of the hub translation, spoke length magnification, and two spoke rotation symmetric spaces. Now consider a sampled m-rep, i.e., a collection of medial atoms to each of which the transformation, relative to its own center, can be applied. This discrete approximation to an m-rep is a Cartesian product of the symmetric spaces corresponding to each atom separately and is thus itself a symmetric space. The set of discrete m-reps of a particular structure made up of n_i internal atoms and n_e end atoms can thus be understood as a curved manifold of dimension $8\,n_i + 9\,n_e$. A point on that surface corresponds to a particular discrete m-rep (see Fig. 8.8).

Why have we gone to the trouble of thinking of a discrete m-rep as a curved, smooth manifold formed by a symmetric space together with a distance function? It is because for such manifolds geodesic calculations are closed-form algebraic operations implemented by two operations between a point $\mathbf{p}$ on the manifold and a point on the linear tangent space at that point. The mapping between the tangent hyperplane and the curved surface is called $exp_{\mathbf{p}}$, and the inverse mapping, between the curved surface and the tangent hyperplane, is called $log_{\mathbf{p}}$. As illustrated in Fig. 8.8, these mappings allow the difference between two m-reps to be computed; and as illustrated in Fig. 8.9. they allow the geodesic to be sampled to form an animation of the deformation between two m-reps. And they allow positions along that geodesic, measured by the fraction of the distance traveled between the end points, to provide an interpolation between two m-reps. While they also allow interpolation among two or more medial atoms, interpolation within a fixed m-rep is more appropriately done (Han et al., 2005a) using interpolation of the medial sheet from the atom hub positions and the sheet normals given by each atom's spoke differences and interpolation of the two respective spoke swings using the mathematics given in Chapter 3. Strictly speaking, neither form of interpolation produces medial atoms in that the atoms' spokes are not orthogonal to the surface over the atoms' ends, but rather they are what Damon calls skeletal atoms (Chapter 3).

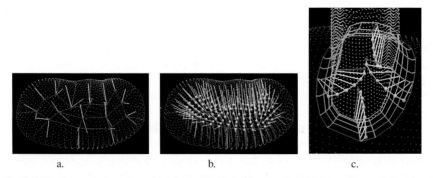

a.   b.   c.

Fig. 8.9 The interpolation of quad-meshes of spokes (Han et al., 2005a). (**a**) The uninterpolated atoms of a single figure kidney object. (**b**) Interpolated atoms for that object. (**c**) For the blend region of a multi-figure object (see Fig. 8.5). The red and green dots show the surfaces of the part of the host figure and the subfigure that are included in the object surface. The red and green curves delimit the region to be blended; on each curve sample points have associated spokes, shown in light blue and pink, on the host figure and subfigure, respectively. The yellow lines show spokes interpolated between the light blue and pink spokes; the end points of these spokes interpolate the blend surface between the corresponding points on red and green curves; these are shown for only four example points. The medium blue curves show similar interpolations between other pairs of corresponding points on the red and green curves

Frequently we are interested in the "shape space" of m-reps, that is, the set of m-reps modulo a global similarity transformation (7 dimensions: translation (3), rotation (3), magnification/contraction (1)). By analogy to shape spaces on flat (Euclidean) manifolds, we consider a space of m-reps that have been corrected by an alignment operation that minimizes the inter-object geodesic distances.

8.7 The Statistical View of Objects

Instead of always talking about an individual entity, such as your liver or your car, as an object, we also frequently speak of an object, e.g., *the* automobile or *the* liver, as a population of instances, either across individuals or within an individual across time. Moreover, even a single object representation may be usefully understood as a center of a population of objects that agree with the representation to within some tolerance. The remaining question is how to represent such probability distributions on m-reps.

The first answer that may come to mind is to find the mean and principal components of the parameters describing an m-rep, since these can be used to parametrize a Gaussian probability distribution. Let us consider even the simplest situation, probabilities on a single medial atom. Because the transformations available include nonlinear ones of magnification and rotation, the linear theory of principal components is not suitable and, if used, generates geometrically illegal objects. However, the theory of principal components has been generalized by Fletcher et al. (2004) to the situation of symmetric spaces including nonlinear transformations. As described in Section 8.6, the mental leap comes from first understanding a geometric entity as a geometric transformation of a base entity (e.g., a medial atom as a transformation of the base medial atom as described above) and then considering it as a point in the of all such geometric transformations. Each point in that space corresponds to a transformation, i.e., to a geometric entity (though some of the points may be geometrically illegal). The idea is to do statistics on the collection of symmetric space points corresponding to a training population or to place probability density measures on the symmetric space.

Let us first consider the symmetric space of translations. Because addition of coordinate values is the operation of the underlying Lie group, the space of translations can be understood as a flat (Euclidean) space of transformations. That is, one can visualize the space as a flat, albeit high-dimensional, surface upon which the difference between two transformations is a vector along the straight line between the points corresponding to the two transformations (Fig. 8.8). The metric on that difference is the Euclidean difference between the points and is thus calculated using the Pythagorean theorem. Principal component analysis is appropriate in such spaces. In such a space the mean of a collection of geometric entities (points) is the point to which the sum of squared distances over the collection is minimum. And the subspace of the first k principal components is that k-dimensional linear

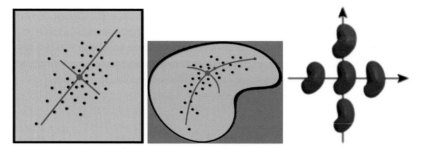

Fig. 8.10 *Left*: A flat symmetric space and *Middle*: a curved symmetric space, with means and principal geodesics. *Right*: The mean kidney m-rep is illustrated at the origin, with the m-reps ± 1 standard deviation from the mean along the two dominant principal geodesics shown along the horizontal and vertical axes

subspace through the mean to which the sum of squared distances over the collection is minimum.

As illustrated in Fig. 8.10, this view can be generalized to the situations where the transformations defining geometric entities are nonlinear. In this case the space is curved, since straight lines between a pair of points achieved by linear interpolation of the parameters describing the points fail to stay on the surface of transformations. For example, the linear interpolation of two 3D rotation matrices or of two unit quaternions (represented as a 4-tuple of coefficients) element-by-element does not yield a 3D rotation matrix or a unit quaternion, respectively. This is because in the rotation Lie group the operation involves multiplication of the matrices or quaternions. However, if closest distance paths on this curved surface of transformations can be defined, i.e., geodesics in the space, the notions of mean and principal components can be generalized. The Fréchet mean of a collection of geometric entities (points) is the point to which the sum of squared geodesic distances over the collection is minimum. And the subspace of the first k principal components is that k-dimensional subspace of geodesic paths through the mean to which the sum of squared geodesic distances over the collection is minimum.

The same idea can be applied to produce statistics on a tuple (Cartesian product) of transformations, i.e., to a sampled m-rep. A strength of this idea is that the different operations on an atom: hub translations, spoke rotations, and magnification of both spokes, can relate in their respectively appropriate ways. This idea can be well illustrated by considering the mean of two sampled m-reps. In producing the mean, the hubs may be translated along the vector in Euclidean 3-space between the two input hubs to the point halfway between them. At the same time, the rotations of a spoke must interpolate as rotations, i.e., on the surface of the sphere in 3-space, to the orientation halfway between the two input orientations. In addition, the lengths (magnifications) of the two spoke pairs needs to be interpolated according to its Lie group operation, namely multiplication. That is, the mean size should be the geometric mean of the two input sizes. This idea is all the stronger when it is applied to the Cartesian product of a set of atoms sampling an m-rep. Then articulations at joints are handled correctly, swellings that are local remain local, etc.

The approach just sketched depends on producing the measures on the Lie group according to which geodesics will be defined. The idea we are using at present is to make equal those differences in the different components of a medial atom that produce equivalent changes in Euclidean position in the ambient 3-space at the implied boundary, i.e., at the spoke ends. We make this choice because so much of the action in image analysis takes place at this boundary. Physical objects are visually sensed at the boundary. Image contrasts typically take place at the boundary. So even though we wish to represent the whole object interior and not just the boundary, the boundary has a special significance. Of course, if in a particular application the properties of interest take place at another place, e.g., some fixed fraction of the way from the medial locus to the boundary, i.e., some fixed value of δ, then the distance r in the following would need to be replaced by δr.

If the spokes have length r, an appropriately small translation of boundary by Δ can be accomplished by moving the hub by Δ, by changing r by Δ (multiplying r by $(1 + \Delta/r)$), or by changing the spoke angle by Δ/r. Thus spoke angle changes of $\Delta\theta$ contribute $r\Delta\theta$ to distances, spoke length magnification by $1 + \lambda$ contribute λr to distances, and translations of the hub by $\triangle\mathbf{p}$ contribute $|\Delta\mathbf{p}|$ to distances.

The means of computing the mean and principal components from a collection of training points on the symmetric space of sampled m-reps can now be specified. Basically, one either does trigonometry on the symmetric space surface itself, or one transfers the points to the tangent hyperplane at the present estimate of the mean via the $log_\mathbf{p}$ function, operates on that surface, and transfers the result back onto the symmetric space surface via the $exp_\mathbf{p}$ function. The second approach can be shown to work for computing the mean, as long as distances on the tangent hyperplane are taken to be the corresponding geodesic distances on the symmetric space surface. As long as the cloud of sample points is tightly clustered, the second approach produces a good approximation to either definition of principal geodesics defined on the symmetric space manifold:

1. The geodesic through the mean that minimizes the sum of squared geodesic distances to the geodesic.
2. The geodesic through the mean such that the points on that geodesic by geodesic projection of the sample point have maximum sum of square geodesic distances to the mean.

The latter two definitions, equivalent in flat space, are not equivalent in any curved space, e.g., the space of tuples of atoms (m-reps). Indeed, the approximation has certain properties preferable to either of the two definitions of principal geodesics defined on the symmetric space manifold, so we use this approximation in computing principal geodesics.

Two issues that arise when doing statistics are preliminary alignment of the training cases and setting positional correspondences among the training cases. We have discussed a method of alignment at the end of the previous section. The issue of correspondence is discussed in Section 8.9.

What advantages for probability distribution estimation on object(s) geometry are provided by analysis of m-reps by geodesic alignment, Fréchet mean computation, and principal geodesic analysis? The following paragraph summarizes some experimental results indicating these advantages. Details can be found in Liu et al. (2008) and Jeong (2008).

Estimation of a probability distribution can be evaluated by the number of modes of variation required to capture a certain fraction of the variance and by the degree to which a test set of represented objects is correlated with the projections of the set onto the shape space provided during that estimation. The latter measure increases with the size of the training set used for the estimation. For any designated level of this measure, two means of representation and probability distribution estimation can be compared by the size of the sample set needed. We have done various comparisons of both types using ellipsoids randomly warped by a composition of bending, twisting, and tapering with m-reps fit to the results, i.e., comprising three nonlinear, independent modes of variation.

The results suggest first, that when local rotations are a significant component of the variation in the population, global analysis via principal geodesic analysis of m-reps may require fewer modes of variation than principal component analysis on boundary point representations. Second, the results show disadvantages of global m-rep analysis to analysis at two scale levels, global and then local by atom on what cannot be described in the stably computable global modes of variation. In this test the single, global scale method requires 4 times the number of samples to give comparable estimation quality than the multiscale method. The multiscale method includes an alignment that depends on knowing normal directions to the medial sheet, information provided stably by the m-rep spokes. A question that remains is whether a multiscale boundary-based representation with not only boundary positions but also orientations given by boundary normals might provide similar advantages.

Other probability distributions and statistical approaches can also be applied on symmetric spaces. Examples are probability distributions produced by Parzen windowing and other Gaussian mixtures and clustering approaches. Statistical approaches include discrimination by support vector machine and related methods, kernel methods, and indeed any statistical method that needs to be applied to objects or shapes.

Statistics may also be calculated on image intensities. However, image intensities, in 3D medical images in any case, only make sense in anatomic, i.e., in object-relative, coordinates. That is, for the intensities across images to correspond, whether position by position or region by region, the image space must be transformed to these coordinates before the statistics is done. This transformation may be within objects and outside near the boundary, or it may also also deal with interstitial regions between objects. Detailing of this approach is left to Chapter 9.

Finally, for deformations that take place within a particular object instance rather than between members of a population, physical models represented by partial differential equations and solved by finite element and other discrete approximations can be aided by m-reps, in meshing (Crouch et al., 2003) and perhaps in solution via eigenanalysis directly from m-reps.

8.8 Discrete M-Reps

To represent a continuous entity such as an m-rep in a computer, one must discretize it. The two most common means of doing this are by sampling or by producing a parametrized representation in terms of basis functions. In this section we discuss m-reps represented by sampling to produce *discrete m-reps*. In Section 8.10 we discuss a parametrized representation formed by m-rep splines.

The medial sheets forming a figure can be sampled by any appropriate sampling scheme: into triangular tiles, quadrilateral tiles, hexagonal tiles, etc. According to the idea that the spoke length r forms a distance metric, the spacing between the tile vertices should be approximately r-proportional. At each vertex is placed a medial atom (Fig. 8.4).

In a continuous m-rep the end curve is formed by atoms whose spokes have come together, i.e., the object angle is 0 (see Chapters 2 and 3). The locus of the spoke ends of these atoms forms a crest on the boundary, i.e., a locus of a local maximum of curvature in the principal direction across the crest. Two problems arise with using such an atom of spoke-multiplicity 2 as an end atom in a discrete m-rep. First, the atoms spokes move towards each other at an infinite rate in the limit as the medial sheet end is approached (Fig. 8.11a), an unstable process. Second, deriving an atom from but a single point of image information is ill-conditioned, and doing this for an atom as critical to the shape as an end atom is therefore ill-advised.

Thus, as introduced in Chapter 1, we invented a new representation called *end atoms* (Fig. 8.11b) for figural ends that was designed to be stable in both of these senses. To avoid the infinitely fast collapse to zero of the angle between the medial spokes, it could allow only a subset of the types of ends generally allowed. At

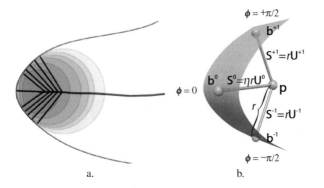

a.   b.

Fig. 8.11 Medial geometry of end atoms. (**a**) In cross-section, the continuous relationship between a point on a medial surface and the points of contact between the disk of principal curvature inscribed at the point and the boundary of the object asymptotes at the crest point: equal steps along the medial surface result in increasing steps along the boundary. (**b**) To account for this asymptotic relationship m-reps describe ends of figures using special end atoms with three spokes that describe the entire end-cap of the figure, symmetric about the crest at $\mathbf{b}^0$

the same time, it needed to be consistent with the description used for the interior portion of the medial manifold. We therefore cut off the interior description while there was still a significant angle between the spokes and insisted that the Blum medial axis of the end portion continue straight from the place where the interior description stopped. That is, the end atom was provided with a new spoke along the bisector of the other two spokes with the new spoke incident to the crest of the implied boundary. By providing this bisector spoke with a specifiable length ηr, we provided a parameter η of crest sharpness additional to the parameters of the interior medial atoms. Even corners can thus be designated, using $\eta = 1/\cos(\theta)$.

The branching structure described in Section 8.5 was also developed to avoid instability in a continuous m-rep. In this representation the parts, called *hinges*, of the ends of protrusion (additive) and indentation (subtractive) subfigure sheets ride on the implied boundary of a host figure. These hinges are represented by designating a sequence of end atoms on the subfigure as hinge atoms. Knowing the hinge atoms' global position, orientation, and size parameters in the figural coordinates of the host boundary allows the subfigure to follow changes of its host figure while remaining appropriately connected to the host. The remainder of the atoms in the subfigure should then change to that member of the principal geodesic space of that subfigure for which the hinge atoms agree most closely, in geodesic distance, with their host-implied positions.

The description of a subfigure riding on a host figure and thus intersecting the host figure requires a means of blending the seam of a subfigure with its host figure and of giving object-relative coordinates within the blend region. An approach consistent with the medial atoms of the host figure and of the subfigure, illustrated in Fig. 8.12, is to cut off the subfigure at a specified value of its u coordinate, produce a hole in the host figure by dilating the subfigure's intersection with the host figure in the (u, v) coordinates of the host figure, associate the positions on the two cut-off curves based on an integrated minimum distance criterion, and between each pair of associated points interpolate the spokes. This interpolation is best accomplished using the ideas in Section 8.6, interpolating the hub position linearly, log of the spoke length linearly, and the spoke orientation along the spherical geodesic.

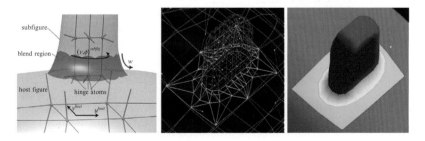

Fig. 8.12 A subfigure and host figure, their relation and blend, and hinge atoms

The blended surface passes over the ends of the interpolated spokes. The degree of dilation and the subfigure cutoff coordinate are parameters of the process.

For speed of computation we presently approximate the surface interpolation via a grossly tiled blend and retiling and interpolative smoothing with normal agreement, using a successive subdivision algorithm described in detail in Thall (2004) (Fig. 8.12). This algorithm is also used to quickly approximate the boundary tile vertices of the individual figures making up an object. However, a method based on interpolating the medial atom hubs and spokes appears preferable. Such a method was briefly discussed near the end of Section 8.6.

The blend region needs its own figural coordinates. As illustrated in the left panel of Fig. 8.12, the v and ϕ coordinates of the subfigure serve to take one around the blend, δ takes one along the spokes, and we create a w coordinate that follows the interpolation from the subfigure cutoff to the edge of the hole in the host figure.

As an object deforms, not only can the individual figures that make it up change shape, but it is also useful to see subfigures as changing their overall conformation relative to their host. If they rotate or translate or magnify, they need to be seen as rotate or translate relative to the host figure, i.e., in the figural coordinates of the host. This includes not only rotation on the surface of the object but also rotation about a hinge in that surface, hinge atom by hinge atom, with a consequence for the remaining atoms within the subfigure's shape space, as described above. This provides a generalized notion of hinging even when the hinge curve is not straight.

One can handle the inter-object relationships in a similar way (Pizer et al., 2005c; Jeong et al., 2006). For each target object certain medial atoms in neighboring objects, namely those near and thus highly correlated with the target object are designated as neighbor atoms. These neighbor atoms are *augmented* to the target object atoms, and after alignment of these sets over the cases, principal geodesic analysis is applied to this augmented set, i.e., to the union of these atoms. The result is a shape space formed from the mean and set of chosen principal geodesic modes for the augmented set. For any position of the neighbor atoms, the projection of the aligned augmented set onto the shape space followed by restricting the result to the target object is the conditional mean of the target object, given the neighbor atom positions. This conditional mean is the target object value *predicted* by the neighbors. Prediction allows the target object to undergo changes sympathetic to the changes of its neighbors. This predicted value can be geodesically subtracted from the aligned target object itself to give a target object residue. A second principal geodesic analysis on these residues can be computed, yielding a shape space for the target object after prediction by its neighbors and a probability density on that space. Geodesic subtraction of the prediction and projection onto this neighbor effect shape space then yields the part of object variation that can be attributed to its own variation, as opposed to the effects of neighbor variation.

8.9 Correspondence of Discrete M-Reps in Families of Training Cases

The principle expounded in this chapter is that a medial representation is usefully thought of as implying a boundary rather than the reverse. To compute the mean and principal geodesics associated with a family of training cases, each training case is typically given by its boundary. The problem is to produce an m-rep for each case such that the figural topology and number of medial atoms per figure are fixed and such that the individual atoms are in good correspondence across the cases. This issue of establishing correspondence is always of concern with discrete representations, but it is particularly of concern because many conformations of atoms can almost equally well describe the same boundary. One of the major strengths of discrete m-reps is its stability: moderate changes in the medial atoms can lead to only small changes in the implied boundary. But said another way, multiple arrangements of medial atoms can imply the same boundary to within some tolerance. This polymorphism can lead to problems of positional correspondence between different instances of the same object, e.g., by representing different individuals in a population or different states of the same individual.

One way to obtain correspondence is to take the point of view of Taylor, Davies, et al. (Davies et al., 2002) that in a population of all reparametrizations of the members of the population, the one to be chosen is the one which has the tightest probability distribution. That is, degradation of correspondence is assumed to broaden the probability distribution. This point of view requires the notion of orbits of individual instance, that is, the subspace of all representations on the symmetric space that lead to the same boundary (perhaps to within some tolerance). Finding the best representative of each training case's orbit leads to a time-consuming computation.

A good step towards achieving correspondence at low cost can be obtained by depending on m-rep's edge fitting in a predictable fashion into a crest of the training case's boundary and then regularly spacing the atoms. The regular spacing can be obtained by including a measure of irregularity in an objective function to be minimized as the m-rep is fit to the object boundary. The measure of regularity that we use is the sum of the square deviations of each atom from the average of its neighbors, where the average and the deviation is computed via geodesic distances.

The objective function that we minimize to fit an m-rep of a given topology into an object is a weighted sum of the following terms.

1. The irregularity measure just mentioned.
2. The sum of squared distances between the implied boundary's tile vertices and the input object boundary. This distance function can be made large in other objects of a complex, to avoid interpenetration of the regions implied by the respective objects' m-rep sheets.
3. The sum of squared geodesic distances of the medial atoms to the corresponding atoms in a reference m-rep translated so that it shares a center of gravity with the m-rep being fit. (Rotation and scaling is also possible, using the second moments

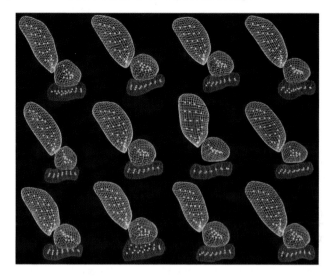

Fig. 8.13 An m-rep fit to 12 training cases of the bladder, prostate, rectum complex with correspondence

about the center of gravity.) Initially, the reference object may be taken as one of the training objects, but after all the objects are initially fit, their mean can form a new reference object, and the fitting can be repeated.

4. The sum, over an optionally chosen set of landmarks on the reference object, of distance squared divided by tolerance squared. This helps with initialization, with getting the m-rep to fit into narrow elongated sections, and with avoiding rotation of the m-rep sheet.

As illustrated in Fig. 8.13, this algorithm works quite effectively, over a variety of complexes of single figure objects. A version for multifigure objects is in trial.

8.10 Continuous M-Reps via Splines or Other Basis Functions

With a parametrization of the medial sheet on (u, v), given the continuous functions of the hub locus $\mathbf{p}(u,v)$ and the spoke length value $r(u,v)$, the *envelope equations* for the family of spheres defined by these functions yield expressions restating equations 5 of Chapter 1 and the equations in proposition 7.1 of Chapter 2. The following analytic expressions for the spoke unit vectors $\mathbf{U}(u,v)$ and the spoke ends $\mathbf{b}(u,v)$ result:

$$\mathbf{U} = -\nabla r \pm \sqrt{1 - |\nabla r|^2}\mathbf{N}, \qquad (8.1)$$

$$\mathbf{b} = \mathbf{p} + r\mathbf{U}, \tag{8.2}$$

where $\mathbf{N}$ is the unit normal to the medial sheet and ∇r is the Riemannian gradient of the function r on the medial sheet (the Riemannian gradient describes the direction in which r changes fastest on the manifold $\mathbf{p}$ and its magnitude is equal to the rate of change per unit step in the tangent plane). These are given respectively by

$$\mathbf{N} = \frac{\mathbf{p}_u \times \mathbf{p}_v}{|\mathbf{p}_u \times \mathbf{p}_v|} \quad \text{and} \quad \nabla r = \begin{bmatrix} \mathbf{p}_u \ \mathbf{p}_v \end{bmatrix} \mathbb{I}^{-1} \begin{bmatrix} r_u \\ r_v \end{bmatrix} \tag{8.3}$$

where $\mathbb{I}$ denotes the first fundamental form on the medial surface, given by the outer product of $\begin{bmatrix} \mathbf{p}_u \ \mathbf{p}_v \end{bmatrix}$ with itself. The spokes themselves are simply $r\mathbf{U}$, the object angle $\theta(u, v)$ is given by $\cos^{-1}(|\nabla r|)$, and the frame $F(u, v)$ is made from the vectors $\mathbf{N}$, $\nabla r/|\nabla r|$, and their cross-product.

Recognizing the ability to compute a continuous locus of full medial atoms from the continuous functions $\mathbf{p}(u, v)$ and $r(u, v)$, Yushkevich et al. (2003) proposed B-splines in (x, y, z, r) as a means of producing these continuous functions. Given a mesh of control points for these B-splines, the internal locus of a single-figure m-rep can be calculated. The greater challenge lay in forming the end curve and representing branches, where the vector ∇r must satisfy certain equality constraints.

Along the end curve, the equality constraint requires that $|\nabla r| = 1$, thus ensuring that the normal component of the spoke vectors $\mathbf{U}$ vanishes, and the spoke ends meet. In 2D continuous m-reps, Yushkevich was able to repose the problem as a constraint on the control points of the B-spline. In 3-D, however, the number of points at which the constraint must hold is infinite, while the number of control points is still finite; hence, the problem is overdetermined. A solution was obtained by letting the domain of the medial surface definition be the region in (u, v) space bounded by the zeroth level set of $|\nabla r| - 1$. In the B-spline framework, this solution was implemented by forcing $|\nabla r|$ to take large values on the perimeter of the unit square, making sure that $|\nabla r| < 1$ somewhere inside of the unit square, and then finding the level set to define the domain. The shape of the domain can be regulated using the spline control points, but the domain can not be fitted exactly to some prescribed curve. Figures 8.14a–d illustrate how the implicit formulation makes it possible to define medial surfaces on free-form domains. Yushkevich et al. recently developed an alternative method in which the continuous m-rep can be based on any set of basis functions and the boundary of the domain can be explicitly defined; this method treats the end-curve constraint as a boundary condition of a partial differential equation (PDE) that can be solved quickly (Yushkevich et al., 2005).

As illustrated in Fig. 8.15, the control points of the continuous m-rep can be adjusted to optimize the fit of the implied boundary to the boundary of a given binary image. In addition to minimizing the distance to the boundary, the optimization includes penalty and regularization terms that ensure that certain inequality constraints are satisfied and that the parametrization of the medial surface by (u, v) coordinates is more or less uniform. One of the penalty terms prevents

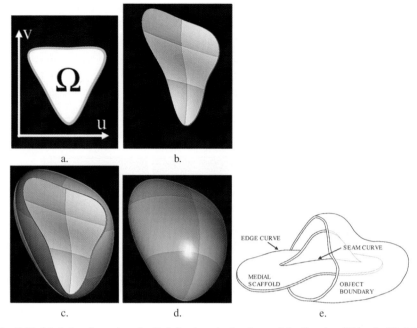

Fig. 8.14 (**a**) A free-form domain Ω defined as the level set of the function $|\nabla r| - 1$. (**b**) A B-spline medial surface interpolated on the domain Ω. (**c**) One side of the boundary defined by the continuous m-rep. (**d**) Both sides of the continuous m-rep's boundary. The constraint $|\nabla r| = 1$ ensures that the two sides form a closed surface. (**e**) The fin-like formation of the 3D Blum medial axes poses challenges for continuous spline-based modeling

the formation of singularities on the implied boundary by ensuring that the Jacobian of the function $\mathbf{U}(u, v)$ is positive.

The resulting continuous m-rep is a Blum medial locus of the object it implies. This contrasts with the locus of medial atoms produced by interpolating a discrete m-rep. That locus is a skeletal surface interpolating its medial atoms, but it is not necessarily a medial surface. The effect is that the discrete m-rep is typically able to fit a particular binary image more closely than the continuous m-rep.

The spline-based cm-rep method described above can model multifigure objects in 2D, but in 3D it is currently limited to single-figure m-reps. The alternative PDE-based method described in Yushkevich et al. (2005) appears to have the same limitation, though it does provide a means of computing the edge of the medial sheet explicitly rather than implicitly. Chapter 3 of the recently completed dissertation by Terriberry (2006) gives a method based on control points and implied control curves that not only provides all medial features explicitly but also supports medial branching. Terriberry's method also provides analytic calculation of various volume and medially implied boundary integrals useful in determining correspondences among a family of m-reps. Therefore, it appears to be an alternative for statistical m-reps worthy of exploration.

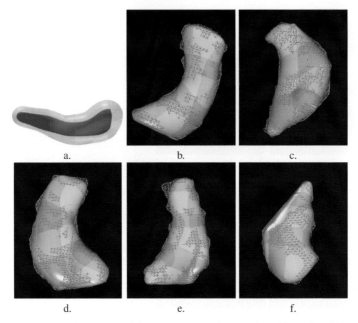

Fig. 8.15 Examples of fitting a continuous m-rep template to manually outlined hippocampus boundaries. (**a**) A hippocampus template shown with a semi-transparent boundary. (**b–f**) Examples of the fitted template's boundary (*shown as green surface*) displayed together with the hippocampus boundary outline (*shown as orange mesh*). The fit is lossy because the continuous m-rep template imposes a simple medial branching topology

Terriberry's cm-reps provide an additional useful capability, calculation of integrals over medially implied object interiors or boundaries or parts thereof by integration over the medial sheet(s) (Terriberry, 2006, Chapter 4). Useful integrals over the object interior are moments of the medially implied object, the volume overlap between an object and a binary image, and various distances between the interiors of two objects. Useful integrals over the object boundary are the ratio of the area of a boundary patch and that of the corresponding medial sheet patch and regional averages of such geometric entities as mean and Gaussian curvature, principal curvatures, and principal directions. In particular, these and other integrals are useful in fitting a cm-res to a binary image or fitting cm-reps to multiple binary images with established correspondences over boundary points.

Both types of integration are done by pulling back the integration to the medial sheet according to the methods discussed in Chapter 3, Sections 4.1–4.4. Recall that the pullback requires calculation of the radial shape matrix S_{rad} at the hub of each spoke involved in the integration. The ability to express S_{rad} as an analytic function of the cm-rep spline coefficients allows the necessary integrations to be carried out using accurate adaptive subdivision numerical integration techniques on the medial sheet.

8.11 Summary and Conclusion

M-reps are designed for deformation by global and local combinations of object interior magnifications, rotations, and translations. The m-rep has the following advantages over alternative representations of objects or object complexes:

1. It inherently allows the deformation of an object to be decomposed into local translations, local twistings and bendings (rotations), and local magnifications.
2. It allows one to distinguish object deformations into along-object deviations, namely elongations and bendings, and across-object deviations, namely bulgings and attachment of protrusions or indentations.
3. It is especially designed to deal with both objects and the surrounding ensembles of objects.
4. It provides an anatomic object and object ensemble based coordinate system in terms of which to deal with the geometry-to-image match and allowing efficient determination of interpenetration of one object with another or a distant section of an object with itself.
5. It directly supports a series of object-based representation at successively smaller spatial scales and thus tolerances, allowing image analysis methods to proceed in an inherently efficient coarse-to-fine manner.
6. It deals with objects in terms of figures, i.e., protrusions and indentations that users frequently think of the objects in terms of and have names for.

A geometry and a statistics for such transformations has been devised, and a physical modeling capability is also partially developed. Among the strengths of this representation, providing efficiency, is a natural multi-scale framework made possible because of the representation's capabilities in providing within object and inter-object geometric neighborhood relationships. These capabilities can be leveraged in a wide variety of applications. Besides the image analysis applications of segmentation and hypothesis testing on shape populations described in Chapter 9, there are many other image analysis applications available and myriad applications in computer graphics and physically based modeling. Among the places of possible use in computer graphics, many of which have been piloted, are animation, texture rendering, image-based rendering, computer games, and computer-aided design. Physically based modeling using the nonlinear basis that m-reps provide is an open opportunity.

Nevertheless, m-reps, as presently designed, force the fitting of a fixed branching structure to objects in a population that may have a more complex or different branching structure. In addition, while the discretely sampled m-reps have an advantage of more tightly fitting objects over the parametrized m-reps, the discrete representation requires a more complex computational infrastructure, which is also more complex than that required by alternative non-medial object representations commonly in use. Also, like the the alternative object representations and associated statistical methods commonly in use, m-reps' statistical analysis is not prevented from including in the domain of random objects those that are geometrically improper, e.g., ones that self-penetrate or have unsmooth boundaries, even if they are less likely than the alternatives to have such improprieties.

Acknowledgment This work was done under the partial support of NIH grants P01 CA47982 and P01 EB02779, NSF SGER Grant CCR-9910419, and a fellowship from the Link Foundation. A gift from Intel Corp. provided computers on which some of this research was carried out. We thank Delphine Bull, Edward Chaney, Guido Gerig, A. Graham Gash, Ja Yeon Jeong, Conglin Lu, Gregg Tracton, and Joshua Stough for help with models, figures, software, and references. We are grateful to J. Stephen Marron, Keith Muller, and Surajit Ray for help with statistical methods.

Part III
Applications

Chapter 9
Statistical Applications with Deformable M-Reps
Anatomic Object Segmentation and Discrimination

Stephen Pizer, Martin Styner, Timothy Terriberry, Robert Broadhurst, Sarang Joshi, Edward Chaney, and P. Thomas Fletcher

Abstract There are many uses of the means of representing objects by discrete m-reps and of estimating probability distributions on them by extensions of linear statistical techniques to nonlinear manifolds describing the associated nonlinear transformations that were detailed in Chapter 8. Two important ones are described in this chapter: segmentation by posterior optimization and determining the significant shape distinctions that can be found in two different probability distributions on an m-rep with the same topology but from two different classes. Both uses require facing issues of probabilities on geometry at multiple levels of spatial scale. The segmentation problem requires the estimation of the probability of image intensity

S. Pizer
Medical Image Display & Analysis Group, University of North Carolina at Chapel Hill, USA,
e-mail: pizer@cs.unc.edu

M. Styner
Medical Image Display & Analysis Group, University of North Carolina at Chapel Hill, USA,
e-mail: styner@cs.unc.edu

T. Terriberry
Medical Image Display & Analysis Group, University of North Carolina at Chapel Hill, USA,
e-mail: tterribe@cs.unc.edu

R. Broadhurst
Medical Image Display & Analysis Group, University of North Carolina at Chapel Hill, USA,
e-mail: reb@cs.unc.edu

S. Joshi
Medical Image Display & Analysis Group, University of North Carolina at Chapel Hill, USA,
e-mail: sjoshi@cs.unc.edu

E. Chaney
Medical Image Display & Analysis Group, University of North Carolina at Chapel Hill, USA,
e-mail: edward_chaney@med.unc.edu

P.T. Fletcher
Department of Computer Science, University of Utah, USA,
e-mail: fletcher@sci.utah.edu

K. Siddiqi and S. Pizer (eds.) *Medial Representations – Mathematics, Algorithms and Applications*.
© Springer Science + Business Media B.V. 2008

distributions given the object description; we describe a way of doing that by an extension of principal component analysis to regional intensity summaries produced using the object-relative coordinates provided by m-reps. Applications of both segmentation and determination of shape distinctions to anatomic objects in medical images are described. Also described is a variant on the segmentation program used in estimating the probability density on an m-rep; this program fits an m-rep to a binary image in a way that is intended to achieve correspondence of medial atoms across the training population.

9.1 Introduction and Statistical Formulation

Both segmentation, i.e., extraction, of objects from images and characterization of geometric differences between classes of objects are usefully accomplished in terms of deformable shape models. In segmentation a geometric model is deformed into the image data, allowing the method to reflect an understanding of what legitimate or typical shapes are. In characterizing the differences between shapes in two different populations, the differences are measured in terms of the deformation from one shape to another. Medial models provide a useful representation of the object or complex of objects that undergoes deformation and of the deformations themselves. Moreover, statistics on medial models are useful for both applications, specifying the typicality of a shape or the population of deformations between shapes in the two classes being compared. Finally, the segmentation application requires not only statistics on the geometry, i.e., on the medial models or their deformations, but also statistics on the image intensities, given a medial model. Because these intensities are best understood statistically in object-relative coordinates, the figural coordinates provided by m-reps are an important means of producing the image intensity statistics.

In Chapter 8 the geometry of discrete m-reps and statistics on these entities were discussed. This chapter discusses the use of these geometric representations and their statistics, as well as the statistics on image intensities in figural coordinates for segmentation of anatomic objects and object complexes. It also discusses the use of these geometric representations and their statistics for statistical shape difference characterization between classes of anatomic objects or object complexes extracted from medical images, e.g., between the hippocampi or lateral ventricles of healthy and schizophrenic individuals as extracted from magnetic resonance images.

In characterizing the difference between two anatomic populations the differences need not only to be specified statistically, but also this specification needs to include *where* the differences are and *what form of deformation* occurs there, for example, whether it is a local twist or a local bend or a local swelling or a local contraction. Also, in segmentation, a coarse-to-fine, i.e., successively more local approach has serious speed advantages for any given quality of segmentation. M-reps with their coordinate systems, their provision of multiscale statistics, and their

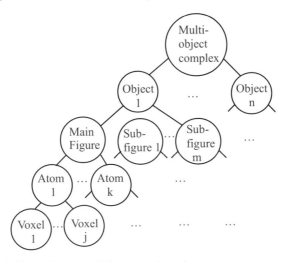

Fig. 9.1 A tree of objects, figures, medial atoms and voxels

medial basis' provision of both local width and local figural orientation are well matched to these needs.

More precisely, as illustrated in Fig. 9.1, consider a tree of geometrical entities such that the discrete m-rep at the root of the tree describes a whole object complex and such that the children of a node describe sub-entities which taken together make up that entity. For example, if the root node describes a complex of objects, its children would respectively describe each object making up the complex. Similarly, if a node describes an object made up of figures, its children would respectively describe the figures making up the object, their children might describe individual medial atoms, and their children might describe sequences of displacement on individual voxels. In each node is a collection of atoms made up of all its children, each atom with a value. The value of a node is the atom values of all the atoms making up that node. Then deforming the entity corresponding to a node deforms all of its sub-entities, and after that deformation we may move on to the sub-entities of the node and deform them further in some order. We refer to these stages at which processing occurs as *scale-levels*.

At each scale-level other than the top of the tree, an entity $\mathbf{m}$ has as set of neighbors $N(\mathbf{m})$, that are at nearby physical positions. It is useful to think of the probabilistic relationship among entities in terms of the value of each child of a node, given the value of that node and the conditional probability of a node given the values of its neighbors. The former describe inter-scale-level differences, and the latter describe inter-neighbor differences, i.e., differences across position. This view allows us to think of the problem with a Markov random field formulation in both the scale and positional dimensions.

That is, if the m-rep $\mathbf{n}$ is a child sub-entity of an m-rep $\mathbf{m}$, and $\mathbf{m} \rightarrow \mathbf{n}$ is the value that $\mathbf{n}$ takes as a result of the deformations of its ancestors and most recently as a result of its parent $\mathbf{m}$, we wish the conditional probability of the deformation describing the difference $(\mathbf{m} \rightarrow \mathbf{n}) \ominus \mathbf{n}$, given the parent node $\mathbf{m}$, where the symbol

$\ominus$ denotes the geodesic path between its two operands. Similarly, if $< N(\mathbf{n}) >$ describes the prediction of $\mathbf{n}$ based on its neighbors, we wish the conditional probability, $p(\mathbf{n}\ominus < N(\mathbf{n}) > |N(\mathbf{n}))$, of $\mathbf{n}\ominus < N(\mathbf{n}) >$ given the neighbor nodes $N(\mathbf{n})$. Because the essence of geometry is that entities are locally correlated, a thesis that for medial atoms of various anatomic objects is supported by our data, it is reasonable to condition $(\mathbf{m} \to \mathbf{n})\ominus\mathbf{n}$, on only the parent node $\mathbf{m}$ and not on ancestors more distant in scale, and it is reasonable to condition $\mathbf{n}\ominus < N(\mathbf{n}) >$ on its immediate neighbors $N(\mathbf{n})$ and not on more distant entities.

In the work described here, we simplify the probabilistic formulation even further. We assume that $(\mathbf{m} \to \mathbf{n})\ominus\mathbf{n}$ is statistically independent of $\mathbf{m}$ and that $p(\mathbf{n}\ominus < N(\mathbf{n}) > |N(\mathbf{n}))$ can be broken up into two factors, one describing the change independent of its neighbors and the other describing the interrelationship of it with its neighbors. Breaking things down according to this Markov formulation allows a segmentation or hypothesis test with final locality such that the total number of primitives at that level of locality is M (e.g., there are M voxels in the objects being segmented or at which shape differences are being tested) to operate in $\mathcal{O}(M)$ time rather than the $\mathcal{O}(M^2)$ that are required when the relation of every primitive with every other one must be dealt with.

The geodesic differences between m-reps used in the foregoing formulation are in the same symmetric space as the subtrahend and the minuend. That is, the, the geodesic differences of a collection of medial atoms is the collection of differences of the corresponding atoms, and the difference of two atoms is the Cartesian product of the corresponding components, as illustrated by the difference between interior slab atoms in the following:

1. The difference of the hub positions, which like a hub position itself is a vector in $\mathcal{R}^n$.
2. The "difference" of the spoke lengths, which is the ratio of these lengths giving the magnification of one into the other, and thus like a length itself is a scalar in $\mathcal{R}^+$.
3. The "difference" of each spoke position on the unit sphere S^2 with the corresponding spoke's position on S^2, which can be understood as a position on S^2. There are difficulties with differences of angle differences associated with having to specify a reference angle; these will not be further discussed here.

As a result, statistics on such geodesic differences can be accomplished by the same methods of computing means and principal geodesics described in Chapter 8.

Finally, consider the probabilities on differences of m-reps that are the target of statistical characterization of inter-class differences. These differences of m-reps are again in the same symmetric space as the subtrahends and minuends. One requires methods of hypothesis testing that yield the significance of distinctions in probability distributions in this symmetric space and, as well, the location of such significant changes, for various levels of locality.

In Section 9.2 we introduce segmentation via posterior optimization of deformable m-reps with an overview of the approach. We find that two log probability densities are needed, one measuring the geometric typicality of an m-rep and the other measuring the match between the m-rep and an image. In Section 9.3 we discuss how to train the first probability density, given binary images of sample objects, and how to measure this geometric typicality on any m-rep, given this training. In Section 9.4 we discuss estimating the probability density on image intensities given a medial model and how to measure this probability density on any target image. In Section 9.5 we conclude our discussion of segmentation by specifying the segmentation scale at the smallest scale level, that of the voxel, followed by the excellent results obtained using our multi-scale method using the geometric and intensity probabilities. In Section 9.6 we discuss means of hypothesis testing based on m-reps for statistical characterization of shape differences between populations of objects or object complexes. Section 9.7 gives some examples of results using this method. In Section 9.8 we discuss the apparent strengths and weaknesses of the medial methods we propose for the segmentation application and characterization of shape differences application, as compared to alternative object representations. In that section we also discuss work that remains in both these methods of application of m-reps and in the formulation of m-reps themselves and their statistics.

9.2 Segmentation by Posterior Optimization of Deformable M-Reps: Overview

Published studies by others and our own research results strongly suggest that segmentation of a normal or near-normal object (or objects) from 3D medical images in all but the simplest cases will be most successful if it uses (1) knowledge of the geometry of not only the target anatomic object but also the complex of objects providing context for the target object and (2) knowledge of the image intensities to be expected relative to the geometry of the target and contextual objects.

We use the general segmentation approach already shown by others to lead to success ((Cootes et al., 1993; Staib and Duncan, 1996; Delingette, 1999), among others; also see (McInerny and Terzopoulos, 1996) for a survey of active surfaces methods), namely deforming a geometric model by optimizing an objective function that includes a geometry-to-image match term which is constrained by or summed with a geometric typicality term. In this approach a model of the object(s) to be segmented is placed in the target image data and undergoes a series of transformations that deform the model to closely match the target object.

In computer vision an important class of methods uses explicit geometric models in a Bayesian statistical framework to provide *a priori* information used in posterior optimization to match the deformable shape models against a target image. Using this approach, we start from a statement of the segmentation objective as finding the most probable conformation of the target object(s) $\mathbf{m}$ given the image I, i.e., of computing $\text{argmax}_{\mathbf{m}} p(\mathbf{m}|I)$. Here $\mathbf{m}$ is the geometric representation of the target object(s), in our case the tree of medial atom meshes that comprises an m-rep, and I is a tuple formed by a 3D array of image intensities. The probability density $p(\mathbf{m}|I)$

is frequently called the *posterior density*, so the method is called one of *posterior optimization* (Duda et al., 2001).

By Bayes rule, $\text{argmax}_{\mathbf{m}}p(\mathbf{m}|I) = \text{argmax}_{\mathbf{m}}[\log p(\mathbf{m}) + \log p(I|\mathbf{m})]$. Thus the geometric typicality term ideally measures the logarithm of the so-called *prior* probability density, the probability density that the candidate geometric entity exists in the population of objects, as described in Chapter 8. And the geometry-to-image match term ideally measures the logarithm of the so-called *likelihood*, the probability density that the target image values, relative to the candidate geometry, would arise in the population of images from that modality. As a fundamental means of obtaining efficiency, we optimize such an objective function for successively smaller spatial tolerances (spatial scales), where each of the spatial scale levels are object-relevant: the object complex, the object, the slab (or tube) figure, the figural section, and the voxels not only interior to the object(s) but also the voxels between them, which we call *interstitial* voxels.

The success of the deformable shape models posterior optimization approach depends on the object representation, i.e., the structural details and parameter set for the deformed model, as well as on the form of the objective function. The most common geometric representation in the literature of segmentation by deformable models is made up of directly recorded boundary locations, sometimes called *b-reps* (Cootes et al., 1993; Kelemen et al., 1999), also see papers surveyed by (McInerny and Terzopoulos, 1996). Our m-reps representation (Fig. 9.2), principal geodesic analysis to produce its statistics, and the associated segmentation method use a medial representation intended to produce improved and/or more efficient segmentations for the reasons given in Chapter 8, Section 11. The most relevant of these advantages for this application are the efficient training of the prior it provides, its ability to provide a coordinate system in which to describe intensities probabilistically, and its inherent multi-object, multi-scale nature, which leads to effectiveness and efficiency of segmentation of single or multiple objects. However, small indentations and protrusions of anatomic objects are impractical to model medially. Our approach to solving this problem is to implement a non-medial voxel stage described in Section 9.5.1.

M-reps, combined with the voxel-level representation, provide their advantages over other deformable object representations at the expense of a level of complexity that required the development of special theoretical underpinnings, software,

Fig. 9.2 M-rep modeled kidney with its medial mesh, a liver model that is made from two figures, one for each lobe, and male pelvis model made from multiple objects (two bones, bladder, rectum, prostate). The kidney model also shows the underlying representation of a sampled medial surface and a tiled boundary

and validations. Largely automatic segmentation by large to small application of deformable m-reps has been implemented in software called *Pablo* (Pizer et al., 2005b) that accomplishes 3D segmentations in a few minutes. Software for building and training models has also been developed. The methods underlying this software and its abilities are the subject of Sections 9.2–9.5.

The next two sections give a more specific picture of Pablo's method (Section 9.2.1) and operation (Section 9.2.2).

9.2.1 Segmentation Method: Posterior Optimization for Multiscale Deformation of Figurally Based Models

Our method for deforming a model into image data typically begins with a manually chosen initial positioning of the mean model, frequently via choosing a few rough landmark positions. The segmentation process then follows a number of stages of segmentation at successively smaller levels of scale. The spatial tolerance of the resulting segmentation can be large at the largest scale level but decreases as the scale gets smaller.

As illustrated in Fig. 9.3, at each scale level, i.e., level of the tree shown in Fig. 9.1, the same log prior + log likelihood objective function is optimized by geometrically transforming the entities at that scale level, using a transformation

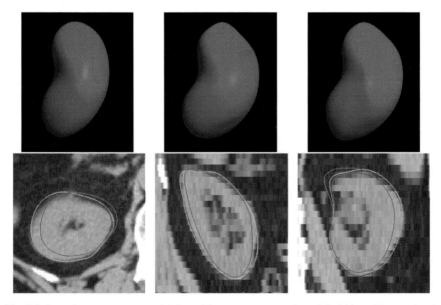

Fig. 9.3 Stage by stage progress of deformable m-rep segmentation of the kidney. Top: rendered 3D view, after model alignment via landmarks, the figure stage, and the figural section (atom) stage. Bottom: results on axial, sagittal and coronal CT slices. Each image compares progress through consecutive stages via overlaid curves: magenta—aligned position; green—post object stage; red—post atom stage

global to the respective entity. Thus, at the largest scale level, the object ensemble stage, the whole object ensemble undergoes a global transformation. At the next smaller scale level, each object making up the object ensemble separately undergoes a transformation global to it. And as the computation moves to successively smaller scale stages, successively smaller entities making up the entities at the next larger scale level, namely figures, subfigures, and medial atoms, are optimized with a transformation global to each of them. The series of optimizations concludes with a small relocation of all of the voxels in the image being optimized.

At all of these scale levels, we follow the strategy of iterative conditional modes, so the algorithm cycles among the component entities in random order until the group converges.[1] For example at the figural atom stage, the algorithm cycles through the atoms in random order.

At each scale level larger than the voxel scale level, the geometric transformation of the entity is made up of a typically deterministic similarity transformation and a maximum posterior warp. The similarity transform, a translation, rotation, and uniform magnification, aligns the entity to neighboring entities of the same type (objects to neighboring objects, medial atoms to neighboring atoms), except it aligns to landmarks at the largest scale. The warp is formed from a few principal geodesics (see Chapter 8) of the deformations of that entity experienced in the training data. At the voxel scale level, the optimization is over displacements per voxel of only a few voxel widths. The result is that we typically optimize 6 or fewer parameters per entity, providing efficiency and convergence of the segmentation at that scale level.

At each scale level we use the conjugate gradient method to optimize the log prior + log likelihood objective function. The log prior metric is detailed further in Section 9.3. As detailed in Section 9.4.2, we have implemented a way of computing the log likelihood that measures the geometry-to-image match based on probability densities on intensity distribution features in various figural-coordinate-specified regions inside and outside of the object (Fig. 9.4) such that each region is expected to be a constant mixture of tissue types (Broadhurst et al., 2005).

9.2.2 Segmentation Method: User Operation

M-rep-defined objects can be viewed as a boundary mesh (at any of a number of vertex spacing levels), a rendered surface, a collection of points at the aforementioned boundary vertices, or a medial atom mesh. Most users find the first two of these the most useful. Images are normally viewed in a tri-orthogonal display, with the three possible slice directions fixed to the cardinal within-image and cross-image slice directions given by the stored target image. The displayed object can be presented together with the intensity display (see Fig. 9.3). Moreover, we also provide

---

[1] The convergence properties are shared with all iterative conditional modes methods and are based on the underlying Markov random field. In practice, convergence always occurs, but sometimes the convergence is to a local maximum of the objective function rather than the desired global maximum.

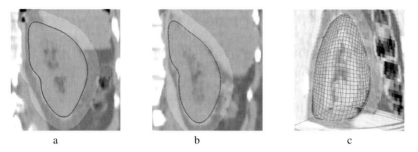

a   b   c

Fig. 9.4 Boundary-relative regions used for measuring geometry-to-image match to a kidney. (**a**, **b**) Example from two different patients displayed in 2D cuts. The kidney interior region is portrayed in blue, and the kidney exterior region is portrayed in orange. (**c**) A mesh showing in 3D the m-rep implied boundary of the kidney, and the kidney interior and exterior regions in two orthogonal cuts through the 3D image

a boundary display mode on the displayed slice, in which the 3D object does not appear but the curve of its intersection with the displayed slice(s) is displayed on that slice (those slices).

Using these viewing mechanisms, the user either chooses the location of pre-selected landmarks in the target image, which is then used as the basis of an Procrustes initialization of the model, or he or she manually initializes the chosen model by placing it in an initial position relative to the 3D image (for example, see Figs. 9.3-bottom row, 9.4c, and 9.9-bottom left). The initialization transform derived from the landmarks is frequently a similarity transform, but we have found it also useful that this landmark-based transform optimize in the shape space of the principal geodesics of the object with a data-match term given by the sum of squared model landmark to image landmark squared distances, with each squared distance divided by its tolerance squared.

The landmarks on the model are chosen as a specified spoke end. These landmarks appear as colored spots on the base model in the display space. These landmarks can also be used for editing an m-rep in the middle of the optimization process or as another term in the geometry-to-image match.

The user is also given control of the values of the weights controlling the strength of the geometric typicality term in the objective function, relative to the geometry-to-image match term. However, since the two terms are now both Mahalanobis distances, the default weight of unity needs seldom be changed.

9.3 Training and Measuring Statistical Geometric Typicality

To be able to measure a log prior, one needs a parametrized function that one can evaluate with any m-rep for the desired object as the argument. Section 9.3.1 describes the means for training the parameters of this prior probability density on m-reps that is then used to measure geometric typicality of any candidate m-rep appearing in the optimization of the log posterior. This training of the prior is done by principal geodesic analysis of m-reps fit to binary images extracted from training greyscale images. Section 9.3.1 describes both the fitting of m-reps to binary images and how principal geodesic analysis is used at multiple scales to produce the prior probabilities needed for the various scale levels. Section 9.3.2 describes the means for measuring the log prior at multiple scales needed in the multiscale segmentation procedure.

9.3.1 M-Rep Model Fitting and Geometric Statistics Formation

Model-building must designate the figures making up an object or object ensemble, give the size of the mesh of each figure, and give the way the figures are related. It must also specify each medial atom in the model forming the mean object or object ensemble and the variability of these at many scale levels. Illustrated in the panels of Fig. 9.5 are m-rep models of a variety of anatomic structures that we have built.

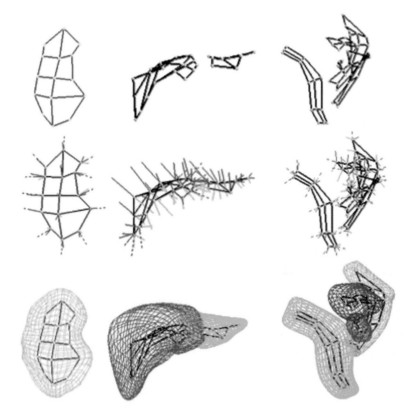

Fig. 9.5 M-reps for a kidney, a liver, and a male pelvis. *Top row*: mesh of atom hubs; *middle row*: mesh of medial atoms (including spokes); *bottom row*: the implied boundaries shown with atom mesh(es)

In the following we sketch our model building procedure, leaving the details of how we meet this challenging goal to other papers (Merck et al., 2006).

Because an m-rep is intended to allow the representation of a whole population of an anatomic object across patients, we build it based on a significant sample of instances of the object. Typically we use some tens of instances, say 50. For each instance we begin with both a 3D binary image representing the interior of the object, typically manually segmented, and an associated 3D greyscale image (CT or MRI or another modality).

Styner et al. (2003a) describe a tool for producing m-rep models from such binary image samples, based on the principle that effective segmentation depends on building a model that can easily deform into any instance of the object that can appear in a target image. We can use this tool to compute the set of appropriate figures at a given level of approximation from a training population, or we can choose the figures based on anatomic expertise to correspond to named anatomic structures. The tool measures the level of approximation in the figure computation step via error in volume overlap (typically 98%). In either case, given the figures, the tool chooses the number of atoms in each figure as the minimal number that can fit every training instance to a given error measured by the mean absolute distance of the surfaces (typically 5% of the average radius).

More recently we have completed a stable web-sharable tool called *Binary Pablo* for fitting an m-rep model to each member of a collection of binary images and deriving the Fréchet mean and principal geodesic modes and variances (Merck et al., 2006). Once a base model is generated, we use Pablo to deform it into the binary segmented training images. The program optimizes an objective function that has an "image match" term giving an average distance between the boundary implied by the m-rep and the binary image boundary, and three geometry terms: (1) giving an average squared-distance between each atom and the geodesic average of its neighbors, thus producing a regular mesh of atoms; (2) discouraging folded objects by penalizing rS_{rad} eigenvalues $\varepsilon 1$ (see Chapter 3); (3) giving a squared-distance from a reference m-rep. The sharable version only operates for single-figure objects, but versions that fit m-reps to multi-figure objects and to multi-object complexes are available in our research toolkit.

Given the m-rep models for all the training cases (Fig. 9.6), we use a tool initially developed in Dam et al. (2004) and further developed by Lu (Lu et al., 2003) to compute the mean model and the principal-standard-deviation-weighted principal geodesics describing its variability. This tool uses the method of principal geodesic statistics on symmetric spaces described in Chapter 8, Section 7. As with linear statistics, each principal geodesic has an associated variance, and moving along that geodesic gives a principal mode of variation of the population of m-reps.

The statistics at one scale level need to describe the variability of the geometric entity at that scale level after the variability at the larger scale levels has been accounted for and after alignment to neighboring entities has been done. Description of this residue statistics, based on the theory of Markov random fields, is given in Lu et al. (2003).

Fig. 9.6 *Left*: A subset of our population of training kidneys. *Right*: the mean of the population and the mean $+/-1$ standard deviation in each of the first two principal geodesic modes

With these means and a number of principal geodesics chosen to capture some fraction of the variance at that scale level, deforming a geometric entity at that scale level involves aligning the object to its neighbors and then computing the coefficients of the principal geodesics of the deformation of that entity.

9.3.2 Measuring Statistical Geometric Typicality

The geometric typicality that we wish to use is $\log(p(\mathbf{m}))$, or in the case that $\mathbf{m}$ has neighbors $N(\mathbf{m})$, $\log(p(\mathbf{m}|N(\mathbf{m}))$. But except for an additive constant and a constant multiplier of -0.5, when the principal geodesic analysis given in Section 8.7 is used, the log probability density in the symmetric space at any scale level is just a Mahalanobis distance in a tangent space to that symmetric space. Thus, when optimizing in the space of principal geodesics, we are optimizing over the weights a_i of the projections $\mathbf{v}^i$ of the unit-variance principal geodesics onto the feature space tangent plane at the mean. For any value of these a_i, and given the variances σ_i^2 of the principal geodesics in that tangent plane that are derived in the principal geodesic analysis, the Mahalanobis distance of $-\sum_i a_i^2$ forms the geometric typicality measure.

As discussed earlier, at all scale levels but the global one this geometric typicality metric of the relevant geometric entity needs to reflect its shape properties but also its relation to immediately neighboring peer entities. This can be accomplished with principal geodesics that were computed with augmenting atoms in adjacent objects or figures (see Chapter 8, Section 7).

Two special neighbor relations deserve comment. One is the non-interpenetration relation among very nearby (possibly abutting) objects (see the male pelvis in Fig. 9.2). Not only the correct relative position, orientation and size need to be reflected in the geometric typicality, but also an interpenetration of the figures needs to result in a low geometric typicality. The second neighbor relation of note is that between a protrusion or indentation *subfigure* to the "host" figure on or into which it sits (see liver in Figs. 9.2 and 9.5) or the relation between an object and a nearby, possibly abutting, object. In Chapter 8 we argued that the subfigure should ride on the boundary implied by the host's representation and be known in the figure-relative coordinates of the host. The augmentation idea mentioned as applying to nearby objects uses a similar concept. Thereby we can make measurements of typicality in terms of the position of the subfigure (or related object) relative to the host, the orientation of the subfigure relative to the host, and the size of the subfigure relative to the host. When slight modifications of the hinge atom relationship are created due to motions in symmetric spaces not maintaining the relationship of hinge atoms to their host figure boundary, we find success in simply projecting the hinge atoms back onto the host boundary along host surface normals (interpolated spokes).

9.4 Training and Measuring Statistical Geometry-to-Image Match

Methods for training and measuring a probability density on image intensities must do so in a way respecting correspondence of locations across the population. There is much good work on correspondence, e.g., (Davies et al., 2002; Yushkevich et al., 2005), but here we suggest that correspondence be obtained through object-relative coordinates (Fig. 9.7). For m-reps that means that the figural coordinates provided by $\mathbf{u} = (u, v, \phi, \tau)$ within figures (see Chapter 8, Section 3) and by $\mathbf{u} = (v, w, \phi, \tau)$, within inter-figural blend regions (see Chapter 8, Section 8) provide the means of correspondence. More precisely, intensity statistics are done with respect to $I(\mathbf{u})$.

Recall that within an object main figure and within a subfigure outside of the blend region, (u, v) measures relative location along the medial sheet, ϕ expresses which side of the medial sheet the location is or at the end where in the transition between the sides the location is, and τ gives the fraction of the distance along the spoke from the medial end to the boundary end. For interfigural blend regions between a subfigure and a host figure v and ϕ are the cross-figure figural coordinates of the subfigure and $w \in [-1, 1]$ moves along the blend from the curve on the subfigure terminating the blend ($w = -1$) to the curve on the host figure terminating the blend ($w = +1$). Section 9.4.1 describes the computation transforming between Euclidean coordinates $\mathbf{x}$ and figural coordinates $\mathbf{u}$. Between objects one must interpolate between the figural coordinates of the nearby objects. The means

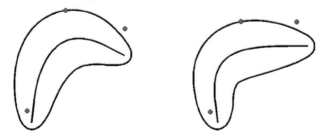

Fig. 9.7 Correspondence over deformation via figural correspondence. In each pair of corresponding marked points, the two points have the same value of the figural coordinates $\mathbf{u} = (u, v, \phi, \tau)$

of this interpolation is still a subject of research, but one of the options is described in Section 9.5.1.

We have used two basic methods that go from m-reps and associated greyscale images to geometry-to-image match functions on an image given an m-rep. The method we used first (Stough et al., 2004) was based, like that of the active shape method of Cootes et al. (1993), on normalized correlation between cross boundary intensity profiles and template profiles determined in training. However, in our method the template in each profile was chosen from among a limited number of possibilities chosen by clustering profiles during training, and values needed in the normalizations of the target profiles at each boundary vertex were also determined during training, thus stabilizing the normalization. Both normalized correlation methods produce a log probability density only under the poor assumption that the profiles are uncorrelated and that the tissue mixture at a voxel in the template can be expected to be precisely the same as that in the corresponding voxel in the target image. To overcome the first weakness Ho (2004) argues for improvements based on multiscale profiles, produced by a variant on Gaussian weighting across but not along the profiles.

Either variant of this profile match method can be expected to achieve less success than our new method, which is designed to produce log probabilities without these faulty assumptions. Our experiments on kidney segmentation, sketched in Section 9.5.2 and detailed in (Broadhurst et al., 2006), showed the new method to give better results in practice. Thus we describe only the new method, which generates a log likelihood on discrete quantile functions from the intensities in regions relative to the m-rep. It is detailed in Section 9.4.2.

9.4.1 Transforming Between Figural and Euclidean Coordinates

The geometry-to-image match term in the objective function requires object-relative image positions $x(u)$ to be computed in large number. Thus, interpolation within $I(x)$ must be very efficient. In Pablo at present, this transformation $x(u)$ is done through the mechanism of subdivision surface methods (Thall, 2004), as described below. Han is developing a more accurate method based on the interpolation of medial atoms (see Chapter 8) and is seeing how to make it adequately speedy.

In the subdivision surface method we interpolate the boundary first and consequently can interpolate medial atoms at any position on the sheet of atoms. The implied boundary is computed from the set of atom spokes connected into quadrilateral and triangular tiles both within figures and in interfigural blend regions (Figs. 8.3 and 8.13). The boundary interpolation is accomplished by a variation of the very efficient Catmull-Clark subdivision (Catmull and Clark, 1978) of the mesh of polygonal tiles. Thall's variation (Thall, 2004) of Catmull-Clark subdivision produces a limit surface that iteratively approaches a surface interpolating in position to spoke ends and with a normal interpolating the respective spokes. That surface is a B-spline at all but finitely many points on the surface. The program gives control of a tolerance on the normal and on the closeness of the interpolations.

The resulting B-spline allows the computation of both boundary positions **b** and boundary normals **U**, which are spoke directions there. Interpolating the medial radius r as well as u and v at such boundary positions allows the computation of $\mathbf{x}(\mathbf{u}) = \mathbf{b} + (\tau - 1)\,\mathbf{U}$.

Points **x** can also be given a figural coordinate **u** by finding the figural coordinates of the closest medially implied boundary point, using the boundary normal or the gradient of the distance function as the spoke direction, and calculating τ from the intersection of this spoke with the sheet of hubs. This calculation, however, is fraught with danger, since the boundary may be inadequately smooth.

9.4.2 Geometry-to-Image Match via Statistics on Discrete Regional Quantile Functions

9.4.2.1 Conceptual Basis for Statistics on Intensity Quantile Functions

Any efficient geometric description does not capture all there is to say about the biology or physics of the individual being modeled. Thus for a given medially specified object or complex of objects, the variation between different images of the same class of object frequently is due not only to intensity noise but more so to the variation of the materials of which the object are made and of the variation across the object of the weights of those materials making up the materials mixture. Thus in medical images there is variation across patients of the locations of specified tissue types within and between their respective objects. This suggests that point-by-point correspondence as provided, for example, in the active shape models and active appearance models of (Cootes et al., 1993, 1999), where the probability densities are on corresponding intensity values, be replaced by probability densities on regional collections of intensities, ignoring the particular locational correspondences within these regions. In particular, this suggests probabilities on intensity summaries, such as histograms, of regions expected to have uniform mixtures of tissue types.

Our regional intensity summary based match method (Broadhurst et al., 2005; Pizer et al., 2005a) uses a region inside the object and a region outside the object (Fig. 9.4) and sometimes subregions of these regions defined according to figural geometry.

The feature space formed by using the bin counts of histograms of intensity provides a poor basis for probabilistic analysis. The weakness is exemplified by the fact that the average of two unimodal histograms in this form would be a bimodal histogram, rather than a unimodal histogram whose center is between the two being averaged. In the following we argue that instead representing the regional intensity collection by the curve of intensity values versus quantile (regional intensity quantile function, or *RIQF*) allows an effective log probability density to be calculated by factor analysis. Also, histogram bin counts as features suffer from quantization effects, i.e., binning errors, while discrete RIQFs do not since no arbitrary bin boundaries are selected.

The RIQF of an intensity distribution i can be shown to be the inverse of the cumulative distribution function I of i. Discretely sampling the RIQF yields the *discrete RIQF (DRIQF)*. The DRIQF is an n bin quantile function where each bin j, representing $1/n$ of the probability distribution area, stores its average image intensity i_j. Considering these values in sorted order, the DRIQF for region k can be represented as a vector $i^k = (i_1^k, \ldots, i_n^k)$. Computing this vector requires partial sorting of the list of N intensities in the region, taking $O(Nlog(n))$ time.

The effectiveness of using standard statistical tools to construct a probability distribution of RIQFs depends on the fact that the space of RIQFs has several known linear properties related to Euclidean distance and thus mean and principal components. Euclidean distance between RIQFs corresponds to the Mallows distance (Mallows, 1972; Levina and Bickel, 2001) between the corresponding probability distributions, defined as follows. For two continuous one-dimensional distributions with cumulative distribution functions Q and R, and RIQFs $q = Q^{-1}$ and $r = R^{-1}$, respectively, the Mallows distance between them is defined as the Minkowski L_2 norm between q and r:

$$M_2(q,r) = \left(\int_0^1 |Q^{-1}(t) - R^{-1}(t)|^2 dt \right)^{1/2} = \left(\int_0^1 |q(t) - r(t)|^2 dt \right)^{1/2}.$$

The Mallows distance can be shown to measure the work required to change one distribution into another by moving probability mass, i.e., the Earth Mover's distance between the corresponding probability distributions, intuitively a good measure of difference between RIQFs. For DRIQFs q and r, the Mallows distance is defined as the L_2 norm of the vector difference between q and r:

$$M_2(q,r) = \left(\frac{1}{n} \sum_{j=1}^{n} |q_j - r_j|^2 \right)^{1/2}.$$

Location and scale changes to any probability distribution, or changes in any affine combination of the DRIQF values, are linear in the space of DRIQFs. Several families of common continuous distributions, including Gaussian, uniform, and exponential distributions, are parameterized by location and scale parameters. Thus, DRIQFs of each of these families of distributions exist in a two-dimensional linear subspace. Also, the Euclidean average (or any linear combination) of a set of DRIQFs from one of these families of distributions results in a DRIQF contained within the family and having means and standard deviations averaging (or correspondingly linearly combining) the respective means and standard deviations. For example, the Mallows distance between two Gaussian distributions $N(\mu_1, \sigma_1^2)$ and $N(\mu_2, \sigma_2^2)$ is $\sqrt{(\mu_1 - \mu_2)^2 + (\sigma_1 - \sigma_2)^2}$. The average in this space of a set of RIQFs corresponding to Gaussian probability densities is a Gaussian with a mean and standard deviation equal to the average mean and standard deviation of the set of Gaussians. However, a weakness of the space is that for probability distributions composed of a mixture of multiple underlying unimodal distributions, changing the mixture amount is a nonlinear operation.

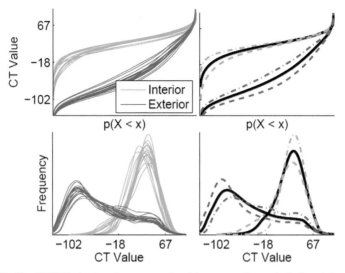

Fig. 9.8 Bladder DRIQFs (*top*) and corresponding histograms (*bottom*). *Left*: training samples; *right*: learned mean and ± 2 standard deviations along the first principal direction

The consequence of the foregoing is that analysis of regional intensity distributions can be captured by linear statistics on their DQRIFs, which can efficiently capture variation in location and scale change. The method is not effective for dealing with multimodal probability distributions with widely separated peaks and varying interpeak separations, but our experience is that it works well for unimodal probability distributions and even multimodal probability distributions whose peaks are not widely separated.

DRIQFs of interior and exterior regions of the bladder in 15 CT images are shown in Fig. 9.8. The first two principal directions of variation of the interior and exterior regions capture 95% and 97% of the variation, respectively. DRIQFs of subregions can also be constructed; this example and this discussion are only in terms of interior and exterior regions. In this example, the contribution of each voxel to the DRIQF is Gaussian weighted by its distance to the surface. This allows narrow regions to be defined that have larger capture ranges and smoother objective functions during segmentation than equivalent non-weighted regions. In each region, gas and bone intensities have been automatically removed from the distribution using a threshold, and the probability of each is independently measured. The Mallows distance is sensitive to the variation in these intensities due to their extreme intensity values compared to the differences in fat and tissue intensities.

9.4.2.2 Training Probability Densities on Regional Intensity Quantile Functions

The probability densities on DRIQFs that we use are estimated by principal component analysis of the DRIQFs, taken as feature vectors. Then the geometry-to-image match of the DRIQF obtained from a target image region is a Mahalanobis distance based on this principal component analysis. In the following we detail the estimation of this Mahalanobis distance function.

We model the variation in our DRIQF feature space as a multivariate Gaussian distribution. The dimension of this space is equal to the number of bins used to represent a DRIQF, which is typically 200. This often results in a high dimension, low sample size situation, which prevents us from estimating a full rank multivariate Gaussian model. Therefore, we use principal component analysis to estimate a low dimensional subspace, typically of dimension 2–4, in which we build a Gaussian model. We then measure the expected distance to this subspace by summing the remaining eigenvalues since during segmentation we expect to estimate the probability of many image regions that are not typical of the training regions. Thus, the final Gaussian model for each region is of dimension equal to the number of eigenmodes plus one.

The geometry-to-image match function is the resulting Mahalanobis distance function. Intuitively, the Mahalanobis distance of a target DRIQF is equal to the Mallows distance between the probability distribution corresponding to the target DRIQF and that corresponding to the mean RIQF, modified by the standard deviations in each direction of the Gaussian model. Thus the Mahalanobis distance is a natural enhancement of the Mallows distance that accounts for the variability in the training set.

The training data on which the principal component analysis is done is formed as follows. For each training case we have a greyscale image, a binary image, and an m-rep fit to the binary image as discussed in Section 9.3.1. Voxel correspondences specified by m-rep based figural coordinates (Section 9.4.1) allows us to compute the set of DRIQFs for each object-relative image region across the training images. When determining if a voxel belongs in a region, we initially use the binary image, not the m-rep, to label voxels as being inside or outside the object. This allows us to define mean DRIQFs that correctly provide references for the Mahalanobis distances used to form the geometry-to-image match. These DRIQFs do not, however, measure the expected variation of the actual training segmentations. Therefore, we estimate the covariance of the DRIQF in each region from the DRIQF values based on m-rep region labeling minus the already computed respective mean DRIQF, which is based on binary labeling.

9.5 Pablo Details and Results

9.5.1 The Voxel-Scale Stage of Segmentation

After all of the stages of segmentation that modify the medial atoms, an m-rep has undergone transformation from the beginning model (typically the mean of the global object complex or object). Figural coordinates allow this transformation to be understood as a diffeomorphism within all of the objects making up the complex represented by the m-rep. This warp can be interpolated into a chosen portion of 3-space including the complex, including the interstitial space between multiple objects or figures, if the complex is made up of more than one figure. A further finer scale transformation on that portion of 3-space can then be determined.

We interpolate the transformation from the objects to the surrounding 3-space, as follows. Each implied boundary position of the m-rep is understood as the tip of a particular m-rep spoke, either one of the basic representation or one interpolated from it. That spoke is from a medial atom $\mathbf{m}(1)$ at particular figural coordinates that allow it to be associated with a corresponding atom $\mathbf{m}(0)$ in the original m-rep model. Paths $\mathbf{m}(t)$, $0 \leq t \leq 1$, in the abstract space of atoms between the original atoms and the corresponding transformed atoms can be found according the mathematics in Chapter 3, Section 3.3, such that at every t the m-rep is unfolded and thus the continuous transformation of m-rep interiors is diffeomorphic. These paths can be sampled in t to produce a path of the corresponding spoke ends, and this sequence of positions can be used as a boundary condition in a landmark interpolation method. For example, one can use the thin-plate spline interpolation (Bookstein, 1991) on each of the corresponding successive pieces of the paths of all of the spoke ends. If the interstitial transformation was not diffeomorphic, as when objects slid along each other between individuals, an interpolation that allowed such transformations would need to be used.

We determine the fine scale warp to be composed with the transformation interpolated from the medial transformation using the fluid-flow warp method of (Miller et al., 1999). If the final map might not be diffeomorphic, as when regions of gas formed or were lost in the rectum or when tumors existed in the particular patient but the model was of well patients, then a warp method that permitted such situations would need to be applied.

The approach of computing a small scale space warp to be composed with a medially determined warp has the following advantages over computing the whole deformation as a space warp from an atlas. Optimizing large scale deformations is likely to be heavily affected by local minima, and in any case it is very likely to be slow as result of having to work over the combinatorially related, many small voxels. Indeed methods that have attempted to compute such warps have found it necessary to begin the process by preceding the voxel-scale warp by applying larger scale transformations such as ones based on manually chosen landmarks (Christensen et al., 1997). Using medial transformations to provide the large scale warp has advantages of being automatic, of using object-based correspondences, and of dividing itself into multiple scales, e.g., global to the object complex, object by object (with sympathetic inter-object relations), figure by figure (with sympathetic inter-figure relations), and medial atom by medial atom (reflecting inter-atom relations). Using these many scale levels produces both a much improved likelihood of convergence to the global optimum and qualitatively improved speed.

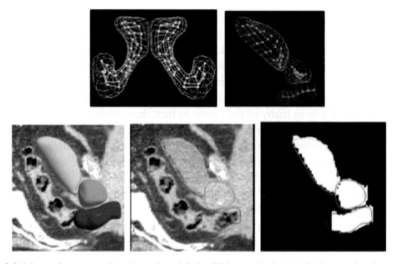

Fig. 9.9 M-reps for segmenting the male pelvis in CT images in later radiotherapy fractions. *Top left*: m-rep for pubic bones, used to register the later day fraction images with the first day fraction. *Top-right*: the m-rep for the bladder, prostate and rectum. *Bottom left*: a visualization of the bladder, prostate rectum m-rep's implied boundaries relative to a slice of the associated 3D CT image. *Bottom middle and right*: the segmentation result in a later fraction, shown in one of the image slices first vs. the greyscale CT image and then vs. the human segmentation shown in white

segmentations that have a median, over the cases, of the intersection/average volume overlap to a human segmentation of 93% for both the bladder and prostate and a median, over the cases, of the average closest point distance to the human segmentation of 1.13 mm for the bladder and 0.99 mm for the prostate. The numbers for the prostate, comparing segmentation based on statistics from a human who produced the training manual segmentations to the that human's result in the left-out-case, should be compared to the numbers comparing another observer's manual segmentation of the prostate to that of the training observer in one of the five patients' set of 16 multi-day CTs (Foskey et al., 2005). The agreement of the two humans' segmentations was 81% volume overlap and 1.9 mm average closest point surface separation.

When our segmentation was not as good as we wished, there were two explanations. First, in many of the segmentations of the bladder, a smaller scale refinement was necessary. We expect this to be accomplished when the log posterior optimizing atom stage is applied. Second, in a few cases the bladder initialization based on prostate landmarks was not adequate, but with a bladder-based initialization the segmentation was improved in a majority of cases.

This multi-object segmentation has been adapted for the clinical situation of adaptive radiotherapy by training the object principal geodesics by a pooling of aligned deviations from the mean of other patients. The results, which will soon be published with the details of the method, are comparable to those reported above.

Also, we expect shortly to report results of the atom-stage refinements of these segmentations.

Moreover, we are presently investigating having each object's change at the object scale level be divided into an m-rep change $\Delta \mathbf{m}^{self}$ independent of neighboring objects and an m-rep change $\Delta \mathbf{m}^{ngbr}$ reflecting the effect on the object of changes in neighbor atoms (Jeong et al., 2006). The neighbor-induced change is statistically described using the method of augmented object descriptions and prediction described in Chapter 8, Section 8. $\Delta \mathbf{m}^{ngbr}$ is decomposed as a conditional mean of the object, given designated neighbor atoms in its neighbors, plus a neighbor-effect residue with its own probability density. Probability densities on $\Delta \mathbf{m}^{self}$, on the augmented object, and on the neighbor effect residue are estimated by repetition of successive principal geodesic analyses. In segmentation the posterior is successively optimized with the prior iteratively in succession being the *self* probability density and the *neighbor residue* probability density, respectively. Initial results from statistical analysis on the bladder, prostate, rectum object complex are biologically reasonable, but it remains to test this approach by segmentations that use the self and neighbor residue probability densities.

9.5.2.3 Speed of Computation

The speed of a 3D segmentation on a Pentium IV, 1.7 GHz computer subdivides as follows.

- Preprocessing computations take less than 1 second.
- The largest scale stage (the object complex stage for an ensemble, the object stage for a single multifigure object, the figure stage for a single-figure object) takes a about 5 seconds per iteration and on the average requiring 20 iterations for a total time of about 3 minutes to determine the geometric warp coefficients.
- When the smaller scale medial stages are appropriately re-programmed, the same numbers will apply to each subfigure stage and about double for a full pass through the atoms at the atom stage, modulo the number of iterations required.
- The voxel displacement stage has not been timed, but it is expected to operate in under a minute.

Thus the total time for a kidney segmentation will typically be 7 minutes to segment a single-figure object.

While the method's speed has already benefited strongly from moving much of the computation from the deformation stage to the model building stage, there is still much room for speedup of integer multiples by more medial levels of coarse to fine, by medial deformation measured directly from the atoms without resort to the implied boundary, by having the gradients of the objective function relative to the changing parameters needed by the optimization steps be computed analytically rather than with numerical derivatives (shown in initial tests to more than double the speed), and just by more careful coding.

9.6 Hypothesis Testing for Localized Shape Differences Between Groups

We now focus on the quantitative morphologic assessment of structures between groups of human subjects. Our examples are individual brain structures in neuroimaging. Conventional methods study only volumetric changes, which explain intuitively global atrophy or dilation of structures. On the other hand, structural changes at specific locations are not sufficiently reflected in volume measurements. Statistical shape difference testing has thus become of increasing interest. Its potential to precisely locate morphological changes and to discriminate between groups makes it a good choice for studying pathological morphologic processes due to disease, as well as neuro-developmental processes. For example, we may wish to understand the shape differences in the hippocampus, caudate nucleus, cerebral ventricle complex in the brain between control patients and schizophrenics, or we may be interested in the differences of the hippocampus between 2-year olds who will develop autism and 4-year olds who will develop autism.

We focus in this section on the *hypothesis testing* of *whether* and *where* there are m-rep shape differences between the groups. We will discuss both global tests and truly local tests. Hypothesis testing applications using other medial descriptions have been proposed by Golland et al. (1999) and Bouix et al. (2005a).

We call the group designator C, which is numbered from 1 to the number of studied groups. Each group C_i consists of the objects of a sample of N_i cases. We assume that the objects or object complexes have been aligned across all cases, with the same alignment applied for the cases in both classes. The discrete m-rep objects are described as a tuple of medial atoms. The first idea is either to take all the atoms together and do a global test by studying the multivariate tuple of atoms $\times$ the 8 or 9 parameters per atom. Such a test can be powerful but will fail to localize the differences found to a particular collection of locations (i.e., parameters).

The alternative is to do a local (for a particular parameter of a particular atom) test on each atom parameter, at each position. We will use the term *location* to refer to such a combination of parameter and atom. The first idea might be to design a statistical test separately on any such location, and then to repeat that test over all atoms $\times$ parameters. However, the atoms are all correlated, and the parameter values are all correlated. To avoid unintended looseness in the threshold for rejecting the null hypothesis for any parameter, the threshold for rejection has to be adjusted for each parameter in a way reflecting the correlations. In Section 9.6.1.2 we describe a non-parametric permutation method to deal with this problem for individual parameters.

Section 9.6.2 will then focus on testing the full m-rep atom parameters jointly in symmetric space at a fixed scale. Finally, Section 9.6.2.2 will discuss why even atom by atom testing is not adequately local to the regions determined by the atoms and how to more appropriately handle locality.

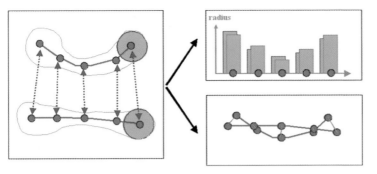

Fig. 9.10 Scalar m-rep shape difference (*schematically in 2D*) of 2 m-rep objects: Differences in the radius (*top graph*) and position (*lower graph*) are studied separately in Euclidean space. The properties express different kinds of underlying processes (*growth vs. deformation*) that are assumed to be statistically independent in the scalar testing

9.6.1 Tests in Euclidean Space

9.6.1.1 Univariate Tests in Euclidean Space

We may test a particular location (see Fig. 9.10). Here we focus on the two parameters, local position and radius (Styner et al., 2003b, 2004) of a particular atom. We first compute the group average objects by averaging the position and radius for each medial atom across each group. The overall average location is then computed as the average over all group average locations and serves as a template for computing univariate shape distance measurements. The signed position and thickness differences to the template are computed separately for the specified atom. The sign of the position difference is computed using the direction of the template medial surface normals.

Global shape analysis is computed by analyzing summarizing features such as the mean, median or other quantile measurements of the local differences across each object by standard statistical hypothesis tests. The choice of the feature evidently influences the outcome of the tests. The statistical tests mainly include parametric mean difference tests based on the Students-t distribution, and non-parametric mean difference tests, as well as parametric analysis of variance tests (ANOVA).

Local shape analysis does not need a summarizing feature as it is a truly local test. It is computed by testing each medial atom independently with a standard statistical hypothesis test. This results in a significance value (P-value) for each parameter and medial atom. We can represent this significance in a 3D visualization using color and size of spheres at the atom positions of the overall average object. This visualization, called a medial statistical significance map, allows one to locate significant shape differences between the groups in an intuitive but not truly local fashion (see Section 9.6.2.2). However, this raw significance map is incorrectly optimistic in regard to false-positive error rate because the atoms as well as the individual parameter values of a single atom are correlated, leading to the *multiple comparison problem*, a topic of active research in the field of shape analysis (Worsley et al., 1996; Nichols and Holmes, 2001).

The raw significance map can be corrected for this multiple comparison problem using a uniformly sensitive, non-parametric permutation test approach (Pantazis et al., 2004) described in the next section. This results in a corrected significance map. In contrast to the raw significance map, which is a quite optimistic estimate of the real significance map, the corrected significance map is a somewhat pessimistic estimate, as discussed in the next section.

9.6.1.2 Multivariate Permutation Tests in Euclidean Spaces

The permutation tests we describe here localize regions (atom indices or parameters thereof) in objects that exhibit statistically significant morphological variation among two population groups while controlling the risk of any false positives, as long as the object features are in a Euclidean space. We find local thresholds that control the false-positive error rate and at the same time achieve uniform sensitivity among all locations.

We assume we have two groups of local parameter sets, group A and group B. Each parameter set represents either shape parameters or differences of shape parameters. We want to test the two groups for difference in the means at each location. Permutation tests are a valid and tractable approach for such an application. Our null hypothesis is that the distribution of the parameter set at each element is the same for every subject regardless of the group. Permutations among the two groups satisfy the exchangeability condition, i.e., they leave the distribution of the statistic of interest unaltered under the null hypothesis. Given n_1 members of the first group a_k, $k = 1 \ldots n_1$ and n_2 members of the second group b_k, $k = 1 \ldots n_2$, we can create $M \leq \binom{n_1+n_2}{n_2}$ permutation samples. A value of M from 20,000 and up should yield results that are negligibly different from using all permutations (Edgington, 1995).

The complex set of steps needed to test the null hypothesis that the two groups have the same probability distributions is illustrated in Fig. 9.11. We take the reader through this process step by step. For each permutation sample j, we compute a difference metric T_j between the groups, with elements T_{ij}. For univariate Euclidean parameters the absolute distance between the means of the groups is often used:

$$T_{ij} = \left| \hat{\mu}_{aij} - \hat{\mu}_{bij} \right| \tag{9.1}$$

where i is the location index, j the permutation index. If we wish to sense locations at which differences of collections of parameters at the locations are signficant, we can use difference metrics for multivariate, Euclidean or non-Euclidean parameters, as long as the difference metric itself is in Euclidean space, such as the multivariate Hotelling T^2 test statistic for the collection: $T^2 \propto D^2 = (\hat{\mu}_a - \hat{\mu}_b)^T \hat{\Sigma}^{-1} (\hat{\mu}_a - \hat{\mu}_b)$, where $\hat{\Sigma}$ is the pooled sample covariance. In $\mathbb{R}^n$ this statistic is invariant to coordinate transformations and is uniformly the most powerful test with this property

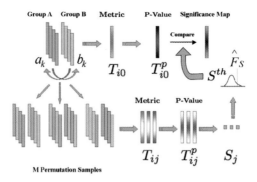

Fig. 9.11 Illustration of the permutation scheme. In the bottom row we create M permutation samples from the original data. We let j index the permutations and let i index the locations. For each permutation and location we compute the group difference metric T_{ij}, which is probability-normalized into T_{ij}^p. The data is then summarized across all locations to create the conservative summary statistic S_j over all locations. The empirical distribution of S_j, called $\hat{F}_S$ is used to define a global threshold S^{th} which for each location is applied to the probability-normalized test statistic obtained from the division to be tested, into groups A and B

(see Anderson, 1958 for a derivation). We cannot use this statistic directly on the multivariate combination of all atoms and parameters due to its inability to provide sensing of location.

In Fig. 9.11 it is assumed that we are given a target division of the data into two groups, A and B. To achieve uniform sensitivity across all locations, the parameter (or group of parameters) value T_{ij} at each location is first transformed to a uniformly distributed probability density value on [0,1], making all locations comparable. This is applied both to the test grouping, producing T_{i0}^p, and as illustrated in the bottom row of Fig. 9.11, it is also applied to the random permutations derived from the union of groups A and B, producing the T_{ij}^p. We can compare T_{ij}^p for each parameter i within each permutation j to produce a conservative summary statistic S_j for each permutation. Across the permutations the distribution of this summary statistic produces a common threshold S^{th} for each of the respective probability-normalized local parameters T_{i0}^p, as illustrated in the top row. The justification and specification of this scheme now follows.

The conservative summary statistic that we use for each permutation is the smallest probability density value over all locations i. We may then use the empirical distribution of this conservative summary statistic to extract thresholds that control the false positives to a desired level.

This method depends on having a form of normalization in the statistic T_{ij} that makes the locations comparable. A suitable normalization scheme is based on computing p-values, i.e., at each spatial location we compute the empirical distribution across permutations and then replace the statistic T_{ij} for each permutation sample with its p-value T_{ij}^p. The normalized metric T_{ij}^p is then guaranteed to have a uniform distribution on [0,1] under H_o for each i.

We can use the normalized data to define a local threshold map that controls the false-positive error rate to a desired level, say $\alpha = 5\%$, when applied to the original data. If the conservative summary statistic of the local parameters is $S_j = min_i\{T_{ij}^p\}$ over all locations i and $\hat{F}_S$ is the empirical cumulative distribution function of S, the appropriate global thresholds for a level α test would be $\hat{F}_S^{-1}(\alpha)$. For example, if we choose a threshold that leaves 5% of the area of the empirical distribution on the left side of S_j, then we have 5% probability of one or more false positives across all locations. This threshold S^{th} can be directly applied to T_{i0}^p (the statistic formed by probability-normalizing the original data with permutation index $j = 0$). Since the statistic T_{ij} is normalized separately for each location i, the same S^{th} corresponds to different values of local thresholds $p_i^{-1}(S^{th})$ of the unnormalized statistic T_{i0} at different locations.

Due to the use of the minimum p-value statistic across the whole surface, this correction scheme is focused only on controlling the rate of false positives at the given level α (commonly $\alpha = 0.05$) across the surface. No similar control of the rate of false negatives is available with this scheme. As the local significance level correctly controlled for false negatives can be anywhere between the raw p-value and the p-value corrected with our scheme, this corrected significance map is a pessimistic estimate of the true significance map.

9.6.2 Tests in Symmetric Spaces

The ideas in the previous section must be generalized to the non-Euclidean feature spaces appearing in m-reps and their symmetric space. We can derive permutation tests for equality of means of two groups using elements of the symmetric space. The sample means of each group, $\hat{\mu}_a$ and $\hat{\mu}_b$, can be computed using the techniques described in Chapter 8. Replacing T_{ij} from (9.1) with

$$T_{ij} = d(\hat{\mu}_a^*, \hat{\mu}_b^*) \tag{9.2}$$

yields a natural extension of local tests to symmetric spaces.

This provides a way to produce tests for a single aspect of the m-reps, such as position or radius of a particular atom, independently of the others, but typically we require a multivariate test using all of the parameters of one or more atoms simultaneously. We cannot fall back on Hotelling's T^2 test because it applies only to the linear case. Instead we can apply a transformation that forms new features from marginal probabilities, handling differing degrees of variability or correlation and making the test independent of magnification.

9.6.2.1 Global Multivariate Permutation Tests in Symmetric Spaces

We must now generalize the desirable properties of Hotelling's T^2 test to a nonparametric, nonlinear setting. One seemingly attractive option is to perform statistics in the tangent plane as is done with principal geodesic analysis, since its linearity means Hotelling's T^2 test can be applied directly. However, with two samples, the question that arises is *which* tangent plane, since there is a different one around each sample's mean, and there may be no unique map between them.

The other conceptual problem is that if one follows geodesics past the *cut locus*—the set of points where two or more geodesics cross—then points on the manifold no longer have a single well-defined representative in the tangent plane. Instead of addressing these problems, we take a more general approach, which only requires that our objects lie in a metric space.

Our approach is based upon a general framework for nonparametric combination introduced by Pesarin (2001). The general idea is to perform a set of partial tests, each on a different aspect of the data, and then combine them into a single summary statistic, taking into account the dependence between the variables and the true multivariate nature of the data. When performing the partial tests, we require that each distribution has the same structure around the mean—equivalent to the assumption of a common covariance required by Hotelling—and test for a difference of means. More precisely, following the idea described in the previous section, we map each feature to its marginal probability and use these probability values as features. The following two sections describe the details.

Partial Tests. We compute test statistics T_{ij} as before, where as before i indexes the model parameters and j is the permutation index. We now turn to the case where we have Q test statistics: one for each of the parameters in our shape model. Let $\mu_{a,i}$ and $\mu_{b,i}$ be the means of the ith model parameter for each group. Then we wish to test whether any hypothesis $H_{1,i} : \{\mu_{a,i} \neq \mu_{b,i}\}$ is true against the alternative, that each null hypothesis $H_{0,i} : \{\mu_{a,i} = \mu_{b,i}\}$ is true. The partial test statistics $T_{ij}, i \in 1 \ldots Q, j \in 1 \ldots M$ are defined analogously to (9.2).

It can be shown that each of our mapped features T_{ij}^p has the properties of being significant for large values, consistent, and marginally unbiased, as defined in (Pesarin, 2001). Given that, Pesarin shows that a suitable combining function (described in the next section) will produce an unbiased test for the global hypothesis H_0 against H_1.

Since each of our tests are restricted to the data from a single model parameter and we have assumed that the distributions around the means in each group are identical, they are marginally unbiased. We cannot add an explicit test for equality of the distributions about the mean, as then the test for equality of means would be biased on its outcome.

To illustrate these ideas, we present a simple example, which we will follow through the next few sections. We take two samples of $n_1 = n_2 = 10$ data points from the two-dimensional space $\mathbb{R} \times \mathbb{R}^+$, corresponding to a position and a scale parameter. The samples are taken from a multivariate normal distribution by exponentiating the second coordinate, and then scaling both coordinates by a factor of ten. They are plotted together in Fig. 9.12a. They have a common covariance (before the exponentiation), and the two means are slightly offset in the second coordinate.

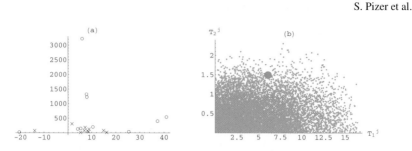

Fig. 9.12 The observed data and test statistics for our simple example. (**a**) shows the distribution of our two samples, with ×'s for the first and ○'s for the second. (**b**) shows the distribution of the partial test statistics under permutation. The large dot indicates the location of the observed data point

We construct $Q = 2$ partial test statistics using (9.2) for each coordinate, and evaluate them using Monte Carlo simulation with $M = 10{,}000$ permutations.

The results are shown in Fig. 9.12b. The first partial test value lies in the middle of the distribution, while the second lies near the edge. However, the scale of the first test is much larger, because no logarithm is involved in its metric.

Multivariate Combination. Given the partial tests from the previous section, we wish to combine them into a single test, while preserving the underlying dependence relations between the tests. This is done in the following manner. We apply the same M permutations to the data when computing each of partial tests, and we then compute a p-value statistic, T_{ij}^p as in Section 9.6.1.2. It is critical to use the same permutations for each partial test, as this is what captures the nature of the joint distribution.

We now wish to design a combining function to produce a single summary statistic, T_j', from each p-value vector $\boldsymbol{T}_j^p$. For one-sided tests, this statistic must be monotonically non-increasing in each argument, must obtain its (possibly infinite) supremum when any p-value is zero, and the critical value S'^{th} must be finite and strictly smaller than the supremum. If these conditions are satisfied along with those on the partial tests from the previous section, T_j' will be an unbiased test for the global hypothesis H_0 against H_1 (Pesarin, 2001).

Our combining function is motivated by the two-sided case (with signed distances), where we can use the Mahalanobis distance. Thus we need to transform the uniformly distributed p-values to a random variable that is normally distributed with mean zero and standard deviation 1. This is straightforwardly accomplished by applying the inverse of the cumulative density function for that Gaussian after subtracting $\frac{1}{2M}$. The extra $\frac{1}{2M}$ term keeps the values finite when the p-value is 1, and it is negligible as M goes to infinity. That is, we compute a U_j vector for each permutation, where $U_{ij} = \Phi^{-1}(T_{ij}^p - \frac{1}{2M})$, $j \in 1 \ldots M$, and Φ is the cumulative distribution function for the standard normal distribution. The map via the p-values and the Φ function gives the statistics known distributions that are directly comparable.

Arranging the U_j vectors into a single $M \times p$ matrix $\mathbf{U}$, we can estimate the covariance matrix $\hat{\Sigma}_U = \frac{1}{M}\mathbf{U}^T\mathbf{U}$ and use the Mahalanobis statistic: $T_j' = (U_j)^T \hat{\Sigma}_U^{-1} U_j$.

In the event that the data really is linear and normally distributed, the $\hat{\Sigma}_U$ matrix converges to the true covariance as the sample size goes to infinity (Pallini and Pesarin, 1992), making it asymptotically equivalent to Hotelling's T^2 test. Even if the sample size is small, the matrix Σ_U is well-conditioned regardless of the number of variables, since it is the covariance over the M permutations.

Typically, our distances are not signed, so we are interested in a one-sided test. In this case, we use the positive half of the standard normal cumulative distance function, $U_{ij} = \Phi^{-1}(1 - \frac{1}{2}(T_{ij}^p - \frac{1}{2M}))$ and assume the U_j distribution is symmetric about the origin. This assumption, however, implies that the covariance between $U_{i_1 j}$ and $U_{i_2 j}$ is exactly zero when $i_1 \neq i_2$. The diagonal entries of $\hat{\Sigma}_U$ are 1 by construction, so $\hat{\Sigma}_U = I$, the identity matrix. The fact that the p-values of the partial tests are invariant to scale obviates the need for arbitrary scaling factors. Thus, our one-sided combining function is

$$T'_j = (U_j)^T \cdot U_j. \qquad (9.3)$$

The normality of the partial test statistics is not required. Also, even though the marginal distributions of the U_j vectors are normal, the joint distribution may not be. Therefore, we must use the empirical distribution of T'_j in order to compute the final p-value of the global test: $T_0'^p$. It is this nonparametric approach that corrects for correlation among the tests, even without explicit diagonal entries in the covariance matrix.

We return to our example from the previous section. The U_j vectors are plotted in Fig. 9.13a, along with the $\alpha = 0.05$ decision boundary, and our sample is shown to lie outside of it. As can be seen, equal power is assigned to alternatives lying at the same distance from the origin in this space. Figure 9.13b shows this boundary mapped back into the space of the original p-values.

The entire procedure is very similar to procedures used in correction for multiple tests described in the previous sections. However, instead of trying to find a local

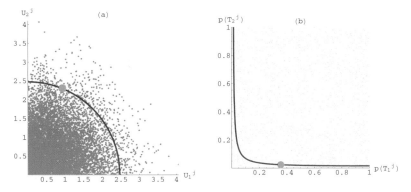

Fig. 9.13 The empirical distribution of our example plotted against the decision boundary at $\alpha = 0.05$. **(a)** The distribution of the U_j vectors, where the cutoff is a circle centered around the origin. **(b)** The distribution of the original p-values with the decision boundary pulled back into this space. The observed value is shown as the large dot in both plots

threshold for each test individually, we carve out a region of the multivariate T_{ij}^p space that contains some particular fraction, e.g., 5%, of the data to label as significant. We lose the ability to say *which* test is significant but gain power in the cases where multiple statistics independently signal significant differences.

9.6.2.2 Local Multivariate Tests in Symmetric Spaces

A test on all of the geometric primitives (e.g., medial atom) taken together is not truly a large scale test, for it confuses correlation with spatial scale. A test on each geometric primitive is not truly a small scale test, for it will respond equally well to a variation with large spatial scale as to one with a small scale. The Markov assumption on geometric neighbors allows the separation of scales by removing the correlation of neighboring elements from an element. In particular, if we can estimate the best predictor of a primitive by its neighbors and subtract that predictor from the primitive, the resulting residue provides the entity to test for significant variation *at the specified locality*.

This idea can be used for primitives such as objects or figures, but we are presently working to test it at the scale of the medial atom. Using the ideas in Section 9.1, the hypothesis testing would thus be done on each geodesic difference of the interpoland from the atom. However, we are still working on this form of local test, so the following section simply tests the atoms, one by one, rather than their residues.

9.7 Applications of Hypothesis Testing to Brain Structure Shape Differences in Neuro-Imaging

This section presents two application of hypothesis testing of m-rep objects. In the first application, scalar hypothesis testing of individual medial parameters was employed (see Section 9.6.1.2) for analyzing hippocampal shape in schizophrenia. In the second application, hypothesis testing in the symmetric space (see Section 9.6.2) was employed for analyzing ventricular shape in healthy twins and in schizophrenia.

9.7.1 Hippocampus Study in Schizophrenia

In the study presented in this section, we investigated the shape of the hippocampus structure in the left and right brain hemisphere in schizophrenic patients (SZ, 56 cases) and healthy controls (Cnt, 26 cases). The hippocampus is a gray matter structure in the limbic system and is involved in processes of motivation and emotions. It also has a central role in the formation of memory. Hippocampal atrophy has been observed in studies of several neurological diseases, such as schizophrenia, epilepsy, and Alzheimer's disease. The goal of our study was to assess shape changes between schizophrenic patients and the control group.

The subjects in this study all have male gender and the same handedness. The two populations are matched for age and ethnicity. The hippocampi were segmented from inversion-recovery-prepped SPGR MRI datasets (resolution: $0.9375 \times 0.9375 \times 1.5$ mm) using a manual outlining procedure based on a strict protocol and well-accepted anatomical landmarks (Duvernoy, 1998). The segmentation was performed by a single clinical expert (Schobel et al., 2001) with intra-rater variability of the segmented volume measurements at 0.95. Spherical harmonic (SPHARM) coefficients were computed using a sampling of 2,252 points, and the results were normalized via a rigid-body Procrustes alignment and a scaling to unit volume. The m-rep model was built on the aligned full population including the objects of all subjects on both sides, with the right hippocampi mirrored at the interhemispheric plane prior to the model generation. The resulting m-rep model has a single figure topology and a grid sampling of 3×8 medial atoms, in total 24 atoms. The range of the average distance error between the fitted m-rep's boundary and the original boundary was between 0.14 and 0.27 mm (mean error 0.17 mm), less than half the voxel size of the original MRI, so the medial shape analysis should capture the shape changes in the image data.

The template for the medial shape analysis was determined by the overall average structure. As the two populations are not equal in size, we computed the overall average as the average of the population averages. Due to age variation in both populations, the shape difference values were corrected for age influence, using a linear least squares model.

The global shape analysis in Table 9.1 shows a strong trend in the m-rep position analysis on the left side. The m-rep thickness analysis is significant for neither hippocampus. This suggests a deformation shape change in the hippocampus between the schizophrenic and the control group. The results of the m-rep position analysis shows a stronger significance than the SPHARM-PDM analysis that was also carried out. Additionally to the mean difference, several quartile measures (Median, 75% and 95%) were analyzed and produced structurally the same results.

The m-rep local position shape analysis (Fig. 9.14) yields significant changes that are in roughly the same position, mainly in the hippocampal tail, as shown by SPHARM-PDM shape analysis and by distance maps of the averages. No significance was found in the local m-rep thickness analysis. Similar to the outcome of the global analysis, the local m-rep position analysis shows a stronger significance

Table 9.1 Results of global shape analysis (average across the surface/medial manifold): Table of group mean difference p-values between the schizophrenic and control group (*: significant at $\alpha = 0.05$ significance level)

Global analysis	M-rep thickness	M-rep position
Left	$p = 0.722$	$p = 0.0513$
Right	$p = 0.751$	$*p = 0.0001$

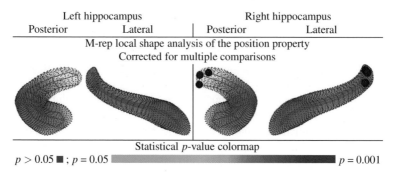

Fig. 9.14 Statistical maps of the local shape analysis from posterior and lateral views, corrected for multiple comparisons. The m-rep analysis shows the statistical significance at each medial atom using both the color and the radius of spheres placed at the atom positions. The main area of significance is located at the hippocampal tail. The corrected results are overly pessimistic

than the SPHARM-PDM analysis. The local shape differences are mainly located at the right hippocampal tail, with near significance in the left hippocampal tail. By inspecting the average structures of the two groups, we further find that the hippocampal tail region of the control group in our study is more bent than the one of the schizophrenic group.

9.7.2 Lateral Ventricle Study of Healthy and Schizophrenic Twins

The data for these experiments comes from a twin pair schizophrenia study conducted by Weinberger et al. (2001). High resolution ($0.9375 \times 0.9375 \times 1.5 \, \text{mm}^3$) MRIs scans were acquired from three different subject groups: 9 healthy monozygotic twin pairs (MZ), 10 healthy dizygotic twin pairs (DZ), and 9 monozygotic twin pairs with one twin discordant for schizophrenia and one twin unaffected (DS). See Fig. 9.15 for some examples. A fourth group of 10 healthy non-related subject pairs (NR) was constructed by matching unrelated members of the two healthy groups. All four groups were matched for age, gender, and handedness. A tenth healthy, monozygotic twin pair was discarded due to possible brain shape changes attributed to major head trauma suffered by one of the twins in a car accident at age seven. A tenth twin pair discordant for schizophrenia was discarded due to hydrocephaly in the unaffected twin.

The left and right lateral ventricles were segmented using automatic atlas based tissue classification (van Leemput et al., 1999) and 3-D connectivity. An automatic morphological closing operation was applied to ensure a spherical topology. An area-preserving map was used to map them to a sphere, after which they were converted to a spherical harmonics representation (SPHARM) (Brechbühler et al., 1995). Correspondence on the boundary was established using the first order harmonics (Gerig et al., 2001). Point Distribution Models (PDMs) were constructed by

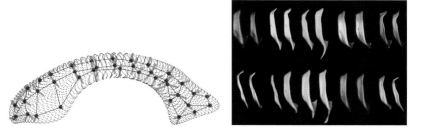

Fig. 9.15 *Left*: An example m-rep of a left lateral ventricle. The mesh vertices and off-shooting spokes make up the medial atoms. The shape the m-rep was fit to is shown as a point cloud surrounding it. *Right*: Ventricle pairs from five monozygotic twin pairs (*top*) and five dizygotic twin pairs (*bottom*)

uniformly sampling the boundary at corresponding points. The m-rep models were constructed using a robust method that ensures a common medial topology (Styner et al., 2003a). For our data, this consists of a single medial sheet with a 3×13 grid of medial atoms, which provides 98% volume overlap with the original segmentations.

From this data set, we wish to determine if the twin pairs that were more closely related had smaller variations in shape. We also wish to see if the shape variations between the discordant and the unaffected twins in the schizophrenic pairs is similar to the normal variation between healthy monozygotic twins. For this purpose, we use the partial test statistics:

$$T_{ij} = \frac{1}{n_1} \sum_{k=1}^{n_1} d(a_{ki}^{1*}, a_{ki}^{2*}) - \frac{1}{n_2} \sum_{k=1}^{n_2} d(b_{ki}^{1*}, b_{ki}^{2*}). \tag{9.4}$$

Here (a^1, a^2) form the twin pairs for one group, while (b^1, b^2) form the twin pairs for the other. The partial tests are applied separately to all three components of the medial atom location, $\mathbf{x}$, as well as the radius and two spoke directions. This gives six partial tests per medial atom, for a total of $p = 3 \times 13 \times 6 = 234$, much larger than the sample size. Each is a one-sided test that the variability in group 2 is larger than that in group 1.

For consistency with previous studies (Styner et al., 2005), all shapes were volume normalized. After normalization, we also applied m-rep alignment, as described by Fletcher et al. (2004), to minimize the sum of squared geodesic distances between models in a medial analog of Procrustes alignment. First, the members of each twin pair were aligned with each other, and then the pairs were aligned together as a group, applying the same transformation to each member of a single pair.

In order to ensure invariance to rotations, we had to choose data-dependent coordinate axes for the $\underline{\mathbf{x}}$ component of each medial atom. Our choice was the axes which diagonalized the sample covariance of the displacement vectors from one twin's atom position to the other at each site. While this had some influence on the results, the general trend was the same irrespective of the axes used.

Table 9.2 p-values for paired tests for the difference in the amount of shape variability in groups with different degrees of genetic similarity. Results from our method are in the first two columns, while results from a previous study (Styner et al., 2005) are in the last two for comparison. Groups are: monozygotic (MZ), monozygotic twins with one twin discordant for schizophrenia (DS), dizygotic (DZ), and non-related (NR) (∗: significant at the $\alpha = 0.05$ significance level).

	Our study		Boundary study (Styner et al., 2005)	
	Left	Right	Left	Right
MZ vs. DS	0.12	0.38	0.28	0.68
MZ vs. DZ	∗0.00006	∗0.0033	∗0.0082	∗0.0399
MZ vs. NR	∗0.00002	∗0.00020	∗0.0018	∗0.0006
DS vs. DZ	∗0.020	∗0.0076	0.25	0.24
DS vs. NR	∗0.0031	∗0.00026	∗0.018	∗0.0026
DZ vs. NR	0.16	0.055	∗0.05	∗0.016

For each pair of twin groups, we generated $M = 50,000$ permutations, and computed their p-value vectors. Following Section 9.6.2.1, these were mapped into U_j vectors, from which the empirical distribution of the combined test statistic T'^k from (9.3) was estimated, producing a single global p-value.

The results are summarized in Table 9.2. For comparison, we list the results of a previous study which used a univariate test on the average distance between corresponding points on the PDMs (Styner et al., 2005). While we note that the significance of a p-value on an experimental data set is not a useful metric for comparing different methods, it is interesting to see the differences between the two. Our tests give a consistent ranking: MZ $\approx$ DS $<$ DZ $\approx$ NR, which is fully transitive. The boundary study, however, finds a significant difference between DZ and NR, but fails to identify the difference between DS and DZ.

We also performed local tests, to identify specific medial atoms with with strong differences. A multivariate test was conducted using our procedure on the 6 components of each atom, and the results were corrected for multiple tests using the minimum p-value distribution across the shape described in Section 9.6.1.2. The results are shown in Fig. 9.16.

9.8 Discussion and Future Work

9.8.1 Are M-Reps Effective?

The main objective of this chapter was to describe m-reps based methods for 3D medical image segmentation and for statistical characterization of differences of anatomic shapes seen in populations of medical images. M-reps have been used both to capture knowledge of object geometry and to give a basis of the positional correspondences needed in training and measuring geometry-to-image match log probabilities. As well, they have allowed efficient, multiscale operation in both training the probabilities and applying them. It has been our expectation that they provide

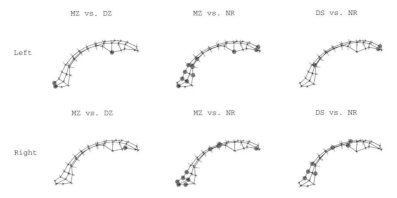

Fig. 9.16 Results for local tests for the difference in shape variability in groups with different degrees of genetic similarity. Atoms with differences significant at the $\alpha = 0.05$ level are shown in a larger size. Tests not shown had no significant local differences

more stable estimates of modes of variation and the associated principal variances for a given number of training samples than alternative bases for geometric statistics, and we have some early results suggesting that this is the case, but this is yet to be proven.

In addition, much more than other geometric representations, m-reps have provided a means of yielding probability distributions whose samples were very unlikely to be geometrically improper, avoiding illegal interpenetrations and creases. Checks on geometric propriety via S_{rad} has avoided creasing or near creasing and the improved estimates of boundary normals without lowering the tightness of boundary fits to training binary images. DRIQF statistics based on these fits have led to improved segmentations.

The success of m-reps as an object representation designed for statistical uses should be judged by the success of the applications. Within the class of deformable models methods that might be considered to provide comparable segmentation accuracy, robustness and low interaction requirements, m-reps based segmentations are among the speedier.

In terms of accuracy and robustness, the 3D segmentation method based on m-reps have produced single-figure object, viz. kidney, segmentations that are competitive with human manual segmentation and are, to our knowledge, the most accurate kidney segmentations reported in the literature. The same can be said of the initial multi-object segmentations of male pelvis objects in CT images using intra-patient statistics, though given the serious challenge of this problem, further work must be done before the method can be tested on many patients and its results compared to the results of alternative methods for segmentation of these objects. Moreover, while the apparatus for segmentation of multi-figure objects exists and has been tried on simple test cases, real application and testing of such segmentation is yet to be done.

Of course, when comparing m-reps to other object representations that are being used for segmentation via deformable models, the issue is not simply whether m-reps are as good or better than the alternatives, but whether they are enough better to justify the complexities of the medial representation. Controlled, quantitated validation on a variety of objects by multiple methods in competition needs to be carried out before this can be judged.

We are in the process of making the following improvements to our deformable m-reps segmentation method and software:

1. Sensing and reporting locations on the segmented object that do not have the expected level of geometry-to-image match, so that the user can take actions of relocating that object section and then restart the segmentation.
2. Bringing to routine usability a posterior optimizing atom stage as well as the option of computing a small scale diffeomorphism both in the target object(s) and in the interstitial space between objects in place of the small scale boundary displacement.
3. Developing a form of our software intended for clinical use and thus being as automatic as possible, and making all interactions in clinical terms.

The m-rep hypothesis testing tools have been applied to several studies in neuro-imaging and have shown to provide meaningful results. The main advantage of our m-rep hypothesis testing tools over boundary based testing tools is the identification of different types of processes using the different m-rep atom properties. This leads to results that are more intuitively interpretable. In several studies of the hippocampus, the caudate and the lateral ventricles, we have shown that the overall results correlate well between medial and boundary description, but also that our m-rep analysis is able to capture additional information not seen in the boundary analysis.

Our current hypothesis testing tools are based on a true multivariate permutation test approach for hypothesis testing in direct products of metric spaces. The resulting test does not require a priori scaling factors to be chosen, and captures the true multivariate nature of the data. It is well-defined even in the high-dimensional, low-sample size case. The method has been developed for m-reps, though it is suitable for any type of metric data, including potentially categorical data. An important area for future research is the design of suitable partial tests to use in each space. Because they cannot be broken into smaller pieces than a single component of the direct product, the distance to the mean and similar tests are limited in the types of distributions they can describe. For example, the distance from the mean can only characterize an isotropic distribution on the sphere. This would allow us to relax our assumption of identical distribution about the mean.

Even though our hypothesis testing tools have matured to a degree that they can be employed routinely in neuro-imaging studies, there are several limitations to our current tools making the following enhancements to our methods necessary:

1. Developing a combined analysis of multiple objects in order to capture correlated differences of the shape in neighboring brain structures such as the lateral ventricle and the caudate.
2. Enhancing the analysis scheme to incorporate several layers of scale starting at the global multi-object scale down to the local single atom scale.

3. Incorporating statistical models of patient covariates such as gender, age and medication in the permutation test algorithm. The current method incorporates covariates by correcting atom parameters independently using least-squares linear regression.

9.8.2 Other M-Rep Uses and Properties

In a separate paper (Crouch et al., 2003) we have shown how the space parametrization provided by m-reps also allows the interior of the object to be divided into mesh elements useful for efficient mechanical modeling of intra-patient motion of anatomic structures due to such interventions as intrarectal imaging probes. The measures of mechanical energy computed in this approach could be used for segmentation of a patient whose segmented m-rep from an earlier (e.g., planning) image can be used as the model for a segmentation in a later (e.g., intra-treatment) image.

M-reps provide one means of modeling objects and collections of objects; boundary representations (b-reps) are a common alternative means of such object modeling. They share the limitations of all object modeling methods, namely that a single object model will not serve for a class of objects with mixed topologies at the figural level. However, because they explicitly model the interfigural relations, they have special weaknesses when these relations are variable over the population of objects. For example, an m-rep for a right kidney and a separate left kidney will not perform well for a horse-shoe kidney, in which the kidneys are joined. For such mixed classes, a separate m-rep is required for each exemplar. Another issue shared with other object models is instability for nearly spherically or circularly symmetric objects. In such cases the nearly degenerate geometry creates computational instabilities in discriminating among the three major axes which in turn can cause an m-rep to "flip" during deformation in the image data in an unstable manner. However, m-reps share with other object models the particular strength of resolving these orientational instabilities via the relations among objects.

M-reps' special abilities relative to b-reps derive from their explicit representation of object orientation changes such as twisting and bending and of object size changes such as widening and narrowing. Thus statistics on rectal widenings due to gas, on the variability in the relative pose of the two lobes of the liver, and on the orientation of the bladder relative to the prostate are very effective in m-reps terms. The limitations not of m-reps by themselves but of m-reps with statistics come in situations when the orientational or magnificational relationships are very variable. Thus, like b-reps m-reps are well suited to complex slabs and tubes such as the cerebral cortex or the intestine, and both are well suited to intra-patient variations of these structures over time. But because in the population of humans the variability of the folding structure of the cerebral cortex is high and the variability of the curvature of the intestine is high, statistics on m-reps is a weak tool over that population for these structures.

Because m-reps represent the interior of objects, they lose their effectiveness in image situations where only one side of an object appears in an image, and they have weakness relative to b-reps in situations where one side of an object boundary is statistically stable but the other side has great variability. In that situation b-reps can ignore the unstable or unimaged side, whereas m-reps inherently must represent both sides together.

M-reps allow statistics by providing a fixed topology of sheets and their branching. As presently designed, populations that are not well modeled by fixed topology m-reps together with voxel scale refinements will require a different geometric representation.

Acknowledgments This work was done under the partial support of NIH grants P01 CA47982 and P01 EB02779 and a grant by the Stanley Foundation. A gift from Intel Corp. provided computers on which some of this research was carried out. We thank Xifeng Fang, Qiong Han, Ja Yeon Jeong, Joshua Levy, Conglin Lu, Derek Merck, Joshua Stough, Gregg Tracton, Guido Gerig, A. Graham Gash, and Delphine Bull for help with models, figures, software, and references. We are grateful to J. Stephen Marron, Keith Muller, and Surajit Ray for help with statistical methods, to Jeffrey Lieberman, Julia Fielding, and Valerie Jewells for medical images, and to members of the UNC Neuroimage Analysis Laboratory and Julian Rosenman for manually segmenting images.

Chapter 10
3D Model Retrieval Using Medial Surfaces

Kaleem Siddiqi, Juan Zhang, Diego Macrini, Sven Dickinson, and Ali Shokoufandeh

Abstract Graphs derived from medial representations have been used for 2D object matching and retrieval with considerable success (Pelillo et al., 1999; Siddiqi et al., 1999b; Sebastian et al., 2001). In this chapter we consider consider the use of graphs derived from medial surfaces for 3D object matching and retrieval. The medial reprsentation allows for a qualitative abstraction based on a directed acyclic graph of components and also a degree of invariance to a variety of transformations including the articulation of parts. The formulation discussed in this chapter uses the geometric information associated with each node along with an eigenvalue labeling of the adjacency matrix of the subgraph rooted at that node. Comparative retrieval results are presented against the techniques of shape distributions (Osada et al., 2002) and harmonic spheres (Kazhdan et al., 2003b) on 425 models representing 19 object classes. These results demonstrate that medial surface based graph matching outperforms these techniques for objects with articulating parts.

K. Siddiqi
School of Computer Science & Centre for Intelligent Machines, McGill University, Canada,
e-mail: siddiqi@cim.mcgill.ca

J. Zhang
School of Computer Science & Centre for Intelligent Machines, McGill University, Canada,
e-mail: juan@cim.mcgill.ca

D. Macrini
Department of Computer Science, University of Toronto, Canada,
e-mail: dmac@cs.toronto.edu

S. Dickinson
Department of Computer Science, University of Toronto, Canada,
e-mail: sven@cs.toronto.edu

A. Shokoufandeh
Department of Computer Science, Drexel University, USA,
e-mail: ashokouf@cs.drexel.edu

K. Siddiqi and S. Pizer (eds.) *Medial Representations – Mathematics, Algorithms and Applications.*
© Springer Science + Business Media B.V. 2008

10.1 Introduction

The problem of object recognition is one of significant interest to the computer vision community. It relates to the process of searching a database of models so as to efficiently retrieve instances that are similar to a particular exemplar. The general problem is difficult because objects can undergo signficant deformation and articulation while retaining their identity. The challenge is to come up with representations that provide a degree of invariance under such transformations and which allow for matching algorithms to be applied. The medial models discussed in this book provide a particularly attractive choice because they allow for a reduction of the topology of an object to a graph indicating the relationship between its parts and sub-parts, where each node carries detailed geometric information. As a consequence, medial graphs have been used for 2D object matching and retrieval with demonstrated success in handling part articulation and deformation.

Broadly speaking, there are three classes of existing techniques: (1) graph edit distance based approaches (Sebastian et al., 2001, 2004), (2) sub-graph isomorphism approaches (Pelillo et al., 1999) and (3) graph-spectral approaches (Siddiqi et al., 1999b).[1] These three approaches share the common view that a representation of the 2D medial axis as a graph of components can handle differences in part structure as well as differences in part shapes. The part structure is reflected by the medial axis branching structure and hence the connectivity of the graph, while part shape is reflected in the geometric information associated with each branch. We begin with a brief overview of these classes of methods. Details of each method are presented in the associated references.

10.1.1 Graph Edit Distance Approaches

Graph edit distance approaches assume that similar objects have similar (but not necessarily identical) part structures and part shapes. The essential idea is to use a prescribed set of edit operations to transform one graph in to another. Each of these operations is assigned a cost, and the distance between two objects is determined by the lowest cost set of edit operations between their underlying graph representations. Whereas the notion of edit distance faces a serious issue of computational complexity for arbitrary graphs, polynomial time algorithms have been developed for the case of shock graphs (see Chapter 2) (Klein et al., 2001), which are attributed tree structures. An example of this approach is illustrated in Fig. 10.1, which is adapted from (Sebastian et al., 2004). In this particular example the edit operations take one graph to another by assigning costs to allowed medial axis transitions, as described in Chapter 2. For these costs to be useful in practice, they must take into account

---

[1] A fourth category for matching medial representations has been developed in detail in Chapter 9, but it assumes that candidate objects can be described by a fixed m-rep topology. Thus, this type of method is less applicable to the problem of 2D or 3D object retrieval, where some variation in part structure for objects within the same category is expected to occur.

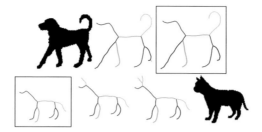

Fig. 10.1 (Adapted from (Sebastian et al., 2004).) Examples of the optimal deformation path between two shapes represented at the extremes of a sequence. The sequence shows operations (symmetry transforms) applied to the medial axis, and the resulting intermediate shock graphs. The boxed shock graphs, which have the same topology, are where the deformation of the two shapes meet in a common simpler shape

both medial axis branching structure (graph connectivity) and medial axis geometry (node attributes).

10.1.2 Subgraph Isomorphism Approaches

Subgraph isomorphism approaches seek to find maximal common subgraphs between two candidate graphs. In the context of matching 2D medial graphs, these approaches have been developed in (Pelillo et al., 1999), using the version of the shock graph developed in (Siddiqi et al., 1999b). The essential idea is to convert the maximal common sub-tree problem into a maximum clique problem on an association graph, and to solve the latter combinatorial problem by converting it to a related optimization problem. The optimization problem is solved by using discrete or continuous time replicator equations. These are differential equations developed in mathematical biology, which have the advantage that they are straightforward to simulate numerically. In these approaches part geometry can be considered in the form of attributes on nodes, leading eventually to a generalization of the maximum clique problem to a maximum weighted clique problem. However, the similarity between part structures, which is reflected in the connectivity of the association graph, is essentially separated from the similarity between part geometries, which is reflected in the weights on association graph nodes.

10.1.3 Graph Spectral Approaches

In graph spectral approaches the essential idea is to create a low-dimensional vector that reflects the topology of the graph. In the context of matching 2D medial directed acyclic graphs (DAGs), one such measure proposed in (Siddiqi et al., 1999b) is based on efficient techniques for computing the sum of the eigenvalues of the adjacency matrix of the DAG. This approach allows geometric similarity between nodes, which may be interpreted as the similarity between part shapes, to be combined with a topological signature vector that captures an object's overall part structure. This combination is then used for both matching and indexing (Shokoufandeh et al., 1999, 2005).

The subject of this chapter is the application of medial graph matching to 3D object recognition. With regard to the choice of method, the sub-graph isomorphism approach carries the disadvantage that for graphs with a large number of nodes and possibly complicated topology, computational efficiency becomes an issue. The graph edit distance approach is an attractive one to pursue, given an appropriate measure of edit distance in 3D. To our knowledge creating such a measure, a challenging task, has not yet been reported. Therefore, in this chapter we focus on an extension of the third approach based on graph spectra.

10.2 3D Model Retrieval

With an explosive growth in the number of 3D object models stored in web repositories and other databases, the computer vision and computer graphics communities have begun to address the important and challenging problem of 3D object retrieval and matching. Although this problem traditionally falls in the domain of computer vision research, it is also of interest to those interested in applications in the areas of solid modeling and computer-aided design. Recent advances include query-based search engines (Funkhouser et al., 2003) which employ promising measures including spherical harmonic descriptors and shape distributions (Osada et al., 2002). Such systems can yield results on databases including hundreds of 3D models, in a matter of a few seconds.

Thus far the emphasis has broadly been on the use of qualitative measures of shape that are typically global. Such measures are robust in the sense that they can deal with noisy and imperfect models, and at the same time they are simple enough so that efficient algorithmic implementations can be sought. However, an inevitable cost is that such measures are inherently coarse and are sensitive to deformations of objects or their parts. As a motivating example, consider the 3D models in Fig. 10.2. These four exemplars of an object class were created by articulations of parts and changes of pose. For such examples, the very notion of a center of mass or a rigid reference point (Alt et al., 1994), which is crucial for the computation of descriptions such as shape histograms (sectors or shells) (Ankerst et al., 1999) or spherical extent functions (Vranic and Saupe, 2001), can be nonintuitive and arbitrary. In fact, the centroid of such models may actually lie in the background. To complicate matters, it is unclear how to obtain a global alignment of such models, and hence signatures based on a Euclidean distance transform (Borgefors, 1984; Funkhouser et al., 2003) have limited power in this setting. As well, measures based on reflective symmetries (Kazhdan et al., 2003a), and signatures based on 3D moments (Elad et al., 2001) or chord histograms (Osada et al., 2002) are not invariant under such transformations.

Fig. 10.2 Exemplars of the object class "human" created by changes in pose and articulations of parts (top row). The medial surface (or 3D skeleton) of each is computed using the algorithm of (Siddiqi et al., 2002) (bottom row). The medial surface is automatically partitioned into distinct parts, each shown in a different color

The computer vision community has grappled with the problem of *generic* or category-level object recognition by suggesting representations based on volumetric parts, including generalized cylinders, superquadrics and *geons* (Binford, 1971; Marr and Nishihara, 1978; Pentland, 1986; Biederman, 1987). Such approaches build a degree of robustness to deformations and movement of parts, but their representational power is limited by the vocabulary of geometric primitives that are selected. Motivated in part by such considerations, there have been attempts to encode 3D shape information using probabilistic descriptors. These allow intrinsic geometric information to be captured by low dimensional signatures. An elegant example of this is the geodesic shape distribution of (Hamza and Krim, 2003), where information theoretic measures are used to compare probability distributions representing 3D object surfaces. In the domain of graph theory there have also been attempts to address the problem of 3D shape matching using representations based on Reeb graphs (Shinagawa et al., 1991; Hilaga et al., 2001). These allow for topological properties to be captured, at least in a coarse sense.

An alternative approach is to use 3D medial loci. As pointed out by Blum, this offers the advantage that a graph of parts can be inferred from the underlying local mirror symmetries of the object (Blum, 1973). A formal abstraction of this type based on the generic singularities of the grassfire flow in 3D has already been discussed in Chapter 2 (see also Leymarie and Kimia, 2003). To further motivate this idea, consider once again the human forms of Fig. 10.2. A medial surface-based representation (bottom row) provides a natural decomposition, which is largely invariant to the articulation and bending of parts.

In this chapter, we build on the technique to compute medial surfaces covered in Chapter 4 (see also Siddiqi et al., 2002) by proposing an interpretation of its output as a directed acyclic graph (DAG) of parts. We then use refinements of algorithms based on graph spectra (Shokoufandeh et al., 2005) to tackle the problems of indexing and matching 3D object models. Graph matching algorithms have already shown promise in the computer vision community for category-level view-based object indexing and matching using 2D skeletal graphs (Siddiqi et al., 1999b; Shokoufan-deh et al., 1999; Pelillo et al., 1999; Sebastian et al., 2001). They have also been demonstrated in the context of matching 3D object models with tubular parts, using a centerline approximation of the 3D skeleton (Sundar et al., 2003). We demon-strate their significant potential for medial surface-based 3D object retrieval with experimental results on a database of 320 models representing 13 object classes, including exemplars of both rigid objects and ones with significant articulation of parts. Comparative results using the information retrieval notion of *precision ver-sus recall* demonstrate that this method significantly outperforms the techniques of shape distributions (Osada et al., 2002) and harmonic spheres (Kazhdan et al., 2003b) for objects with articulating parts.

10.3 Medial Surfaces and DAGs

A number of algorithms for computing 3D medial loci and related representations have been covered in this book. These include the average outward flux-based and object angle extension skeletons of Chapter 4 and the related algorithms based on continuous properties of the Euclidean distance function; the methods based on dig-ital distance tranforms of Chapter 5; methods based on Voronoi diagrams (Chapters 6 and 7); methods for computing m-reps (Chapter 8) and methods which combine constructs from computational geometry with wavefront propagation such as the shock-scaffold technique (Leymarie and Kimia, 2003). See also the applications discussed in Chapter 11. For several of these algorithms the segmentation of the 3D skeleton into its constituent medial manifolds remains a challenge. In this chapter we choose to employ the method of Chapter 4 since it has the advantage that the dig-ital classification of Malandain et al. (1993) allows for the taxonomy of generic 3D skeletal points (Giblin and Kimia, 2004) to be interpreted on a rectangular lattice, leading to a graph of parts.

Under the assumption that the initial model is given in triangulated form, we begin by scaling all the vertices so that they fall within a rectangular lattice of fixed dimension and resolution. We then sub-divide each triangle to generate a dense inter-section with this lattice, resulting in a binary (voxelized) 3D model. The average outward flux of the Euclidean distance function's gradient vector field is computed through unit spheres centered at each rectangular lattice point, using the algorithm of Chapter 4 (Section 2.3). As explained in that chapter, this quantity has the property that it approaches a negative number at skeletal points and goes to zero elsewhere (Siddiqi et al., 2002), and thus it can be used to drive a digital thinning process. Fur-thermore, the limiting average outward flux values for the case of shrinking discs reveals the object angle, and thus this quantity may be viewed as a type of *flux invariant* for both obtaining the medial locus and for determining the geometry of the bounding surface implied by it (Dimitrov et al., 2003; Dimitrov, 2003). The thin-ning process has to be implemented with some care, as described in Chapter 4, so that the topology of the object is not changed. As mentioned above, this process uses the digital classification of points due to Malandain et al. (1993) to label points on the digital medial locus according to the A_k^m classification given in Chapters 1 and 2, i.e., as surface points, rim points, junction points or curve points. See also Table 4.3 of Chapter 4. We refer to this as a *medial surface* representation. This suggests the following 3-step approach for segmenting the (voxelized) medial surface into a set of connected parts:

1. Identify all manifolds comprised of 26-connected surface points and border points.
2. Use junction points to separate these manifolds, but allow junction points to belong to all manifolds that they connect.
3. Form connected components with the remaining curve points, and consider these as parts as well.

This process of automatic skeletonization and segmentation is illustrated for two object classes, a chair and a human form, in Fig. 10.3.

We now propose an interpretation of the segmented medial surface as a directed acyclic graph (DAG). We begin by introducing a notion of *saliency* which captures the relative importance of each component. Consider that the envelope of maximal inscribed spheres of appropriate radii placed at all skeletal points reconstructs the original object's volume (Blum, 1973). The contribution of each component to the overall volume can thus be used as a measure of its significance. Since the

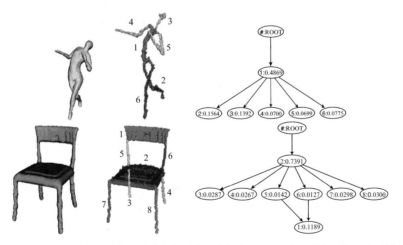

Fig. 10.3 A voxelized human form and chair (*left*) and their segmented medial surfaces (*middle*). A hierarchical interpretation of the medial surface, using a notion of part saliency, leads to a directed acyclic graph DAG (*right*). The nodes in the DAGs have labels corresponding to those on the medial surface, and the saliency of each node is also shown

spheres associated with adjacent components can overlap, an objective measure of component j's saliency is given by

$$Saliency_j = \frac{Voxels_j}{\sum_{i=1}^{N} Voxels_i},$$

where N is the number of components and $Voxels_i$ is the number of voxels *uniquely* reconstructed by component i.

The above notion is a reasonable choice for a saliency measure in the context of 3D model retrieval, but is certainly not the only one. In fact, more principled saliency measures could be developed by using the metric measure introduced by Damon in Chapter 3, or by computing appropriate boundary and regional integrals via their analogous medial integral versions, as discussed in that chapter. At present the development of such saliency measures, as well as computational approaches to approximiate them, remains the subject of future work.

We now propose the following construction of a DAG, using each component's saliency. Consider the most salient component as the root node (level 0), and place components to which it is connected as nodes at level 1. Components to which these nodes are connected are placed at level 2, and this process is repeated in a recursive fashion until all nodes are accounted for. The graph is completed by drawing edges between all pairs of connected nodes, in the direction of increasing levels, hence avoiding the occurrence of any cycles. However, to allow for 3D models comprised of disconnected parts we introduce a single dummy node as the parent of all DAGs for a 3D model.

This process is illustrated in Fig. 10.3 (right column) for the human and chair models, with the saliency values shown within the nodes. Note how this representation captures the intuitive sense that the human is a torso with attached limbs and a head, a chair is a seat with attached legs and a back, etc. This DAG representation of the medial surface is quite different than the graph structure that follows from a direct use of the taxonomy of 3D skeletal points in the continuum presented in Chapter 2 (Giblin and Kimia, 2004). Our motivation is to be able to exploit the hierarchical structure-indexing and structure-matching algorithms reported in Siddiqi et al. (1999b); Shokoufandeh et al. (2005). However, this conversion can also lead to some limitations; we shall return to a discussion of these at the end of this chapter.

10.4 Indexing

A linear search of the 3D model database, i.e., comparing the query 3D object model to each 3D model and selecting the closest one, is inefficient for large databases. An indexing mechanism is therefore essential to select a small set of candidate models to which the matching procedure is applied. When working with hierarchical structures in the form of DAGs, indexing is a challenging task that can be formulated as the fast selection of a small set of candidate model graphs that share a subgraph with the query. But how do we test a given candidate without resorting to subgraph isomorphism and its intractability? The problem is further compounded by the fact that due to perturbation and noise, no significant isomorphisms may exist between the query and the (correct) model. Yet, at some level of abstraction, the two structures (or two of their substructures) may be quite similar. Thus, our indexing problem can be reformulated as finding model (sub)graphs whose structure is *similar* to the query (sub)graph.

Choosing the appropriate level of abstraction with which to characterize a DAG is a challenging problem. We seek a description that, on the one hand, provides the low dimensionality essential for efficient indexing, while on the other hand, is rich enough to prune the database down to a tractable number of candidates. In recent work (Shokoufandeh et al., 2005) we draw on the eigenspace of a graph to characterize the topology of a DAG with a low-dimensional vector that will facilitate an efficient nearest-neighbor search in a database.

The eigenvalues of a graph's adjacency matrix encode important structural properties of the graph, characterizing the degree distribution of its nodes. Moreover, we have shown that the magnitudes of the eigenvalues are stable with respect to minor perturbations of graph structure due to, for example, noise, segmentation error, or minor within-class structural variation (Shokoufandeh et al., 2005).

We can now proceed to define an index based on the eigenvalues. One simple structural abstraction would be a vector of the sorted magnitudes of the eigenvalues of a DAG's adjacency matrix.[2] However, for large DAGs, the dimensionality of the index would be prohibitively large (for efficient nearest-neighbor search), and the descriptor would be global, prohibiting effective indexing of query graphs with added or missing parts. This problem can be addressed by exploiting eigenvalue sums rather than the eigenvalues themselves, and by computing both global and local structural abstractions (Siddiqi et al., 1999b). Let V be the root of a DAG whose maximum branching factor is Δ, as shown in Fig. 10.4. Consider the sub-

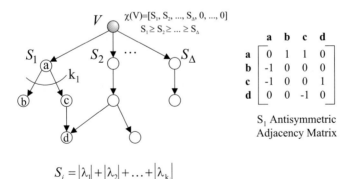

$$S_i = |\lambda_1| + |\lambda_2| + \ldots + |\lambda_{k_i}|$$

Fig. 10.4 Forming a Low-Dimensional Vector Description of Graph Structure. At node a, we compute the sum of the magnitudes of the k_1 largest eigenvalues of the adjacency sub-matrix defined by the subgraph rooted at a. The sorted sums S_i become the components of $\chi(V)$, the *topological signature vector* (or TSV) assigned to V

---

[2] Since the eigenvalues of an antisymmetric matrix are complex we utilize the magnitude of an eigenvalue.

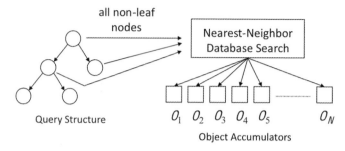

Fig. 10.5 Indexing Mechanism. Each non-trivial node (whose TSV encodes a topological abstraction of the subgraph rooted at the node) votes for models sharing a structurally similar subgraph. Each object accumulator O_i is a bin that stores the number of votes received for object model i. Object models receiving strong support are candidates for a more comprehensive matching process. Adapted from (Shokoufandeh et al., 2005)

graph rooted at node a, the first child of V, and let the out-degree of a be k_1. We compute the sum S_1 of the magnitudes of the k_1 largest eigenvalues of the adjacency sub-matrix defined by the subgraph rooted at node a, with the process repeated for the remaining children of V. The sorted S_i's become the components of a Δ-dimensional vector $\chi(V)$, called a *topological signature vector* (TSV), assigned to V. If the number of S_i's is less than Δ, the vector is padded with zeroes. We can recursively repeat this procedure, assigning a vector to each nonterminal node in the DAG, computed over the subgraph rooted at that node.

Indexing now amounts to a nearest-neighbor search in a model database, as shown in Fig. 10.5. The TSV of each non-leaf node (the root of a graph "part") in each model DAG defines a vector location in a low-dimensional Euclidean space (the model database) at which a pointer to the model containing the subgraph rooted at the node is stored. At indexing time, a TSV is computed for each non-leaf node, and a nearest-neighbor search is performed using each "query" TSV. Each query TSV "votes" for nearby "model" TSVs and these votes are stored in object accumulator bins, with a distinct bin for each object. In this fashion evidence for models that share the substructure defined by the query TSV is accumulated. Indexing could, in fact, be accomplished by indexing solely with the root of the entire query graph. However, in an effort to accommodate large-scale perturbation (which corrupts all ancestor TSVs of a perturbed subgraph), indexing is performed locally (using all non-trivial subgraphs, or "parts") and evidence combined. The result is a small set of ranked model candidates which are verified more extensively using the matching procedure described next.

10.5 Matching

Each of the top-ranking candidates emerging from the indexing process must be verified to determine which is most similar to the query. If there were no noise, our problem could be formulated as a graph isomorphism problem for vertex-labeled graphs. With limited noise, we would search for the largest isomorphic subgraph between query and model. Unfortunately, with the presence of significant noise, in the form of the addition and/or deletion of graph structure, large isomorphic subgraphs may simply not exist. This problem can be overcome by using the same eigen-characterization of graph structure we use as the basis of our indexing mechanism (Siddiqi et al., 1999b).

As we know, each node in a graph (query or model) is assigned a TSV, which reflects the underlying structure in the subgraph rooted at that node. If we simply discarded all the edges in our two graphs, we would be faced with the problem of finding the best correspondence between the nodes in the query and the nodes in the model; two nodes could be said to be in close correspondence if the distance between their TSVs (and the distance between their domain-dependent node labels) was small. In fact, such a formulation amounts to finding the maximum cardinality, minimum weight matching in a bipartite graph spanning the two sets of nodes. In a modification of Reyner's algorithm (Reyner, 1977), we combine the above bipartite matching formulation with a greedy, best-first search in a recursive procedure to compute the corresponding nodes in two rooted DAGs which, in turn, yields an overall similarity measure that can be used to rank the candidate. Details of the algorithm can be found in Siddiqi et al. (1999b); Macrini (2003).

10.5.1 Node Similarity

The above matching algorithm requires a node similarity function that compares the shapes of the 3D parts associated with two nodes. A variety of the measures used in the literature as signatures for indexing entire 3D models could be used to compute similarities between two parts (nodes) (Osada et al., 2002; Ankerst et al., 1999; Vranic and Saupe, 2001; Elad et al., 2001; Kazhdan et al., 2003a). Some care would of course have to be taken in the implementation of methods which require a form of global alignment. In the experiments carried out in this chapter we have opted for a much simpler 1D signature vector, which is based on the use of a mean curvature histogram. The essential idea is to compute a distribution of mean curvature values over all the level sets of the Euclidean distance function within the interior of a part. This is implemented as follows.

First, consider the volumetric part that a node i represents, along with its Euclidean distance function D. At any point within this volume, the mean curvature of the iso-distance level set is given by $\text{div}(\frac{\nabla D}{\|\nabla D\|})$. On a voxel grid with unit spacing the observable mean curvatures are in the range $[-1, 1]$ because the smallest principal curvature that can be measured corresponds to a sphere having radius 1. We compute a histogram of the mean curvature over all voxels in the volumetric part, over this range, using a fixed number of bins N. A mean curvature histogram vector $\mathbf{M}_i$ is then constructed with entries representing the fraction of total voxels in each bin. The similarity between two nodes i and j is then based on an L_2 distance between their mean curvature histogram vectors:

$$Similarity(i,j) = [1 - \underbrace{\sqrt{\sum_{k=1}^{N} [\mathbf{M}_i(k) - \mathbf{M}_j(k)]^2}}_{Distance(i,j)}].$$

By construction, this similarity function is in the interval $[0, 1]$. This measure could be further modified to take into account overall part sizes. In the experiments described in the following section we choose not to do this since our object models have undergone a global size normalization.

10.6 Experimental Results

In order to test our 3D object retrieval algorithms we have used selected models from the Princeton Shape Benchmark (Shilane et al., 2004). This standardized database, which contains 1,814 3D object models organized by class, is an effective one for comparing the performance of a variety of methods including those in (Kazhdan et al., 2003a; Osada et al., 2002; Ankerst et al., 1999; Vranic and Saupe, 2001; Elad et al., 2001). However, this database contains only a limited number of models with articulating parts and hence we have supplemented it with a set of articulated models that we have created. The resulting database, which we call the *McGill Shape Benchmark*, includes 455 exemplars of which we have used 425 in our experiments. The full database can be viewed under *http://www.cim.mcgill.ca/~shape*. The exemplars span 19 *basic level* object classes (hands, humans, teddy bears, spectacles, ants, octopuses, snakes, crabs, spiders, tables, chairs, cups, airplanes, birds, dolphins, dinosaurs, four-legged animals, fish). These classes are divided into two categories, those with significant part articulation, and those with moderate or no part articulation. In our experiments we merge the categories "four-legged" and "dinosaurs", treating them as a single category "four-limbs" Fig. 10.6 depicts 5 exemplars from each of the object classes.

To obtain a fully satisfactory set of exemplars, one would have to sample from a large population of models to be recognized, both with and without articulating parts. We have attempted heuristically to accomplish this in a small way, but carefully achieving that goal is currently beyond the scope of our experimental work. The results which follow must be interpreted with this caveat.

10.6.1 Matching Results

On a large database we envision running the indexing strategy first to obtain a smaller subset of candidate 3D models and to match the query only against these. However, given the moderate size of our database we were able to generate the $425 \times 425 = 180,625$ pairs of matches in a matter of 25–30 minutes

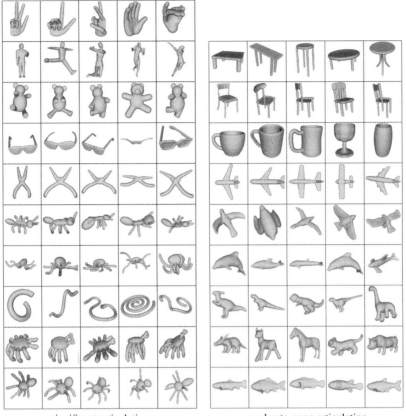

significant articulation   moderate or no articulation

Fig. 10.6 The McGill Shape Benchmark: 5 exemplars are shown from each of the 19 object classes. Exemplars from classes on the left have significant part articulation, whereas those on the right have moderate to no part articulation. The full database of 455 models can be viewed at *http://www.cim.mcgill.ca/~shape/benchMark/*

on a 3.0 GHz desktop PC. We compared the results using medial surfaces (MS) with those obtained using harmonic spheres (HS) (Kazhdan et al., 2003b) and shape distributions (SD) (Osada et al., 2002). For both HS and SD we used as input a mesh representation of the bounding voxels of the voxelized model used for MS. The pair-wise distances between models using harmonic spheres were obtained using Michael Kazhdan's executable code (*http://www.cs.jhu.edu/~misha*) and those using shape distributions were based on our own implementation of the algorithm described in Osada et al. (2002). For this latter implementation we took care to sample points uniformly and randomly on each outward face of each boundary voxel so that the signature curves were faithful. In particular, we were able to reproduce several of the *D*2 shape distributions in Fig. 3 of Osada et al. (2002). The comparisons between the three techniques were performed using the standard

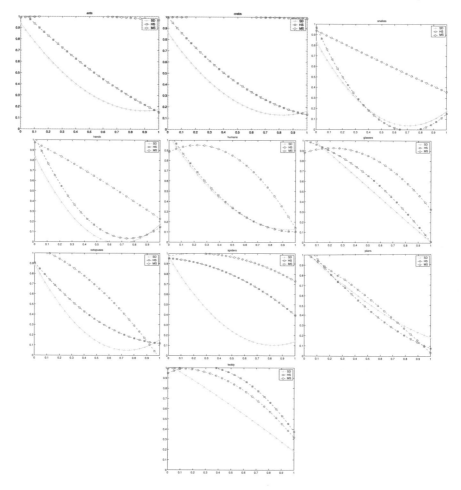

Fig. 10.7 Precision (y axis) versus Recall (x axis): Objects with articulating parts. The results using medial surfaces (MS) are shown with red circles, those using harmonic spheres (HS) with blue squares and those using shape distributions (SD) with green crosses. *Top row*: Ants, crabs, snakes. MS gives superior results. *Second row*: Hands, humans, spectacles. MS gives superior results. *Third row*: Octopuses, spiders, pliers. MS gives superior results. *Fourth row*: Teddy bears. HS gives slightly better results than MS

information retrieval notion of *precision versus recall*, where curves shifted upwards and to the right indicate superior performance.

The results for objects with articulating parts are presented in Fig. 10.7. For the category teddy bears both MS and HS give excellent results. However, for all other categories MS outperforms the other two techniques. For most of these models part structure is largely preserved, but parts articulate and deform. A particularly interesting case is the category snakes, whose exemplars consist of a single tube like

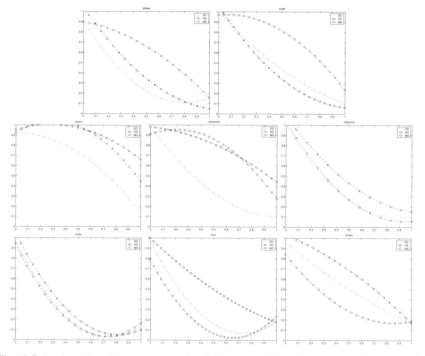

Fig. 10.8 Precision (y axis) versus Recall (x axis): Objects with moderate or no articulation. The results using medial surfaces (MS) are shown with red circles, those using harmonic spheres (HS) with blue squares and those using shape distributions (SD) with green crosses. *Top row*: Tables, cups. MS gives superior results. *Second row*: Chairs, airplanes, dolphins. MS and HS give comparable results for chairs and airplanes. For dolphins HS gives superior results. *Third row*: Birds, four-limbed, fishes. The results are comparable for birds. For four-limbed and fishes MS and SD give superior results

structure that is deformed in a variety of ways, causing significant difficulty for both HS and SD.

Figure 10.8 shows the results for objects with moderate or no part articulation. For categories in the top row MS gives superior results. For categories in the middle row HS and MS give comparable results, with the exception of dolphins for which HS gives superior results. For categories in the third row the results are comparable for birds, but for four-limbs and fish, both HS and SD outperform MS. The HS technique does particularly well on these categories, taking advantage of the pose alignment of the four-limbed models, and the "flat" mass distribution of the fish models. The MS technique would requires a degree of regularization to handle categories with changing part structure; we shall discuss this limitation further in Section 10.7.

computational geometry techniques for computing the Euclidean distance function directly from a mesh, which provides the additional advantage that the points on the shrinking sphere used to measure the average outward flux can be sampled very densely using a coarse-to-fine algorithm (Stolpner and Siddiqi, 2006). It might also be fruitful to explore Voronoi methods, discussed in Chapters 6 and 7, for computing medial surface-based DAGs that could in principle be applied directly to point clouds, provided that the sampling density is high enough (Amenta et al., 2001b) or to use the shock scaffold technique (Leymarie and Kimia, 2003).

With regard to the limitations of the part similarity measure, we expect that the performance of graph theoretic algorithms for comparing medial surface based representations will improve with more discriminating measures, and any one of a number suggested in the literature can be investigated.

The instability for complex objects of the graph structures we compute, as exemplified by the poorer results on the four-limbed animals and the fish, has a variety of aspects. One aspect has to to do with the assumption that we have made in converting a medial surface to a DAG, that an object has a well-defined part hierarchy. Such an assumption can fail for objects which have several main parts of comparable sizes (e.g., a caterpillar). Since this property would in turn be reflected in component parts with approximately the same node saliency, such models could at least be flagged. A second aspect has to do with instabilities in the branching topology of a medial surface based DAG, e.g., the precise manner in which the limbs attach to the torso can change with part deformation and movement. This latter aspect can be dealt with, at least in part, by exploring coarser representations based on the medial surface, e.g., by using Damon's metric measure along with the medial integrals developed in Chapter 4 to develop and incorporate a notion of ligature (Blum, 1973) in 3D. A third aspect has to do with the sensitivity of segmentation techniques that use only digital labelings on a rectangular lattice. These can suffer from discretization artifacts. As mentioned above, we are currently carrying out research in the direction of using computational geometry approaches to apply the average outward flux implementation in 3D directly to a mesh, as well as to refine the sampling (Stolpner and Siddiqi, 2006). Preliminary evidence suggests that the medial surfaces so obtained are more precise and that they may allow for estimates of the differential geometry, specifically the expectation that at medial surface junctions there is a discontinuity in the tangent plane, to be used to improve the segmentation process.

Acknowledgements We are grateful to Sylvain Bouix and Ran Chen for help with the numerical simulations and the preparation of the 3D models. This research was supported by grants from the Natural Sciences and Engineering Research Council of Canada, the Canadian Foundation for Innovation, FQRNT Québec, CITO, IRIS, and PREA.

Chapter 11
From the Infinitely Large to the Infinitely Small
Applications of Medial Symmetry Representations of Shape

Frederic F. Leymarie and Benjamin B. Kimia

Abstract We conclude the book by covering a wide spectrum of applications of medial symmetries of shape from the infinitely large toward the infinitely small. Our journey starts with a dynamic model of the formation and evolution of galaxies. We move on to the description of geographical information at the scale of regions of planet Earth. Next is the representation of cities, buildings, and archaeological artifacts, followed by the perception of gardens and the generation of virtual plants. Having reached the scale of human activities, we consider the perception and generation of artistic creations, the study of motion and the generation of animated virtual objects, and the representation of geometrically complex systems in machining, metal forging and object design. We then move inside the human body itself with applications in medical imaging and biology, followed by the representation of molecular structures. Our final stop is to consider the abstract scale of the perception of visual information.

11.1 Introduction

"Do not disturb my circles." — Archimedes of Syracuse (287–212 BC), last words.

In this chapter we take on a journey covering a full gamut of applications of medial representations of shape, from the large scale features of galaxy formation to the minute description of matter at the molecular level and the abstract scale of thoughts and cognitive processes.

We consider a family of medial representations closely related to and including the medial axes (MA) of (Blum, 1973), the Voronoi diagrams (VD) (Okabe et al.,

F.F. Leymarie
Goldsmiths College, University of London, UK,
e-mail: ffl@gold.ac.uk

B.B. Kimia
Division of Engineering, Brown University, USA,
e-mail: kimia@lems.brown.edu

K. Siddiqi and S. Pizer (eds.) *Medial Representations – Mathematics, Algorithms and Applications*.
© Springer Science + Business Media B.V. 2008

2000), intensity symmetry axes (IAS) of (Gauch and Pizer, 1993), the watersheds of mathematical morphology (Soille, 2004, Chapter 9), the (full) symmetry sets (Bruce and Giblin, 1992), the shock graphs of (Siddiqi et al., 1999b; Sebastian et al., 2004) and medial scaffolds of (Leymarie, 2003), the process inferring symmetries (PISA) of (Leyton, 1992), generalized cylinders and axes (Naeve and Eklundh, 1994), the molecular graphs of (Bader, 1990) and the critical nets of (Johnson et al., 1996).

11.2 Formation and Description of Galaxies

Consider an extremely large universe, possibly infinite in size, filled with an initially homogeneously distributed gaseous medium. As soon as "perturbations" occur in the distribution, the gaseous mass begins to evolve under its own gravity pulls. The morphological pattern toward which this universe will tend to stabilize itself takes the form of a set of highly dense nodes connected by filaments and walls of lesser density, which mimics the MA of the original voids, i.e., of the regions of space initially more empty of galactical matter; these voids can be considered as the "generators" of the final morphology.

When the infinite gaseous medium collapses under its own gravity, the regions that happen to be denser than average contract faster, while regions that are less dense than average expand somewhat faster than the rest of the universe. Thus, matter flows away from the low-density regions and towards the zones of higher density. A consequence of this observation is that the low-density regions become more and more spherical as they occupy more and more volume (Dubinski et al., 1993; Icke and Weygaert, 1987, 1991; Weygaert and Icke, 1989; Weygaert, 2001).

In this model of the formation and evolution of the universe on astronomical scales, it is as if matter flows away from the generators (voids) and gathers along *walls*, i.e., the symmetry sheets for pairs of generators. If matter entering the domain of a new cell passes such a wall, its velocity component perpendicular to the wall is reduced by friction with incoming matter from that symmetric generator. In time, galactic matter flows along these walls to eventually reach *filaments* where three walls intersect, i.e., A_1^3 curves of the 3D MA, and where the density is higher than in the walls. Matter keeps flowing along these filaments to finally end at nodal junctions where four such filaments intersect. These A_1^4 points of the 3D MA "are to be identified with high-density galaxy clusters in the Universe" (Weygaert and Icke, 1989). This dynamic formation constitutes what Icke and van de Weygaert call "the skeleton of the cosmic large-scale mass distribution by tracing the locus of points towards which the matter streams out of the voids" (Weygaert and Icke, 1989) (Fig. 11.1).

An early study of stellar arrangements taking the form of MA or Voronoi tesselations is to be found in the works of Descartes (Descartes, 1664; Okabe et al., 2000). By 1633, Descartes considered "cosmic fragmentation" in the solar system, i.e., the spatial distribution and relative influence of matter, using Voronoi-like

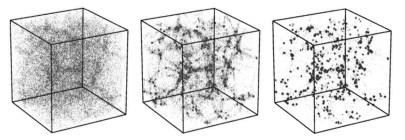

Fig. 11.1 (Adapted from (Weygaert, 2001, Fig. 3), with permission from R. van de Weygaert.) Example of *galactical systems formation* approaching the 3D MA of the original voids. Simulation for a cubic volume delineating a cosmic framework for which about 32,000 galaxies have aggregated

arrangements of solar system bodies—i.e., the sun, earth, moon, and so on—and their surroundings.

11.3 Geography: Topography, Cartography, Networks

In *topography* one is interested in characterizing the shape of a surface modeling a geographic area, usually much smaller than the earth, and such that the earth curvature plays no significant role. Then, the surface can be studied as a Monge patch, a height function of two spatial variables (x,y). Typically, one characterizes such a surface via local and regional features. Local features can be, for example, the singularities of the gradient of the height function: minima (pits or sinks), saddles (passes or relays), and maxima (tops or peaks). This classification by peaks, relays and sinks is the classical "hills and dales" representation of (Cayley, 1859) and (Maxwell, 1870) to characterize the topography of a surface by following where water tends to flow if dropped on the surface. Regional features include the surface lines linking extrema, such as ridges along crest lines, valley lines, and boundaries defined by watershed areas (McAllister and Snoeyink, 1999).

The connective structure of boundaries of watershed regions is discussed in (Nackman, 1984). For a recent survey and contribution on such concepts, see the work of Koenderink and van Doorn on the structure of relief (Koenderink and van Doorn, 1998). The watersheds can be constructed by a "flooding" growth simulating a non-necessarily Euclidean propagation from source generators constrained by the surface topography (Vincent and Soille, 1991; Meyer and Maragos, 1999). These "sources" are usually taken as minima of the intensity function. The flooding concept is similar to the original "grassfire" idea of (Blum, 1967): where fronts of "liquid" from different minima first meet, start tracing a "dam" which we can interpret as an MA sheet. The ribs of the final dams when the entire topography is flooded trace ridges; applying the same process on the inverted image traces valley lines.

One way of representing topographical features is to imagine the height function as a 3D object for which we may compute its symmetry sheets such that their extremities correspond to ridges or valleys. Such an analysis can be applied to any height function, not just in the domain of geography, for example, in image analysis.

Gauch and Pizer defined such a MA for height functions, and called it the *intensity axis of symmetry* (IAS) (Gauch and Pizer, 1993), where intensity refers to their application in the realm of image analysis, where grey level intensity values are taken as heights. In their variant on this concept, the IAS is made by stacking the 2D MA of each slice having as boundary a level curve of the image graph, intensity level by intensity level. A level curve in topography is a closed curve such that all its points have the same height. The succession of such 2D MA when connected along the intensity (height) dimension, creates a set of medial sheets representing a 3D MA of the image landscape. This is a particular type of "3D" MA which is oriented along one direction, here the height or intensity dimension. Gauch and Pizer then extract the ribs of the IAS medial sheets to capture significant ridge and valley lines of the original topography or image landscape. Gauch and Pizer (1993) argues for the superiority, as to which regions are determined, of the IAS–based subdivision over the similar but different ridges computed from watershed boundaries and those computed by the the 3D MA of the inverse of the topographical surface.

In *cartography*, the 2D MA has been used to extract and represent roads, rivers and other elongated structures, from aerial or satellite images, to help in the automatic production of maps. From image analysis and semi-automatic initialization, a road might be represented by its centerline, and tracked from a starting point and initial direction (Airault et al., 1996; Leymarie et al., 1996). The MA is also used for other shape analysis purposes, such as the characterization of river banks and area estimates (McAllister and Snoeyink, 2000). The MA is a "natural" and useful representation because it relates opposite points of river banks, relates centerlines or river networks to the "original river bank data," makes explicit and simple the calculation of surface areas of rivers, and can be used to tie-up the representation of elongated rivers and wider rivers as well as lakes (McAllister and Snoeyink, 2000).

Gold (2000) tackles the problem of *map generalization*, i.e., the production of multiscale versions of maps by simplification (of contours, objects, features). They use an iterative process of *retraction* applied to branches of the MA to simplify it, which in turns smoothes the associated boundaries[1]; they apply their technique to hydrological networks, elevation contours, and cadastral maps (Gold and Thibault, 2001). They also use the MA to tightly couple level heights of a contour map to regenerate terrain models from these contours, where the MA is helpful to preserve the topological relation between nearby level heights, leading to valid local terrain slopes (Thibault and Gold, 2000). Other applications include watershed and flow estimation from river network input, drainage network estimation from basin boundaries, text recognition and placement in cadastral maps.

---

[1] This is closely related to the more general method developed by (Tek and Kimia, 2001, 2003) to structurally smooth 2D contour sets. A 3D analog of this principle has been explored by (Leymarie et al., 2004b).

Cartograms, i.e., geography-related maps deformed to represent statistical information on demographics, epidemiology, agriculture, economics, and so on, can be constructed and visualized using the MA (Keim et al., 2005). In this application the MA becomes a local, somewhat deformable and flexible, 2D Cartesian grid where one axis is provided by the MA itself, and the other is taken as (local) perpendiculars to the MA. These perpendiculars cut polygonal areas of the map (such as State borderlines of the USA); the cut points are used to stretch or compress parallel-wise to the perpendiculars (or along the MA) this area of the map. The MA is particularly useful in this application to keep the deformed map recognizable in shape with respect to the original, untouched one. That is, the overall topology of the MA is left unchanged, and the local deformations preserve main features, such as corners or significant curvature extrema of the map, where MA branches end.

In the area of *wireless sensor networks* an important problem is to plan the routing for static sensor nodes in a geometric space (2D or 3D) with constrained energy supply. The goal is to build a lightweight, efficient structure, constrained by its environment; in particular, geographically close nodes can communicate better. Complex environments, like those in cities, will result in complex shapes of the space where sensors may be deployed, including the presence of "holes", e.g., obstacles like buildings. Connecting a pair of nodes should be done so as to not overload the network by passing by the same part over and over.

Bruck et al. (2005) applies the 2D MA to optimize routing in sensor networks. Given a starting and goal nodes to be connected, they use the MA of the deployment space as a reference to trace a guiding route parallel-wise to the nearest set of MA branches (Fig. 11.2). Compared to other popular approaches, Bruck et al.'s MA method results in comparable average route lengths, but "much better load balancing." Modeling via the MA also leads to good robustness to variations in the network model, a function of the distances between node pairs. Funke (2005) reports on a discrete methods to approximate the MA in sensor networks where nodes have no geometric information. He first evaluates the boundary nodes of the deployment space as well as nodes at the boundary of holes. Then, an approximate distance transform for the deployment space is possible, given the boundary constraints. Bruck et al. (2005) suggests that their MA model of routing would be even more usable if it was extended to the routing constrained on curved (topographical) surfaces under geodesic metrics, as well as to sensors deployed in 3D space with complex geometries (and with 3D obstacles).

11.4 From Urbanism to Architecture and Archaeology

Nearly 50% percent of the world population now lives in cities (closer to 75% in industrialized countries), up from 4% in 1800 and 14% in 1900 (Brand, 2005). The needs for the 3D modeling of large cities to support efficient *urban information systems* are present and growing, in particular in the domains of urban planning and management, civil protection, environment surveillance and crisis mitigation, and

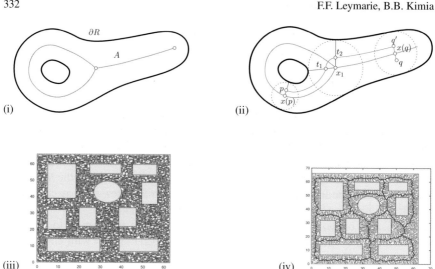

Fig. 11.2 Routing for *wireless sensor networks*; adapted from (Bruck et al., 2005). (**i**) An example of the interior MA of the boundary of a closed region R with two medial vertices. (**ii**) The road system on R for a routing protocol based on the MA, together with a routing from p to q. Two canonical cells C_1 and C_2 may share a common medial vertex but no common chord. (**iii–iv**) Scenario for a university campus where each sensor has a unit normalized coverage radius. (**iii**) A sensor network of 5,735 nodes deployed on a campus. (**iv**) The shortest path forest rooted on its approximate MA

sensor network modeling. One key aspect that requires new technological development is in the management of reconstructed 3D scenery in a geographical context (Leymarie and Gruber, 1997). This in turn calls for a representation of 3D data permitting spatial queries going beyond today's essentially 2D GIS (Geographical Information Systems) (Leymarie, 1997).

Spatial queries in a 3D graph structure have been proposed by (Lee, 2001, 2004; Kwan and Lee, 2005) to be based on the "straight" MA, a simplified medial axis restricted to the description of polygonal layouts. One advantage of abstracting 3D volumes via a medial graph structure is to allow for a hierarchical organization of the data at multiple scales, which is supported by studies on human abstraction of geographic space (Car, 1997). In the system of Lee et al., a series of 2D MAs for the communication network, e.g., the hallway structure, augmented to capture all useful horizontal connectivities, e.g., between rooms, are stacked-up and connected vertically to capture the spatial relationships between 3D entities of an entire building. Each augmented 2D MA represents a level in a building. This use of stratified 2D layouts abstracted by a graph structure, through the MA, similar to the IAS concept of Gauch and Pizer to model topography (see Section 11.3), is favored as (i) it naturally captures the important topological structures of human-made buildings, and (ii) it permits easy extension of existing 2D-based GIS systems. This 3D GIS has been augmented to integrate a ground transportation system together with the hierarchical representation of buildings, in the context of emergency responses in times of crisis (Kwan and Lee, 2005).

In the context of the built environment, spatial analysis is also fundamental to humans to create internal images of the urban area, the city, and the village. The representation of the built structures is understood as "cognitive maps" that are used to plan movement and navigate by different means, such as driving, cycling, walking. In particular, predicting human spatial behaviour in urban environments has been modeled by a *space syntax* based on *axial maps* (Hillier, 1996; Penn, 2003). These maps are constructed from "lines" derived from the main means of communication between or within buildings, such as roads, alleyways, and corridors. An axial map consists of a minimal set of axial lines that preserves the connectivity of the space, such that every axial line which may connect a pair of otherwise unconnected axes, is included (Turner et al., 2005). The concept of an axial line itself comes under various definitions, from the straight centerline of a communication pathway, i.e., its straight MA, to lines connecting vertices of "obstacles", e.g., corners of buildings, thus delimiting "isovists", i.e., the free space that can be seen from a vantage point (Batty and Rana, 2004).

An axial map for a given cityscape is then used to measure certain spatial properties such as "connectivity" and "integration" (Hillier et al., 1987). Connectivity is a local measure computed for a particular axis based on the number of other connected axial branches. Integration is a more global measure which evaluates how many axes need to be traversed to reach a particular goal; this can also be understood as the (topological) depth in the graph one needs to traverse to reach a certain goal from a starting point.

According to Le Corbusier, *architecture* is "based on axes" (Corbusier, 1985). Axes are defined by walls, corridors, lighting, and the spatial layout of other design elements. "Good architectural design thus enables the observer to extract relevant spatial information" (Werner and Long, 2003). This "architectural legibility" or "intelligibility", i.e., the capacity of a space to give clues to the understanding of the whole system (Hillier, 1996), is particularly relevant in wayfinding in a building. The complexity of an axial system for a building can be evaluated in terms of linearity (how straight a path is), connectivity (as for the space syntax above), and consistent alignment with respect to main reference axes, e.g., when moving from floor to floor (Werner and Long, 2003).

Another major issue in architecture is to capture *form and function* in a uniform, integrated framework. Typically, form is encoded in sets of connected floorplans and profiles delimiting the outlines of main walls, doors, stairways, bathrooms, windows, etc. Function encodes the integrated relationships of these different elements: not only their respective position and label, but how they interact, e.g., how a door relates one room to another and how a staircase loops around a set of walls. In architecture, "axial systems" capture the "symmetry between objects" (Pranovich, 2004; Pranovich et al., 2005). As previously illustrated, applying the 2D MA to floorplans and profiles offers the possibility to capture the tracing of symmetric outlines—as the envelopes of maximal contact disks centered on the MA or via sweep functions.

Furthermore, it also captures how various elements are interconnected, and how movement through the intermediary created space is possible.

The 3D MA offers the additional possibility of capturing how volumes are created both within and outside the architectural structures, specifying their form, volume and topology. The 3D MA also captures the detailed traces of ridges and associated main central axes of many structural elements used to construct the basic frames of buildings, including walls, floors, ceilings, stairs, ramps, and so on. This is useful for structural analysis as well as for shape rendering in graphics packages based on defining central axes, and sweeping profiles (the generalized cylinder representation).

A *generative* theory of shape by (Leyton, 2001, Chapter 15) has been applied to architecture, where axes of symmetry (in 2D or 3D) are information carriers summarizing the structure of a building. One of Leyton's key results is to provide a perceptually and mathematically coherent history for the elaboration of complex architectural spaces.

In *archaeology*, in addition to the above issue of representing form and function in a uniform coherent way, one is faced with the challenge of capturing significant descriptors of shape for the multitude of fragments associated to the original objects (Leymarie et al., 2001). The goal here is (1) capturing shape features so as to posit valid or approximate global alignment, and (2) extracting regional or more local information permitting fine matches between pairs of fragments at their breaks, i.e., along the ridges and corners where they were initially separated by fracture.

In recent studies on systems for the automatic re-assembly of archaeological artifacts from their fragments (Ucoluk and Toroslu, 1999; Leito and Stolfi, 2000; Cooper et al., 2002; Papaioannou and Karabassi, 2003), the main theme is to rely on a fine description of surface breaks or their bounding fracture curves. One uses the 3D MA for the interior of a fragment to specify the most significant ridges indicative of where the surface initially broke (Fig. 11.3).

Cooper et al. (2002); Willis and Cooper (2004a,b) uses a two stage strategy when matching pot sherds. The first stage relies on a probabilistic method to postulate a

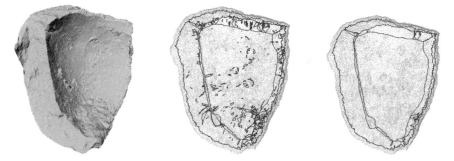

Fig. 11.3 (Adapted from (Leymarie et al., 2004b).) From the fine detailed laser scanning of a pot sherd (left) to the initial computation of the medial scaffold (graph representation of the 3D MA, middle), to the simplified scaffold (right) making explicit the most significant surface ridges (blue rib curves of ridges), useful to determine fine matches between pairs of sherds along their breaks. Red curves indicate axial MA curves, where three (or more) medial sheets meet

potential axis of revolution for each fragment, hence specifying the orientation of a sherd. Once an approximate orientation is found for a fragment, other fragments with compatible positioning can be looked at for close fine matches in a second stage. Willis et al. use a coarse sampling of the most significant ridges indicative of surface break curves to perform close matching of pairs of fragments.

In addition to the 3D geometric information these recent systems rely upon, surface and image or color textures, when available across the break curves, should also prove useful to augment the robustness of the matching. Here the 2D MA can be used in a process of graph alignment or completion for the fine adjustment of matching pairs of sherds. This is particularly useful for those smaller and flatter pieces that show weaker reliable 3D information, for example when the break curves are straight or the interior of break surfaces has little distinctive local structure.

11.5 From Garden Layouts to the Genesis of Plants

When contemplating a garden, we often select what we feel are better viewpoints to admire the structure and layout of the landscape, its plants, flowers, rocks, sculptural elements, and so on. Recently, it was shown by van Tonder *et al.* that certain famous 15th century Japanese garden layouts can be modelled by an approximate 2D MA which represents a perceptual (visual) tension flow field (van Tonder et al., 2002; van Tonder and Lyons, 2005; van Tonder, 2006a,b).[2]

The design of the Ryoanji garden has been a long lasting mystery; van Tonder *et al.* have shown how by using the rock and plant structures of a garden as the generators of a propagation like Blum's grassfire, an approximate oriented flow field they call the *Hybrid Symmetry Transformation (HST)* (van Tonder and Ejima, 2003) indicates the best viewing locus.

The dichotomous tree structure of the empty spaces between the five rock clusters of the Ryoanji garden is elucidated by the HST, showing various properties specific to this structure (Fig. 11.4.A2). It clarifies the hierarchical branching pattern, strict branching rule, approximate uniformity of branch nodes, consistent sloping of each branch towards the viewing verandah (bottom of Figs. 11.4.A1 and A2), convergence of the empty space onto the classical viewing location of the garden, and visual balance of the global structure. The branching structure appears to follow the rule of approximate self-similarity of a fractal which is often suggested as a model for vegetation and plants (West et al., 1999; Ferraro et al., 2005): e.g., it repeats locally in the left most rock cluster of the design (Fig. 11.4.A1).

The approach enables a new level of formal comparison between different Japanese dry rock gardens, and even between Japanese gardens and their counterparts from various locations around the world (van Tonder and Lyons, 2005). Awareness of MA structure in Japanese gardens motivated new attempts to reconstruct

---

[2] Readers familiar with the Japanese language should note that the acronym MA (medial axis) is not to be mistaken with the Japanese term "ma," denoting the interval between "things," spatial or temporal, although the two *appear* to be related.

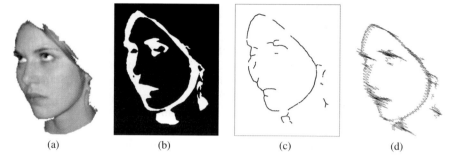

Fig. 11.5 Example of MA *based sketching* (adapted from (Tresset and Leymarie, 2005)). (a) Automatically segmented face region, Maeliss, 2003. (b) Binary map representing two grey levels after *k*-means clustering after blurring. (c) Approximate MA map of binary map in (b). (d) Example of sketching executed using (c)

(Deussen et al., 1999) uses the 3D MA to provide the main axial directions from which to derive perpendicular intersection planes to draw hatching strokes on the object's surface.

In *sculpting* (Hatcher et al., 2005) has developed a concept for biomimetic sculpture, i.e., sculpture that either mimics living organisms (animals, plants) or incorporates in its form the influence of the environment, e.g., sun exposure, rain, wind, or the proximity of humans. The resulting sculptures take the form of layered scaffoldings constituting a volumetric matrix into which smaller sculptural elements can be integrated. The layering approximates a discrete 3D wave propagation.

The 3D MA proves useful in two ways when simulating this biomimetic sculpting process (Fig. 11.6). First, given an initial layer, e.g., giving a description of a natural form, the MA indicates the singularities for the growth and can therefore be used to automatically either stop propagation or slow it down when near these. This is necessary to obtain a matrix that can be built and can sustain its own weight. In particular, minimal angles on the final scaffolding nodes need to be imposed. Second, a regular mesh for the 3D MA sheets can itself serve as a set of initial layers to be grown to generate biomimetic scaffoldings.

11.7 Motion Analysis, Body Animation, Robotics

The motion analysis of human or animal (articulated) figures has been modeled by dynamic 2D MA's, i.e., MA's derived from the outlines of figures in video or snapshot sequences. This representation in a skeletal form of moving figures has its roots in the works of Etienne-Jules Marey (circa 1880), when the first application of rapid photography was made to study the motion of horses, humans, etc. (Braun, 1992, p. 83).

In computer graphics, for visualization purposes such as in *animation* and the simulation of movement, articulated figures have been summarized by various

Fig. 11.6 A biomimetic sculpture using the 3D MA (with permission from Brower Hatcher and Karl Aspelund of Mid-Ocean Studio, Providence, RI, USA). A toy bear was laser scanned at coarse resolution to get a 3D solid rendering. The resulting wire-mesh was used as a source of propagation in 3D, where self-intersections are automatically detected and avoided using the 3D MA (Hatcher et al., 2005)

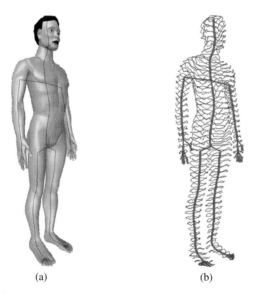

(a)   (b)

Fig. 11.7 (Adapted from (Lazarus and Verroust, 1999), with permission from the authors.) (**a**) Man body modeled with 15,000 faces and with the "source" of the flooding, which creates level-sets, being taken at the top of the head. (**b**) A "center" of each level set is selected and linked to its predecessors as to minimize the path-length to the source

types of skeletons where articulations correspond to junction nodes of the MA. For example, (van Overveld and Wyvill, 1997) uses 3D skeletons approximated with polygons and line elements; these are also used to re-render the shape (as it is animated) by "extruding" the surrounding surfaces associated to each polygonal elements of the skeleton. Unwanted blending and bulging are avoided with this approach. Verroust and Lazarus (2000) uses a level-set method to construct a 3D tubular MA, where a solid shape is flooded from various "sources," and where the center of each (flooding) level set is tracked at successive layers to obtain a final curve-like 3D skeleton (Fig. 11.7). This 3D curve skeleton is then used for animation and shape deformations. A related approach is proposed by (Wade and Parent, 2002), who build a control skeleton for inverse kinematics ("IK skeleton") by successive discrete thinning applied to the voxelization of a 3D solid. In yet another related technique, (Teichmann and Teller, 1998) uses the 3D Voronoi diagram, which is then pruned to obtain an approximation of the 3D MA upon which a network of virtual springs are attached for animation purposes.

In *motion planning*, the outline of the environment, such as the floorplans of buildings, the layouts of streets, or any workspace with obstacles, are used as constraints under which one needs to find (possibly optimal) paths leading from an initial configuration to a goal configuration. The notion of an optimal path involves a trade-off between the total path length, its clearance from obstacles, and other factors including travel time and energy required by a robot to execute its plans. Bounding volumes such as spheres or convex hulls are often used to ensure a minimum clearance to obstacles and to simplify the computation of collision detection (Cohen et al., 1995; Hubbard, 1996; Foskey et al., 2001). The Generalized Voronoi diagram (GVD), which is closely related to the Voronoi methods for computing the MA discussed in Chapters 6 and 7, has been proposed as a basis for motion planning (O'Dunlaing et al., 1983; Wilmarth et al., 1999; Guibas et al., 1999; Holleman and Kavraki, 2000; Garber and Lin, 2002) and collision detection algorithms (Ehmann and Lin, 2000). The MA can also be relaxed toward the visibility graph which provides shortest linear paths which compromise between path length, smoothness and clearance (Wein et al., 2005).

When considering articulated objects or robots with n degrees of freedom typically modelled as Euclidean translations and rotations at the joints, a practical method considers first a set of approximated loci sampling a high-dimensional configuration space. Such discrete paths can be efficiently computed and usually ensure some minimum clearance, but they often are far from being optimized in length, smoothness, and maximum clearance. However, an approximate path can be improved by "retracting" it to the MA of the workspace (Geraerts and Overmars, 2004). Geraerts and Overmars (2005) further propose finding "ridges" along this MA to maximize clearance to obstacles. A "ridge" corresponds to a path on the MA of the workspace such that the radius function is kept locally maximal. For example, for a 3D workspace, this will involve navigating via axial curves, where multiple MA sheets intersect, or along MA sheets following a gradient descent method.

11.8 Machining, Metal Forging, Industrial Design, Object Registration

The MA implicitly defines object offsets, which are required in *numerical tool machining* applications for milling, tuning, punching and drilling (Held, 1991, 2001). In addition, it has been used for surface meshing (Hoffmann, 1995; Amenta et al., 1998b), volume meshing and finite element analysis (Price and Armstrong, 1995, 1997; Patrikalakis and Maekawa, 2002; Wolter et al., 2004), dimension reduction and detail suppression (Sheehy et al., 1996; Leymarie et al., 2004b), as well as shape morphing (Blanding et al., 2000) and haptic exploration of surfaces (Okamura, 2000, 2003). The ability of the MA to implicitly represent the distance of a point to a (complex polygonal) shape is useful (Fig. 11.8) in dynamic path modification for rendering, collision and self-intersection prevention, tolerance verification, visibility computation, accurate motion dynamics and 3D path planning.

In the simulation of the *flow of materials*, the MA can be used to rapidly predict changes in patterns for complex dynamically changing geometries. For example, the 2D MA has been used in metal forging to model the hot metal material constrained to move in an enclosed space. Cavities, which are invaded by the flowing metal, keep changing shape as pressure is applied (Wienstroer and Mathieu, 2002). At each iterative steps of a reverse process simulation, the 2D MA of cross sections of the metal in that space is used as an accurate and efficient deformation predictor.

The MA or some equivalent skeletal representation can be used as a tool for *shape interrogation* (Patrikalakis and Maekawa, 2002), reconstruction, modification and design, and even as a basic element for building intuitive new interfaces for *shape modelers* (Armstrong et al., 1998; Blanding et al., 1999; Wolter and Friese, 2000; Wolter et al., 2004) (Fig. 11.8). Using a 3D skeleton as the underlying shape

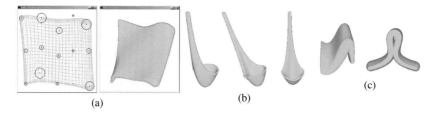

(a)   (b)   (c)

Fig. 11.8 *Modeler* based on the notion of 3D MA (with permission from Franz-Erich Wolter). The boundaries of the generated objects are the envelope surface of maximal spheres whose centers are located on the MA surface. These examples illustrate that it is possible to create and modify the shape of solid objects in an intuitive and systematic manner by modifying the maximal disc radius function associated with the initial MA. (**a**) Growing or shrinking of the maximal disc radius function results in the solid's growing or shrinking, respectively. (**b**) A "spoon"-like solid with concave and convex bounding surfaces. (**c**) Self-intersections occur if the maximal disc radius is not properly controlled during the construction of the envelope surface supposed to bound the newly designed solid. A careful treatment of the 3D MA permits controlling and avoiding this problem

representation gives the designer greater capacity to impose in unison geometric and topological constraints on a shape model, to attach functional representations to different parts of the model, and to specify non-uniform material distribution via offsets (Thompson, 2000).

The application of the MA to graph-based object recognition has already been discussed in Chapter 10. A related problem is the alignment or *registration* of two objects using hierachical representations derived from medial loci. The essential idea is to use edit costs associated with the generic transitions discussed in Chapter 2 to remove less significant object features (Tek and Kimia, 2001; Leymarie et al., 2004b). The coarser representation that survives captures the more salient object parts and is used to align two objects (or distinct instances of the same object) via a matching process on their medial graphs. As an example Fig. 11.9 illustrates the application of a medial scaffold to register two distinct scans of a 3D object (Chang et al., 2004).

11.9 Medicine and Biology

One of the first intended domains of application for the MA was in the areas of medicine and biology. Blum conceived of the 2D MA as a natural descriptor for shapes such as cells and body tissues, as well as entire anthropomorphic bodies or sub-parts, such as the arm, hand and fingers. His main manuscript was published in the Journal of Theoretical Biology (Blum, 1973). We mention below some of the main areas of application of the MA in medicine and biology illustrative of the possibilities it offers.

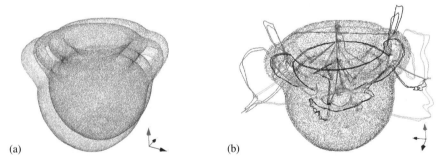

(a)   (b)

Fig. 11.9 Adapted from (Chang et al., 2004)). This figure illustrates how the medial scaffold can be used for registering two scans of a 3D object obtained at the same resolution (20,000 points) by two distinct operators. (**a**) An archaeological pot under two distinct scans. Due to range scanner capability, the data has holes. (**b**) The registration result obtained by aligning the underlying medial scaffolds

11.9.1 Object Segmentation from Images

Segmentation of anatomic objects from medical images by posterior optimization of m-reps has been discussed at length in Chapter 9. (Sebastian et al., 2003) has introduced the idea of using the 2D MA to modulate the "competition" between nearby deformable contours when segmenting noisy image data with tissues coming in close proximity. The 2D MA positioned as an inter-region skeleton acts a "predictor of boundaries" in a feedback loop with the deformable curve models. Each curve model is grown and deformed from an initial "seed" region (either manually or automatically). This method leads to improved bone segmentation of the wrist and spine regions (Crisco et al., 2003). Images of bones are often hard to segment because of (i) the varying densities observed in the bone structure (the outer layer or cortical bone being denser than the spongy bone tissue it encases), and (ii) the close proximity of separate bones part of closely integrated articulated structures, where successive or coupled bones form complex 3D architectures. Sebastian et al. (2003) use the result of segmenting successive CT scans to isolate profiles, which can then be connected to reconstruct in 3D each carpal bone of the wrist.

11.9.2 Path Planning and Virtual Endoscopy

In a way similar to the motion planning problem of robotics discussed in Section 11.7, the 3D MA of solid objects representing tissues can be used to generate central paths to help simulate navigation in the human body without approaching too closely the tissue boundaries. This is helpful for example to plan probe insertion or visualize the 3D navigation through vascular regions, the colon, the lungs, etc. (Zhou et al., 1998) computes a discrete 3D MA from a voxelization of medical image data (CT, MRI). They then compute various paths along the resulting 3D MA following voxels with local maximal medial radius. A similar method based on the 3D MA is used by Paik et al. for virtual endoscopy where a central path is obtained more directly and robustly by the use of a shortest path constraint between a source and sink for each branch to be navigated along (Paik et al., 1998) (Fig. 11.10). Kaufman et al. improve on the detection of sources and sinks and the use of distance fields to smooth centerline paths for virtual camera navigation (Bitter et al., 2001; Wan et al., 2001).

11.9.3 Morphometry, Branching Tree-like Structures

Based on the belief that brain morphometry understanding is the key to the study of "neuroanatomical structures which may be preferentially modified by particular cognitive skills or diseases," (Mangin et al., 2004) have been developing an approach to the study of brain data (typically obtained from MRI scanners) emphasizing the

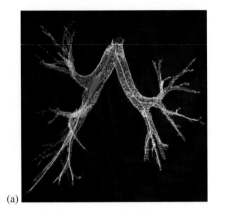
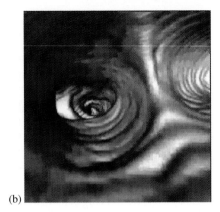

(a)   (b)

Fig. 11.10 (Adapted from (Paik et al., 1998), with permission from D. Paik.) (**a**) Bronchi and virtual camera pose near a branch along a computed path. (**b**) A shot from the virtual camera in the virtual bronchoscopy sequence produced for the bronchi in (a)

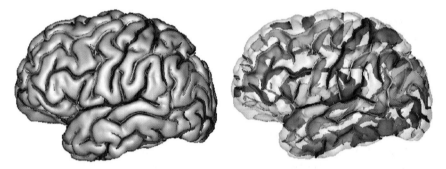

Fig. 11.11 (Adapted from (Mangin et al., 2004, Fig. 4), with permission from J.-F. Mangin.) Cortical sulci are extracted by computing a 3D MA using an erosion process on the cortical surface

morphometry of large structures, in particular the intricate crevices delimiting the 3D pattern of folds in the the cortex, called *sulci*. Their methodology depends on the use of the 3D MA. Using erosions, a discrete thinning process from Mathematical Morphology (Serra, 1982; Soille, 2004), they compute a discrete 3D skeleton of the cortex (Fig. 11.11). To obtain good localization of the ribs of the skeleton in the depths of the sulci by reflecting the influence of various "materials" and thicknesses, the erosion is performed at varying speed. Maintenance of the spherical topology of the cortex is imposed in the erosion process, filling in possible holes in the original dataset. Ribs of the computed 3D MA are then used to create a 3D graph representation of the sulcal medial loci.

A fundamental geometric structure found in all pluricellular living organisms takes the form of *space-filling tree-like branching shapes* that support both plants—roots, trunk, branches and leaves—and animals—vascular blood systems, bronchial tree of the lungs, epithelial ducts of the prostate gland, ureteric tree of the kidneys, etc. For example, a normal human kidney contains between 300,000 and 1 million "nephrons," tips of branches of the ureteric tree (Cullen-McEwen et al., 2002). A practical way to study such complex branching structures is to represent them as 3D graphs via a 3D MA. Jeulin et al. use a 3D thinning technique to capture the fine branching structure of mouse kidneys imaged with laser scanning confocal microscopy (Cullen-McEwen et al., 2002; Cain et al., 2005). They use the resulting 3D MA to measure the lengths of individual ureteric branches and the total length of the ureteric tree. (Chaturvedi and Lee, 2005) also use a 3D thinning process to obtain a 3D skeleton of the airway structure for the lungs of small animals. This skeleton is then used to represent the bronchial tree structure which serves in morphometric studies over multiple generations. As discussed in Chapter 5, Sanniti di Baja et al. use a version of the 3D thinning of volumetric images to obtain a 3D curve skeleton of the blood vascular system from angiographic images.

11.9.4 Growth, Form Genesis

In Chapter 9, Styner discusses the use of the 3D MA for understanding the changes in the forms of brain structures associated with neurodevelopmental or neurode-generative diseases. The 3D MA can also be used to study growth problems in embryology. For example, Mangin et al. propose that further understanding shall be gained by "studying the cortical folding process during brain growth" (Mangin et al., 2004). Durikovic et al. use the MA to represent the brain of an embryo in the womb, from key frames of a video sequence (Czanner et al., 2001), where the MA is interpolated between key frames to produce additional renderings, by re-growing the shape according to the associated radius function of the MA in order to obtain smoother visual effects when producing the final animation. They use a similar approach to study other organs' growth, such as for the stomach and the intestine (Durikovic and Czanner, 2001). They combine the MA with an L-system (typically used to model plants) to capture the topology of the entire digestive system represented as a tree structure (Durikovic, 2004). Jeulin et al. study the growth of kidneys' ureteric trees modeled by a 3D MA in a population of mice. The length distributions of the ureteric trees for different growth periods indicate "the existence of a programmed pattern of ureteric branching and growth" (Cullen-McEwen et al., 2002).

11.9.5 Deformation and Motion of Cells

Leymarie and Levine (1993) form a medial descriptor of live cells to represent the deformation of their boundary, and in particular to predict growths of pseudopods of neutrophils, known to use these to move in their environment. For this purpose they use growing and shrinking "dynamic skeletons" formed using the 2D PISA (Leyton, 2001), where the tip of significant MA branches are constrained to be attached to the associated local curvature extrema of the cell outline. As a branch of the MA grows (or shrinks), their system makes the prediction that the region at the tip of this branch (at the front, if growing, or behind, if shrinking) can be used to restrict the tracing of deformations of the boundary of the cell. In particular, this makes the cell tracking system robust to rapid or discrete changes and motions.

11.9.6 Recognition: From Tissues to Cellular Material

As expressed in Chapter 6, for the purpose of the object recognition of body tissues, the 3D MA has been proposed as a potentially useful representation. Using the medial scaffold graph representation of the 3D MA discussed in Chapter 10, steps toward this goal have been achieved by Chang et al. (2004). Attali and Montanvert (1997); Ferley et al. (1997) characterize the shape of complex elements such as a vertebra or a chromosome via a 3D graph structure built from a Voronoi diagram.

Beil et al. study dynamics of the keratin filament network part of the cytoskeleton of epithelial cells. Such a network structure is key in the formation of a scaffold defining the shape and mechanical properties of cells (Beil et al., 2005). Beil et al. map the 2D MA of electron microscopy images of regions of human pancreatic cancerous cells to a graph structure. Such graphs can then be studied under deformations, e.g., during cell migration.

Bajaj et al. (2003) propose to process cellular material from molecular tomographic imaging to study, characterize and recognize "cellular machines" built from hundreds of individual proteins. Different structural arrangements of these proteins lead to different "machines" that can efficiently carry out their physiological functions. The 3D MA is used to simplify the data while retaining important structural features useful to compare different multi-protein cellular complexes. Based on the notion of a critical graph of a Morse function and on Reeb graphs (Hilaga et al., 2001), which characterize the topology of level sets, Bajaj et al. construct a graph version of the 3D MA by directly linking critical points of the 3D tomographic images seen as density fields. Such a 3D graph can also be used to study and visualize plant and animal viruses from data obtained via electron microscopy (Bajaj, 2007).

11.10 Crystallography, Chemistry, Molecular Design

At scales smaller than what is typical of medical and biological problems, we reach the molecular and atomic structure of matter. At that level, 3D graph structures—which make explicit the topology of the networks connecting atoms to form molecules—are a favored representation. In crystallography they are called "critical nets," while in computational chemistry they are usually called "molecular graphs."

Diffraction data from X-ray crystallography reveals individual atoms in electron density maps. X-rays have the proper wavelength (in the angstrom (rA) range, $\approx 10^{-10}$ m) to be scattered by the electron cloud of an atom of comparable size.

At high enough resolutions, typically for less than 2.5–3.5 rA, a stick model may be directly fit to the 3D data where atoms are well isolated, although this usually involves human interaction. Whether at such high resolution or lower ones, an automatic processing of the data is desirable to identify the "backbone" or 3D skeleton of a potential stick model.

A common practical method to retrieve such structures consists in applying an erosion process to the 3D electronic density field capturing the shape of the space occupied by a molecule or a set of atoms. The 3D skeletonization of electron density maps is a popular method, dating back to the early 1970s and refined since, to automate the tracing of the molecular chains linking different atomic centers (Greer, 1974, 1985; Ioerger and Sacchettini, 2002; Amenta et al., 2002b; Potterton et al., 2004; Gopal et al., 2005). This method first segments a useful volume enclosing the atoms or molecules of interest, which can then be thinned down to retrieve an approximate 3D MA from which the skeleton is derived. In its simplest form, the segmentation is based on selecting an iso-surface of the electronic density field (Fig. 11.12).

Based on the quantum theory for atoms (Bader, 1990), a direct construction of the molecular graph is also possible. Critical points of either (i) the 3D MA obtained from the above thinning process of an iso-surface (Li, 2002), or (ii) a gradient field derived from the electron density map (Leherte et al., 1997; Bajaj et al., 2003), become nodes of the graph structure.

At lower resolutions of more than 2 rA that are common in practice, the points of maximum density do not always correspond to isosurface centers, preventing accurate positioning of the final stick model. A solution to this problem has been proposed by (Aishima et al., 2005) starting from the 3D MA of iso-surface segments of the electron density map (Fig. 11.13). An approximate MA, produced using "simplify" from the *Power Crust* software package of Amenta et al. described in Chapter 7, is thinned down. To create a 3D graph, the result is converted to edges and

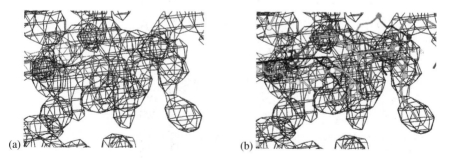

(a)   (b)

Fig. 11.12 3D skeletonization of an electron density map to produce a backbone for a molecular graph; figures provided by Thomas Ioerger et al., similar to results in (Gopal et al., 2005). (**a**) Typical electron density map around a protein; the density map is generated from x-ray crystallography data. (**b**) The 3D structure of the protein at atomic level, derived from the trace points, which are essentially along the MA of the density, and chains that connect some trace points (that are putative carbon-alpha atoms), the "backbone" of the protein is shown, where different inter-atomic bonds are in various colors

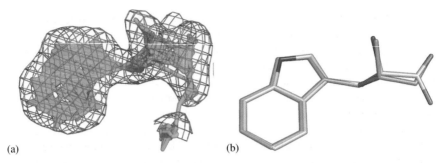

(a)   (b)

Fig. 11.13 (Adapted from (Aishima et al., 2005, Fig. 4), with permission from the authors.) Electron-density map for a tryptophan ligand bound to anthranilate synthase. (**a**) Prior to thinning, the full MA (cyan) is a wide flat shape inside the isosurface (blue mesh). Both the full MA and individual MA segments are too complex for direct graph matching against ligand-coordinate graphs. (**b**) After thinning and conversion to edges and vertices, the resulting model aligns well with the best available ligand model

vertices separated by approximately atom-to-atom distances. Aishima et al. report that this MA-based method is at least as powerful as critical point graphs at high resolution and that it looks promising for applications at lower resolutions.

Once a reliable molecular graph is obtained, various applications are possible, such as pharmaceutical structure-based drug-discovery efforts. In drug design the MA is a suitable substrate for (1) building molecular surfaces and volumes (Chapman and Connolly, 2001), (2) modeling interface surfaces of protein-protein complexes (Ban et al., 2004), receptor sites and the docking of ligands inside protein cavities (Lewis and Bridgett, 2003), (3) identifying geometric invariance among molecules exhibiting similar activity, and (4) mining databases, all key geometric problems in this field (Connolly, 1992; Boissonnat et al., 1994; Lin et al., 1994; Parsons and Canny, 1994; Finn and Kavraki, 1999; Huan et al., 2005).

11.11 Perception and Cognition

In addition to being used in models of human vision, as discussed in Chapter 1, Section 3, a number of researchers have emphasized the role of medial loci in perception and cognition.

Arnheim, a student of the Gestalt school of perception, defines "psychological forces" as an interplay of directed tensions, having magnitude and direction, inherent in any percept. In essence, according to Gestalt theory (Koehler, 1947; Koffka, 1935; Wertheimer, 1923), any percept (visual, auditory, tactile) is represented via a dense vector field. Arnheim characterizes the lines of equilibrium of forces with the notion of "structural skeleton." The percept is seen as a continuous field of forces. It is a dynamic landscape, in which lines (of the structural skeleton) are ridges being the centers of attractive and repulsive forces, whose influence extends through their surroundings, inside and outside the boundaries of a figure. This is the MA in disguise. The structural skeleton serves as a frame of reference by help-ing determine the role of each pictorial element within the balance system of the whole. Arnheim partly justifies the definition of the structural skeleton based on early studies in humans of dynamic sensitivity maps, where a movable dark disk on a white canvas generates preferred directions where to move next: the "directional tendency" cluster along principal axes of the structural skeleton (Arnheim, 1974, pp. 14–15; Fig. 4).

As discussed in Chapter 1, Section 3, Kovács, Julesz and Feher have derived dif-ferential contrast sensitivity maps for 2D shapes. These maps predict where contrast sensitivity improvements, involving higher activity in the primary visual cortex, are maximal, given a contour. The maps are consistent with a medial function called D_ε representing the percentage of boundary points equidistant from the observation point within a tolerance of ε (Kovács and Julesz, 1993, 1994; Kovács et al., 1998).

A deeper understanding of texture perception and texton theory (van Tonder and Ejima, 2000a), as well as of illusory contour perception, has arisen from work in image segmentation on a model called the "patchwork engine," by (van Tonder and Ejima, 2000b). The engine is based on the HST described in Section 11.5, which generates a full symmetry set by modifying the boundary propagation model from Blum's MA to the more general model of geometric optics. Figure 11.14 gives a schematic of the segmentation model, called the "patchwork engine," and some results.

The computational scheme of Kimia et al. based on shock graphs (in 2D) and scaffolds (in 3D) supports the speculation of Kovács et al. that a "sparse skeleton representation of shape is generated early in visual processing" (Kimia,

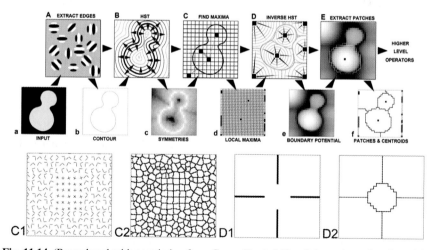

Fig. 11.14 (Reproduced with permission from G. van Tonder) *Top*: Schematic explanation of the Patchwork Engine (van Tonder and Ejima, 2000b). *Bottom*: Forward-inverse HST output segment-ing various types of edge maps (van Tonder and Ejima, 2003). The texture map in C1 is segmented into a cell grid in C2, and the Ehrenstein ("sun-illusion") figure in D1 is segmented as shown in D2 (here computed at low resolution)

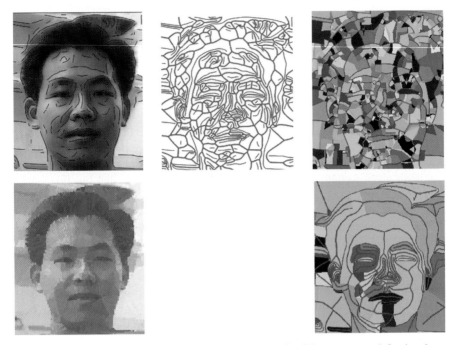

Fig. 11.15 Using the edge map associated with an image, visual fragments are defined and are used as canonical elements for perceptual grouping (Tamrakar and Kimia, 2004). From top-left to bottom-right: (i) a grayscale image with its superimposed contour map, (ii) its shock graph, (iii) the associated visual fragments, (iv) an average intensity reconstruction of the image from the visual fragments, and (v) the resulting visual fragments after applying a sequence of selected transformations mimicking a process of perceptual grouping

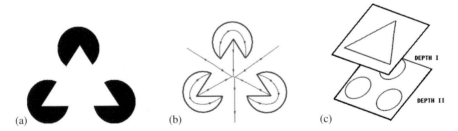

Fig. 11.16 (From (Kimia, 2003)) (**a**) The Kanisza triangle (Kanisza, 1979). (**b**) The medial shock graph arising from the edge map. (**c**) The simultaneous regional encoding of the MA renders it more powerful than curve-based disambiguation. The symmetry transform sequence recovers the illusory triangle at one depth and completes the "pacmen" at another

2003). Kimia et al. extend the traditional model of the MA to represent images, where each MA segment models a region of the image and is called a *visual fragment* (Fig. 11.15). They present a unified theory of perceptual grouping and object recognition where through various sequences of transformations of the MA representation, visual fragments are grouped in various configurations to form object hypotheses, are related to stored models, and are also consistent with the formation of illusory contours (Fig. 11.16) (Tamrakar and Kimia, 2004).

11.12 Conclusion

"To gaze is to think." — Salvador Dali.

We have reviewed a broad selection of the many applications of medial symmetries of shape to demonstrate the power of medial analysis. An exhaustive reporting on all recent applications was not the aim here. Rather, we probed into a number of domains, providing an overview meant to demonstrate how medial symmetries of shape can form the basis of a geometric language powerful enough to capture the representation of a vast array of objects in static or dynamic, evolving form, over a wide range of scales.

Acknowledgements We are grateful to the following people and their collaborators who provided figures, comments, or references for this chapter: Karl Aspelund, Axel Brunger, Ming-Ching Chang, John Dubinski, Jie Gao, Christopher Gold, Kreshna Gopal, Brower Hatcher, Vincent Icke, Thomas Ioerger, Dominique Jeulin, Anxiao (Andrew) Jiang, Daniel Keim, Ilona Kovacs, Francis Lazarus, Jiyeong Lee, Michael Leyton, Hongzhi Li, Ming Lin, Jean-Francois Mangin, David Paik, Christian Panse, Stephen M. Pizer, Noah Raford, Bernhard Rupp, Kaleem Siddiqi, Pierre Soille, Patrick Tresset, Gert van Tonder, Anne Verroust-Blondet, Rien van de Weygaert and Franz-Erich Wolter.

Appendix A
Notation

A.1 Common Notation

p The hub of a medial atom; the center of a sphere bitangent to the corresponding object surface.

r The spoke length of a medial atom; the radius of a sphere bitangent to the corresponding object surface.

$\mathbf{S}^{\pm 1}$ The spokes of a medial atom; a vector from the center of a sphere bitangent to the corresponding object surface to the points of bitangency $\mathbf{b}^{\pm 1}$ on that surface.

b A point on the boundary of an object.

$\mathbf{b}^{\pm 1}$ The points on the boundary of an object that are the points of bitangency of a single bitangent sphere within an object. These points are called *medial involutes*.

$\mathbb{R}^n$ The standard mathematical notation for an n-dimensional Euclidean space.

∇ Gradient, with respect to Euclidean distance or geodesic distance along a surface, as appropriate.

n The normal to an object surface.

A_k^m A label by which singularities are categorized in singularity theory. In this book each medial point is given such a label, where m indicates the number of distinct points at which the bitangent sphere has contact with the object boundary, and the subscript k indicates the order of contact between the ball and the boundary.

$\partial\Omega$ The boundary of an object Ω, also denoted B, on which points are denoted **b**.

$\mathbf{U}^{\pm 1}$ the unit length vectors indicating the directions of the spokes $\mathbf{S}^{\pm 1}$ of a medial atom, i.e., the directions of the vectors from the center of a sphere bitangent to the object boundary to the point of bitangency, i.e., the outward normal **n** to the object surface.

$\mathbf{U}^0$ The unit length bisector of $\mathbf{U}^{\pm 1}$. This vector is in the tangent plane to the medial locus for an interior point of that locus, and it gives the direction of the spoke of multiplicity 2 from an end point of the medial locus.

θ The object angle of a medial atom, i.e., the angle between the medial locus and each spoke.

N The unit vector normal to the medial locus.

K. Siddiqi and S. Pizer (eds.) *Medial Representations – Mathematics, Algorithms and Applications*.
© Springer Science + Business Media B.V. 2008

A.2 Chapter 1

$B \subset \mathbb{R}^n$ A maximal inscribed ball in an object; a sphere bitangent to the corresponding object surface.

Ω The region within $\mathbb{R}^n$ forming an object.

$\mathscr{C}(t,p)$ A curve or surface parametrized by p evolving with time parameter t. That curve or surface follows a grassfire flow from an object boundary to its medial locus.

$\mathbf{n}(p)$ The normal to an object surface $\mathbf{b}$ or $\mathscr{C}(0,p)$ and thus to $\mathscr{C}(t,p)$ if $\mathscr{C}(t,p)$ describes the grassfire flow.

P The center $\mathbf{p}$ and the radius r describing a sphere bitangent to an object surface.

$M(\mathbf{m})$ A medial strength, i.e., medialness, function for the medial atom $\mathbf{m}$ in some image.

A.3 Chapter 2

$\gamma(t)$ A curve in $\mathbb{R}^2$ parametrized by t.

$\gamma(u,v)$ A surface in $\mathbb{R}^3$ parametrized by u,v.

$g(t), g(u,v)$ A distance-squared function used in the description of contact of two curves or surfaces.

(t_1,t_2) A point in the pre-symmetry set for an object in $\mathbb{R}^2$.

D_4 A contact singularity corresponding to points of the boundary that are elliptic points at which the two principal curvatures κ_1 and κ_2 coincide.

r', θ' The derivative of r or θ with respect to arclength along a medial curve.

ψ The angle between two medial axis branches at a branch point.

A.4 Chapter 3

$\mathbf{V}$ The field of spokes on a medial or skeletal surface or curve, also called the radial vector field. $\mathbf{V}$ is denoted by $\mathbf{S}$ in other chapters.

$\mathbf{U}$ The field of unit vectors in spoke directions on a medial or skeletal surface or curve.

$\eta_\mathbf{V}$ The compatibility 1–form on $\mathbf{V}$, which relates the gradient of the radius function with the unit radial vector field $\mathbf{U}$, implying properties of the boundary of a medially implied boundary.

M_{reg} Smooth points of the medial locus M, i.e., all but end (edge) points and branch points.

M_{sing} End points and branch points of M.

$\mathbf{V}(x)$ The spoke(s) emanating from the point x on the medial or skeletal locus M.

$\psi_t(x)$ The point(s) that are the fraction t along spoke(s) emanating from medial or skeletal point x. $\psi_t(x)$ is called the radial map corresponding to fraction t.

$d\psi_t$ The linear operator that summarizes the derivatives of the radial map ψ_t in all tangential directions.

$\mathscr{B}_t$ The curve or surface formed by dilating the medial surface by fraction t of all of its spokes. That is, $\mathscr{B}_t$ is the union of all points $\psi_t(x)$, over all points x on the medial or skeletal locus, for fixed t.

U_{tan} The component of the vector U in the tangent plane to the medial locus.

$S(v)$ The shape operator from differential geometry applied to the direction v. This describes the rate of change of the surface normal per multiple of step size $|v|$ when one walks in the direction v on the tangent plane to the surface.

$proj_V(w)$ The vector resulting from projecting out (removing) from w the component in the direction of the vector V. In the projection considered under the topic of the radial shape operator, V is a vector transverse but not necessarily orthogonal to the medial tangent plane, which has a basis u, v, so if w is written as a linear combination of the u, v, V basis vectors, $proj_V(w)$ results from zeroing the coefficient of V in this linear combination.

$I_{n-1,1}$ The $n \times n$–matrix obtained from the identity matrix by changing the last entry to 0.

κ_r, κ_{ri} A principal radial curvature, one of the eigenvalues of S_{rad}.

κ_E, κ_{Ei} A principal edge curvature, a generalized eigenvalue of the pair of matrices $(S_{Ev}, I_{1,1})$.

r_{ri} A principal radial radius of curvature, i.e., the reciprocal of the radial curvature κ_{ri}.

ρ The component of the spoke direction U in the direction of the normal N to the medial sheet.

S_{med} The ordinary shape operator applied to the medial surface.

H_r The Hessian, i.e., second derivative matrix, of the medial radius r with respect to distance on the medial surface.

∇_v The covariant derivative with respect to the vector v.

$\mathscr{RC}(f)$ The relative critical set of a function. It consists of height ridges, valleys, and connector sets.

μ While in most of the book μ refers to a mean, in Chapter 4 it refers to an eigenvalue of a Hessian.

φ A diffeomorphism on an open subset of $\mathbb{R}^n$.

(M', V'), $\mathscr{B}'$, U' The primed entities refer to the medial description of the object described by the unprimed entities after a diffeomorphism has been applied to the object.

r_1 The medial radius function of a medially described object to which a diffeomorphism has been applied.

σ While in most of the book σ refers to a standard deviation, in Chapter 4 it refers to a scaling factor of the medial radius value corresponding to a diffeomorphism on the object.

Q_φ The radial distortion operator corresponding to a diffeomorphism φ.

$Q_{\varphi v}$ The matrix representing Q_φ in the basis v.

$Q_{E,\varphi}$ The edge distortion operator corresponding to a diffeomorphism φ.

E_φ A distortion operator accounting for the failure of a diffeomorphism φ to send the medial sheet normal to its primed value.

$E_{\varphi \mathbf{v}}$ The matrix representing E_φ in the basis $\mathbf{v}$.

h A multi-valued integrand on a medial surface that has a single value for each spoke. That is, h corresponds to an ordinary function h' on $\tilde{M}$.

ρ The factor relating the Riemannian measure of a medial surface and the medial measure needed in integrals over a medial surface.

$\tilde{g}$ The integrand on a medial surface corresponding to an integral of g on an object region or object boundary.

$m_i(g)(x)$ The i-th radial moment of g along a spoke emanating from x.

$I_\ell(h)$ The ℓ-th weighted integral, i.e., moment, of the multivalued function h.

$\Omega \backslash M$ The set Ω without the points M.

TM The tangent space to the medial curve or surface M at any specified point.

G In most places this symbol refers to the unit vector field generating grassfire flow. In one other place it refers to the metric tensor giving distances and angles on a surface.

$\Gamma_\varepsilon(x)$ A small convex region centered at x.

$vol_k(A)$ The k-dimensional volume of the entity A.

$m_{T\Gamma}$ A medial density in a small region of the medial locus, used in grassfire flux calculations.

M_Γ The smallest medial density for non-edge points on a medial locus.

$\Gamma(M)$ The tree describing the attaching of irreducible medial components to form a medial surface or axis.

A.5 Chapter 4

$\mathbf{W}, \mathbf{W}_0$ An object boundary subject to the grassfire flow.

t The time variable in the grassfire flow. Unlike in Chapter 4, time here is proportional to distance.

$\mathbf{W}(t)$ The object boundary after flow through time t.

$f(\mathbf{x})$ The speed of flow at position $\mathbf{x}$.

$\phi(\mathbf{x})$ A function giving the time t at which the grassfire front reaches the position $\mathbf{x}$.

$\mathbf{q}(t), \mathbf{q}$ A path swept out as a function of t.

$\dot{\mathbf{q}}(t), \dot{\mathbf{q}}$ The velocity, of the path $\mathbf{q}(t)$, i.e., $\mathbf{q}'(t)$.

$\mathscr{L}(\mathbf{q}, \dot{\mathbf{q}}, t)$ The Lagrangian of $(\mathbf{q}, \dot{\mathbf{q}}, t)$.

ψ A functional over the space of curves obtained by integrating the Lagrangian with respect to t.

$\mathscr{H}(\mathbf{q}, \mathbf{p}, t)$ The Hamiltonian of $(\mathbf{q}, \mathbf{p}, t)$.

γ A curve providing an extremum of $\psi(\gamma)$.

$\phi(\mathbf{q}, t)$ The action function, obtained by integrating the Lagrangian over an extremal curve γ.

G The unit vector field corresponding to the grassfire flow.

N_n^* The set $N_n(\mathbf{x})$ with $\mathbf{x}$ removed.

C^* The number of 26-connected components 26-adjacent to $\mathbf{x}$ in the intersection of an object and N_{26}^*.

$\bar{C}$ The number of 6-connected components 6-adjacent to $\mathbf{x}$ in the intersection of an object and N_{18}^*.

$D^\pm\phi$ The directional derivatives of ϕ in the reverse spoke directions.

J Those points on the medial locus whose medial sphere has more than two points of contact with the object boundary.

A.6 Chapter 5

I A binary image, whose pixels or voxels take on the value 0 or 1.

$\bar{I}(\mathbf{x})$ The complement of $I(\mathbf{x})$.

Z^n The set of all n-tuples of integers.

S_o The 8-connected (2D) or 26-connected (3D) set of voxels with value 1 in a binary image.

S_s The discrete skeleton of S_o.

S_c The discrete curve skeleton of S_o in 3D, i.e., an approximation to S_s made of discrete curves.

$\bar{S}$ The complement in I of the voxels in S, i.e., the set of voxels with value 0 in S.

$N(\mathbf{x})$ A set of pixels or voxels that are the neighbors of the pixel or voxel $\mathbf{x}$.

$N_f(\mathbf{x})$ The face neighbors of the voxel $\mathbf{x}$.

$N_e(\mathbf{x})$ The edge neighbors of the voxel $\mathbf{x}$.

$N_v(\mathbf{x})$ The vertex neighbors of the voxel $\mathbf{x}$.

$N_v(\mathbf{x})$ The knight neighbors of the voxel $\mathbf{x}$.

$N(\mathbf{x})$ All neighbors of the voxel $\mathbf{x}$ in a particular metric.

DT A digital image whose voxels hold the shortest distance from that voxel to the set $\bar{S}_o$.

EDT The Euclidean distance transform.

rDT The reverse distance transform.

s_i A seed value for a reverse distance transform.

CMB, CMD, CMO The center of a maximal ball, disk, or object.

SP A saddle-point of a distance transform.

a, b, c The step lengths between edge-neighbors, vertex neighbors, and knight-neighbors, respectively.

D^k A distance transform based on pixel or voxel having k neighbors.

T A threshold for elements of a distance transform.

$X_4(\mathbf{x})$ The crossing number of pixel $\mathbf{x}$, i.e., the number of 4-connected components in $N(\mathbf{x})$.

$C_8(\mathbf{x})$ The connectivity number of pixel $\mathbf{x}$, i.e., the number of 8-connected components in $N(\mathbf{x})$, with one exceptional value.

$d^{\mathbf{y}}I$ The directional derivative image of $I(\mathbf{x})$ in the direction $\mathbf{y} - \mathbf{x}$.

N^k The number of k-connected components of S in $N(\mathbf{x})$.

T_j collects measures of the differences between summary indices (frequently means) of the two respective classes, over the different elements making up the random variable on which an hypothesis is tested.

T_{ij} The i-th element of T_j. It is a metric of the difference in the j-th permutation between the summary indices (frequently means) of the two respective classes for the i-th element of the random variable on which an hypothesis is tested.

T_{ij}^p A test statistic in the form of a probability value.

T_j^p The summary statistic tuple made from the test statistics in the form of probability values, for the j-th permutation.

T_j' A single test statistic for the j-th permutation. This statistic combines the information from the tuple T_j^p.

S^{th}, S'^{th} A threshold on the statistics in the form of probability values.

$H_{0,i}$ The null hypothesis on the i-th element of the multivariate random variable on which a hypothesis is tested.

$H_{1,i}$ The complement of the null hypothesis $H_{0,i}$ on the i-th element of the multivariate random variable on which a hypothesis is tested.

U A matrix of vectors for each permutation which are distributed according to the standard normal distribution.

$\hat{\Sigma}_U$ An estimated covariance matrix for the multivariate normally distributed random variables U_j.

A.10 Chapter 10

V The root node of a DAG.

Δ The maximum branching factor of a DAG.

$\chi(V)$ The topological signature vector assigned to V.

S_i The sum of the magnitudes of the k_i largest eigenvalues of the adjacency sub-matrix defined by the subgraph rooted at the i-th node of a graph.

λ_i The i-th largest eigenvalue of an adjacency matrix or sub-matrix.

k_i The out-degree of a graph node.

O_i The object accumulator for the i-th object model, which is a bin that stores the total number of votes received for that object during the indexing process.

D The Euclidean distance from an object boundary over points in a specified object region.

$M_i(k)$ The k-th element of the mean curvature histogram associated with object model i.

Glossary

Action function The solution to a variational problem where trajectories are found which minimize or maximize a particular quantity. One such example, which arises in geometric optics, is the search for a path between two locations in a medium such that the integral of a quantity along that path is an extremum (an extremal path). In this setting the action function is the integral along the path of the Lagrangian, a quantity that measures the cost of passing through a particular location.

Active shape method A method of segmentation via a deformable model pioneered by Cootes and Taylor. The method represents geometric entities by a tuple of points and limits its solution to a shape space computed via principal component analysis of the point tuples in a training population.

Adjacency matrix An $n \times n$ matrix that contains a complete description of the connectivity structure of a graph having n nodes. With i and j representing the indices of two nodes, a non-diagonal (i, j)-th entry is the number of directed edges from node i to node j. A diagonal entry is the number of self edges (loops) at that node. When studying directed graphs with at most one edge between two nodes i and j, a directed edge from i to j is represented by a 1 for entry (i, j) and a -1 for entry (j, i).

Affine area A measure of the space interior to an object, where the measure does not change when an affine deformation is applied to the space in which the object resides.

Affine deformation A transformation on 2- or 3-space in cardinal coordinates (x, y) or (x, y, z) given by a linear transformation on the coordinates plus an additive translation vector.

Anti-crust This is the dual of the crust. It is used in Voronoi methods for skeletonization to provide an approximation of the medial axis. Also see crust and power crust.

Association graph This is a type of derived graph that serves a purpose in obtaining solutions to the problem of matching two graphs. If node u in the first graph is compatible with node w in the second graph, a node representing the pairing (u,w) is created in the association graph. An edge is then placed between nodes (u,w) and (v,z) in the association graph if the relationship between nodes u and v in the first graph is the same as that between nodes w and z in the second graph. It can be shown that maximal common subgraphs between the two original graphs correspond to maximal cliques in the association graph. Also see maximal/maximum clique.

Augmentation The act of taking the union of the primitives of a geometric entity and a selected set of primitives of the entity's neighbor entities.

Average flux The integrated flow of a vector field through a closed region divided by the area (in 3D) or length (in 2D) of the boundary of the region. The term average outward flux is used when the flow points outwards from the region enclosed.

Average outward flux See average flux.

Average surface distance Between two object boundaries A and B, it is the average of the average distance from A to B over all positions on A and the average distance from B to A over all positions on B.

Basis function One of an indexed set of functions forming a family of functions by taking all linear combinations of the set.

Bayes rule A formula for reversing the order of conditional probabilities using prior probability distributions. The conditional probability $P(A|B)$ or probability density $p(A|B)$ of the random variable tuple A given the random variable tuple B is proportional to $P(B|A)$ or $p(B|A)$ with constant of proportionality given by the ratio of $P(B)$ or $p(B)$ divided by $P(A)$ or $p(A)$.

Bias of an estimator The difference between the correct value of a parameter being estimated and the mean of that estimator over the whole population of training samples.

Binary image An image all of whose voxels or pixels have the value 0 or 1. Typically the binary image represents an object by the set of voxels or pixels with value 1. Sometimes a binary image is called a *characteristic image*.

Bipartite graph A bipartite graph is a graph whose nodes can be divided into two disjoint sets such that there are no edges between any nodes in the same set.

Bitangent sphere A sphere that is tangent to an object boundary at two or more locations or at one location with multiplicity 2 or greater.

Blend A fillet smoothly connecting two cut-off surfaces.

Blend region An object subregion implied by a m-rep connecting each subfigure to its host figure. Its boundary smoothly connects the m-rep implied boundary of the subfigure at a specified cutoff to the host figure surface in which hole has been punched by dilating the subfigure.

Blum medial axis A medial locus for an object. It is a locus of the centers and radii of spheres bitangent to and interior to the object boundary.

Boundary A general term used to mean a curve bounding a region in 2D or a surface bounding a region in 3D.

Boundary evolution A process by which an object boundary is deformed, typically using a partial differential equation that describes the motion of each point in the direction of the normal.

Boundary implied by m-rep The envelope of the spoke ends of a medial representation in the form of an m-rep.

Boundary normal A unit vector orthogonal to the tangent line (in 2D) or tangent plane (in 3D) to an object boundary. In 3D it is frequently computed by normalizing the cross product of the partial derivatives of boundary position with respect to the two variables parameterizing the boundary.

Boundary noise Errors in the positions of a boundary. Often the errors produce small pimples and dimples.

Boundary pre-image The points of tangency on an object boundary of the bitangent sphere centered at a designated medial point.

Boundary sampling

Density. The rate and manner in which points on the boundary are sampled prior to the generation of the Voronoi diagram in Voronoi method-based skeletonization. A number of theoretical results and guarantees are stated in terms of the parameters of the sampling density.

Epsilon. A point set S is said to be an ε-sample of a continuous surface if for every point x on that surface the minimum distance from it to any point in S is at most ε times the distance from x to the medial axis of the surface. Intuitively this means that the sampling density is higher for portions of the surface that are closer to the medial axis.

Poles. When S is a dense sample of a surface, the Voronoi cells tend to be long and skinny and approximately perpendicular to the surface. The poles of a sample $s \in S$ are defined to be the two Voronoi vertices, each at opposite ends of the Voronoi cell, that are furthest from s.

Branch curve The same as fin curve. Also called a Y-branch curve or the junction of figures.

Branch point A location on a medial or skeletal surface where three smooth curves or surfaces come together with discontinuities in their limiting tangent planes.

Branching topology The branching structure of medial curves (2D) or sheets (3D) that is invariant to diffeomorphisms. This structure indicates which sheet (curve) is a branch of which and, for curves, the order along a curve of its branches.

B-rep A representation of an object in terms of its boundary. The representation may be a set of sample boundary points, a continuous function of variables parametrizing the boundary, or a locus defined implicitly as the zero level set of a function on 2-space or 3-space.

B-spline A spline is a continuous function made from polynomial patches and with a specified degree of smoothness at the patch boundaries. A B-spline is specified by control points in such a way that each patch is a linear combination of basis polynomials on the patch parameters, with the control points of a limited number of neighbors of the patch forming the polynomial weights. The spline does not pass through the control points.

Cartesian product of transformations The Cartesian product of sets of transformations $T_i, i = 1, 2, ..., n$ is a set whose elements are an ordered tuple of geometric transformations, with the ith element of the tuple being a member of the set T_i.

Chord A line between two points P and Q on an object boundary.
   Chord residual. For a 2D object the chord residual is the difference between the length of the boundary segment between P and Q and the length of the chord.

Class A family of entities (in this book, geometric entities) defined by some common property of interest. Examples are disease state (e.g., well or ill), sex, and age range.

Classification Applying or generating a method that, given a value of a typically multivariate random variable, determines which of a number of pre-specified classes the value falls into.

Cm-rep See continuous m-rep.

Coarse-to-fine An algorithm that deals first with information at the largest spatial scale, then at a smaller spatial scale, and so on through a sequence of scale-levels.

Compatibility condition See compatibility 1-form. A skeletal structure that satisfies the compatibility condition at all smooth points on the skeletal curve or surface is said to satisfy the partial Blum condition.

Compatibility 1-form A measurement comparing the rate of change of the radial distance (spoke length) with respect to a direction on the tangent to the medial curve or surface with the component of the spoke direction in that direction. The compatibility 1-form being zero is called the compatibility condition. That condition holding at a point on the skeletal curve or surface corresponds to the spokes of the medial atom there being orthogonal to their envelope, i.e., the implied boundary.

Computer vision For some this term is used to describe the discipline of extracting descriptions of the imaged world from images made from reflective light. For others it is used to mean the analysis of any type of image.

Conditional object mean, given neighbors The expected value of the probability distribution of the representation of an object, given the representation of selected primitives in neighboring objects.

Cone of spokes For some objects in 3D, called *tubes*, the medial locus is a space curve. At each point on that curve the spokes from the medial point to points of spherical bitangency form a right circular cone whose axis is tangent to the curve.

Connector curve A curve in the n-dimensional argument space of a function where the function is a relative minimum in a k-dimensional subspace orthogonal to the function's gradient where the Hessian is negative definite, for some $0 < k < n$, and a relative maximum in the complementary $n - 1 - k$ dimensional subspace also orthogonal to the gradient. A series of connector curves for adjacent values of k connect ridge curves and valley curves.

Consistency condition A geometric constraint that a medial axis point must satisfy at certain special points, such as a Y-junction in 2D, where the maximal inscribed disk has 3 points of contact with the boundary.

Contact The touching of two objects, typically a circle (bounding a disk in 2D) and a boundary, or a sphere (bounding a ball in 3D) and a boundary. We may also say that there is contact between the disk (resp. ball) and the boundary, meaning always that they touch at a point of the circular boundary of the disk (resp. ball).

Contact function. A function used to measure the degree of intersection between a curve in 2D (resp. a surface in 3D) and some standard geometric object. For medial axis applications this object is typically a circle (resp. sphere). A parametrization of the boundary is substituted into the equation of the circle (resp. ball). The contact function is a function of one variable for the 2D case and of two variables for the 3D case.

Contact, order or type of. The order of contact measures how closely two objects approximate each other at a contact point. In 2D the order of contact is said to be k if the first $k - 1$ derivatives of the contact function at the point under consideration vanish, while the k-th derivative does not. A change of parameter reduces the contact function to the form x^k; this is also called an A_{k-1} singularity. For $k = 2$ we also say 'ordinary contact' and for $k = 3$ we also say 'osculating contact'. In 3D the contact function (of two variables) is reduced by reparametrization to a 'normal form' which is typically $x^2 \pm y^k$ (analogous to k-point contact and called A_{k-1} where $k = 2$ is 'ordinary contact') or $x^2 y \pm y^3$ (called D_4 contact).

Continuous m-rep A form of m-rep given as a locus of 0-order medial atoms. The locus is provided as a spline of the medial location variables (x, y, z) and the medial radius variable r. It therefore represents the medial variables by polynomials in the patch parameters of the splines.

Contractible region A region without holes or cavities.

Control points Primitives, e.g., locations or medial atoms, that determine an approximating entity, typically as a spline. While control points have the form of a primitive on the resulting geometric entity, those primitives need not fall on that entity but may only be near it.

Coordinate system A mutually orthogonal, right-handed set of unit vectors with an origin.

Core A height ridge of a scalar medial strength function on the variables of a medial atom.

Correction for correlation When the random variable on which hypothesis testing is done has multiple positions, changing the p-value used in hypothesis testing at one position to reflect the fact that in geometry the null hypothesis holding at one position is correlated with it holding at neighboring positions.

Correction for multiple comparisons When the random variable on which hypothesis testing is done has multiple positions, changing the p-value used in hypothesis testing at one position to reflect the fact that the null hypothesis is more likely to be falsely rejected at any of multiple positions than it is at a single position.

Correspondence In a population of geometric entities each described by a collection of discrete primitives, a particular primitive is said to be in correspondence with the others with the same index. The correspondence is improved by moving the primitives along the manifold describing the object, e.g., the boundary or the medial manifold, so as to tighten the probability distribution on the geometric entity while keeping satisfactory coverage of the manifold.

Crack A continuous boundary of a discrete digital object can be defined by considering the vertical and horizontal edges in 2D, or faces in 3D, between background and foreground pixels or voxels. Each such edge or face is sometimes called a crack.

Crest A curve on the boundary of an object in 3D made from crest points. At a crest point the lesser of the two principal curvatures is negative (convex) and is a relative minimum of principal curvature along the integral curve of corresponding principal direction.

Crest curve A curve on a surface made up of crest points.

Crest point A point on a surface where, along an integral curve of principal direction, the curvature is convex and the magnitude of the principal curvature has a relative maximum.

Critical point A critical point of a possibly vector-valued function of one or more variables is a place where the Jacobian matrix does not have full rank. For a scalar function of one or more variables this is a point where the gradient vector is zero.

Crust Under ε-sampling of the boundary to give a set of points, the crust is a selection of Delaunay edges associated with the Voronoi diagram of the set. Its dual, the anti-crust, gives a topologically correct approximation of the medial axis.

CT image A type of medical image produced by computed tomography. The image is reconstructed from an angular sequence of x-ray projections. Typically a CT image is a 3D image produced as a pile of slices.

Cumulative distribution function (CDF) A function of a scalar random variable giving the probability that the variable is less than or equal to the argument of the CDF.

Curvature

Gaussian curvature. A measure on a surface analogous to the curvature of a curve in a plane. It is an infinitesimal measure that describes over what angles the surface normals swing per unit of area about a point on the surface. It is given by the product of the two principal curvatures at the point.

Mean curvature. The average of the normal curvatures over all directions of motion at a point. It is also the average of the normal curvatures in any two orthogonal tangent plane directions.

Normal curvature. For a point on a surface and a direction on the tangent plane to that surface given by a unit velocity vector, it is the component in the velocity direction of the rate of change of the normal.

Principal curvature. The normal curvature in a principal direction.

Curve evolution See boundary evolution.

Cusp A location where the directed tangent to a curve turns 180 degrees. An 'ordinary cusp' can be reduced by change of coordinates to the form (t^2, t^3). In this book cusps are found in reference to the symmetry set in 2D.

Cut locus Per the Wikipedia definition, the cut locus of a point p in a manifold is the set of all other points for which there are multiple geodesics connecting them from p. The cut locus is important in measures involving distance, such as Fréchet means and principal directions, since the distance function from a point is not smooth on its cut locus.

Decision boundary A hypersurface in a feature space used in classification. Points on one side of the surface are taken to be in one class, and points on the other side are taken to be in another class.

Deformable model A model of an object or of a spatial region that changes shape smoothly with changes in its parameters.

Delaunay tetrahedralization The generalization of Delaunay triangulation to 3D. The Delaunay tetrahedralization of a discrete point set is a set of tetrahedra with the points as vertices such that no tetrahedron contains any of the points in its interior. It is formed as the dual of the Voronoi diagram: by drawing straight lines connecting each point to all the neighboring points with which it shares a single Voronoi face. In this book this operation is computed for the set of vertices of a mesh representing an object boundary.

Delaunay triangulation In 2D the Delaunay triangulation of a discrete point set is a set of triangles with the points as vertices such that no triangle contains any of the points in its interior. It is formed as the dual of the Voronoi diagram: by drawing straight lines connecting each point to all the neighboring points with which it shares a single Voronoi edge. Each such connection is called a Delaunay edge.

Diffeomorphism A function from points in a space onto a copy of that space such that is 1-to-1 (invertible) and both the function and its inverse have continuous derivatives of all orders. Intuitively, a diffeomorphism is a smooth, non-folding warp.

Differential geometry The branch of mathematics that studies the shape of geometric domains such as manifolds or stratified sets via derivatives of of position along the domain. Normal directions and curvatures are typical quantities of interest.

Dimension of feature space The number of entries in the tuple designating a point in feature space.

Dimension of shape space If the representation of an object or other geometric entity requires n variables, the dimension of the feature space of all possible values of those variables is n. However, the dimension of actual objects in a population, or at least adequate approximations to them, may form a subspace of much smaller dimension. This subspace is called a shape space.

Directed acyclic graph (DAG) A directed acyclic graph is a graph containing directed edges which has the property that there is no nontrivial path beginning at any node that returns to that node.

Discrete m-rep A representation of an m-rep whose sheets of medial atoms are represented by a grid or chain of medial atoms. Atoms anywhere on the sheet are interpolated from the grid, as are spokes in any inter-figure blend region.

Discrete medial model A medial model made from samples of a continuous medial model.

Discrimination The act of determining the way in which the feature values in one class differ from those in another class, frequently based on the probability distributions on the features in the two classes.

Displacement field A function on an image where at each position a vector is given that indicates a new position onto which the original position is mapped.

Distance function A function on spatial position giving the distance from the closest point in a designated set of positions. Typically the designated set is an object boundary. The distance function may be signed if the distances in the object interior are given a sign opposite from those in the object exterior.

Distortion operator, radial and edge An operator that for all step directions on the medial surface yields the change in the radial or edge shape operator as a result of a deformation of a skeletal structure.

Divergence theorem A theorem relating the integral of the divergence of a differentiable vector field over a region to the flux of the vector field through the region's boundary.

Double of M A medial surface understood as having two pasted together sides. A single spoke (radial vector) emanates from each point on the double of M.

Earth Mover's distance A measure of the distance between two probability distributions given by the physical work required to move the probability mass between the two distributions.

Edge curve A curve on a medial or skeletal surface made from edge points. Along that curve the two radial vectors, i.e., spokes, become equal, and they touch the corresponding object boundary at a crest point.

Edge point A point on the double of a medial or skeletal curve or surface where the curve or surface folds back on itself, i.e., the two sides end. These points can also be understood as points where the medial curve or surface ends. In the medial case the bitangent disk or sphere at those points touches a single object boundary point with a multiplicity 3.

Edge strength A measure at a point on the image, possibly together with a direction, measuring how great the gradient or directional derivative of the image intensity is at the point.

Eikonal flow A non-linear partial differential equation that arises in geometric optics and takes the form $|\nabla \phi| = F(x), x \in \Omega$, with the condition that $\phi = 0$ on $\partial \Omega$. A special case is Blum's grassfire flow, in which case ϕ is the Euclidean distance function to the boundary and $F(x) = 1$.

Elongation A measure of length relative to width.

Ellipsoid The 3D object that has the simplest generic medial locus, namely an ellipse. An ellipsoid has three orthogonal cross-sections, each an ellipse, three orthogonal principal axes, and three corresponding principal radii.

End atom A special form of medial atom in a discrete m-rep. Such an atom is formed from an order 1 medial atom together with a third spoke bisecting the other two spokes.

End curve See edge curve.

Endoskeleton Given an object sampled on a digital raster with boundary represented by the cracks separating the voxels, the endoskeleton is that portion of the skeleton that lies in the object's interior.

End point See edge point.

Estimation of probability distribution Determining a probability distribution on a random variable or tuple of random variables given a collection of training samples of the random variable or tuple.

Euler-Lagrange equation The central equation that arises in the calculus of variations. It provides a way to solve for functions that give extrema of a particular cost functional.

Exoskeleton Given an object sampled on a digital raster with boundary represented by the cracks separating the voxels, the exoskeleton is that portion of the skeleton that lies outside the object.

External medial locus A Blum medial locus of the complement of the interior of an object, i.e., for which all of the bitangent spheres are entirely outside the object interior.

Extraction of objects See segmentation.

Extrema Locations where a function of interest has a relative maximum or a relative minimum.

Euclidean distance function See distance function.

Feature space An abstract space of n-tuples of scalars describing an entity, such as a geometric representation of an object.

Fermat's principle A variational principle in geometric optics that states that when a ray of light passes through a medium it follows a trajectory that minimizes the optical path length.

Figural coordinates Coordinates of points in an object interior or near-boundary exterior that are given in terms of the location on the medial sheet or axis, a choice of spoke, and the relative distance along the spoke from the medial sheet.

Figure In this book, this term is used to refer to a part (or whole) of an object that is represented by a single, i.e., unbranched, medial sheet (in 3D) or curve (in 2D or for a tubular object in 3D).

Fin or fin sheet A section of the medial surface for a 3D object transversely attached to a host section of the medial surface along a curve.
   Fin curve. The curve attaching a fin on the medial surface to its host. At all locations except the curve ends (fin points), the host surface has a discontinuity of its normal. That is, in the plane orthogonal to the curve the fin and the host join in a Y.
   Fin point. An endpoint of a fin curve, where the edge of the medial surface meets the smooth part of the medial surface. Also called a fin creation point.

Fluid-flow warp A means of finding a diffeomorphism on 3-space (or 2-space) that minimizes the sum of a penalty for data mismatch with an atlas and a geometric penalty measured by the energy of viscous flow and modeled by the Navier-Stokes partial differential equation.

Flux invariant When considering the average outward flux of the gradient of the Euclidean distance function through a shrinking disk (2D) or sphere (3D), the limiting value obtained can be viewed as a type of invariant, in the following sense.

It is precisely zero at points not on the medial locus. At points on the medial locus it is non-zero, and its value is a fixed function of the object angle. Hence, at medial points, when combined with the Euclidean distance function, it can be used to reconstruct the points of contact with the boundary.

Form In the usage describing perception, this term describes the spatial conformation of an object, i.e., its shape, location, orientation, and size.

Fourier harmonics Components of a function that are each a constant times a sinusoid at some integer multiple of a base frequency, with some phase. In the representation of boundary of an object of circular topology in 2D, there are two functions: the x position as function of the variable parametrizing the boundary and the y position as function of the variable parametrizing the boundary.

Frame In an n-dimensional space, a set of mutually orthogonal, unit vectors. Conventionally the vectors are indexed in a way that forms a right-handed coordinate system.

Fréchet mean The Fréchet mean of a set of sample points in a feature space is a point in that space that has the minimal sum of squared distances to the members of the set.

Gabor patch A function centered at a point in 2D that is a sinusoid multiplied by a Gaussian.

Gamma-medial axis In Voronoi-based skeletonization the parameter $\gamma(m)$ is twice the object angle at a medial point m. The γ-medial axis is then defined as the subset of medial points with gamma parameter greater than or equal to the constant γ. It is used to develop a scale-invariant approach to the approximation of the medial axis. Also see object angle.

Gauss-Bonnet theorem A formula connecting the Gaussian curvature integrated over the surface and the Euler characteristic (number) of the surface, where the Euler number is a topological property.

Gaussian probability distribution A probability distribution on a random variable of dimension n that has two parameters, the mean μ and the covariance matrix Σ, given by $\frac{1}{[(2\pi)^n det(\Sigma)]^{1/2}} exp(-1/2(x-\mu)^T \Sigma^{-1}(x-\mu))$.

Generative approach A means of defining an entity, in this book typically an object, by describing a means of computing the object boundary or interior.

Generic property Intuitively, a property of a geometric entity is generic if it persists under "almost all" small changes of the entity; and if it fails in some case to persist, almost any arbitrarily small change will yield an entity that has that property.

Generic transition The generic creation and/or loss of a geometric property under a specified family of transformations.

Geodesic center Given two points on a curve, their geodesic center has equal arc-length to both points.

Geodesic difference The difference between two entities in a feature space that may be curved is the shortest distance (geodesic) path between the locations corresponding to the two entities.

Geodesic path A curve in a space with a Riemannian metric such that for all sufficiently close points on the path, the projection of a sufficiently small segment of the path onto the tangent plane at the midpoint of the segment has zero curvature, i.e., is locally approximated to second order by a line segment. For every pair of points in the space, the path of shortest distance is a geodesic path.

Geodesic subtraction The geodesic difference $A \ominus B$ between two points A and B in a feature space is the shortest geodesic path from B to A.

Geometric entity A group of objects, an object, a section of an object's interior or boundary, or a primitive component of an object's representation, such as a medial atom, a boundary vertex, or a voxel.

Geometric illegality A representation of an object that fails to satisfy the geometric requirements of objects. Examples of illegality are local or global self inter-penetration and nonsmoothness.

Geometric medial map A map between geometric properties and loci on the medial surface and geometric properties and loci on the corresponding object boundary.

Geometric transformation A function on all or a part of space that maps each point in that subspace to another point in space.

Geometry-to-image match A measure of how well the image data understood relative to a geometric representation matches that image data in some template.

Graph In certain places in this book a "graph" is an algebraic structure made from nodes and edges connecting specified nodes. In other places in the book a "graph" is a curve or surface giving the value of a scalar function for each value of its arguments.

Graph edit distance This is a cost function assigned to a set of moves such as insertions or deletions of nodes and edges in one graph (or subgraph) to make it match another graph (or subgraph).

Graph matching The problem of finding matches between the nodes of two graphs. The matches are required to satisfy the condition that any two nodes being matched are compatible in terms of their labels and also in terms of their edges to adjacent nodes in the original graphs. Also see maximal common subgraph.

Graph of curvature The curvature value as a function of position along a curve.

Graph theoretic method A method based on the mathematical theory of graphs, which are discrete structures made from a collection of nodes, with certain pairs of nodes connected by edges (arcs).

Grassfire flow The mathematical operation of evolving an object boundary inward at fixed speed along its normals until opposing (fire fronts) quench. This boundary evolution is one means of computing the medial locus. The flow along a normal stops when it reaches the medial axis or surface. Also see eikonal flow.

Grey skeleton This scalar function is used in Chapter 4, and it arises via an extension of the definition of the object angle to all points in the interior of the object, with the exception of medial axis points whose associated maximal inscribed circles touch the boundary at more than 2 points.

Hamiltonian formulation An alternate formulation of a variational problem where the state of the system is described by points and their momenta and where these points move in a $2d$-dimensional phase space. For a given Hamiltonian only one trajectory passes through a point in the phase space. A Hamiltonian formulation can be obtained from a Lagrangian one by applying a Legendre transformation. Also see Lagrangian formulation.

Hamilton's canonical equations When applying the Legendre transform to relate the Lagrangian to the Hamiltonian, the process involves an exchange of roles of velocities with momenta, where the momenta are expressed as appropriate partial derivatives of the Lagrangian. This leads to Hamilton's canonical equations, which describe the evolution of phase space.

Hamilton-Jacobi equation A first-order nonlinear partial differential equation that arises in classical mechanics. In Chapter 4 the action function satisfies a Hamilton-Jacobi equation that relates time derivatives of the action function to the Hamiltonian of the problem. Also see action function.

Hausdorff distance Between two object boundaries A and B, the Hausdorff distance is the larger of the largest distance from A to B over all positions on A and the largest distance from B to A over all positions on B.

Hessian A matrix capturing the second derivative information of a function f at a point. In n dimensions with variables $x_i, i = 1, 2, ...$, it is an $n \times n$ matrix whose ij^{th} element is $\frac{\partial^2 f}{\partial x_i \partial x_j}$.

High dimension, low sample size Refers to statistics done when the number of training samples is not much larger than the inherent dimension of the feature space in which the relevant probability distributions lie.

Hinge atom A medial atom connecting a subfigure to its host figure. Changes in the host figure or motion of the hinges on the surface of the host result in changes in the hinge atoms and consequently changes in the whole subfigure.

Hippocampus A subcortical brain structure. Each human brain has two hippocampi.

Homotopic See homotopy type.

Homotopy preserving thinning See thinning.

Homotopy type Two geometric regions are said to be of the same homotopy type or be homotopic if they can be continuously deformed into each other.

Hotelling T^2 test statistic A generalization of the t statistic to multiple random variables.

Hub The center point of an order 1 or order 0 medial atom. The medial locus is the curve or sheet made from all hubs.

Hypothesis testing For two classes (as considered in this book) a process of comparing probability distributions on a tuple of random variables between classes. For a sample of a particular size, the process operates by determining the probability that the differences seen would happen at random if there were no differences in the classes' probability distributions (the null hypothesis).

   Global test. A hypothesis test using random variables from all parts of a geometric entity.

   Local test. A hypothesis test using random variables from a limited portion of a geometric entity or on limited shape features.

Image A scalar function (the image intensity) on a typically rectangular but possibly infinite region of typically 2-space or 3-space, or such a scalar function on a finite, discrete grid (of pixels or voxels, respectively) within that 2-space or 3-space. In some cases an image may have a tuple of scalar values at each pixel or voxel.

Image analysis Extraction of descriptions of image contents from images.

Image intensities The values taken at the respective pixels or voxels making up an image.

Image match See geometry-to-image match.

Implied or reconstructed boundary An object boundary calculated from the medial representation of the object. It is the envelope of the spokes or of the spheres used in the medial representation. In particular, for a Blum m-rep, the implied boundary is incident and orthogonal to the spokes at their ends.

Indentation figure A region described by an unbranching medial locus to be removed (set-subtracted) from from a medially described host region to produce the description of an object, where the region to be removed includes part of the boundary of the host region.

Indexing In computer vision this term refers to the organization of a large database of models so as to support efficient search, typically in the context of an object recognition or model retrieval task. Given a query object (or model) it is computationally infeasible to match it against all instances in the database, particularly when that database is large. The idea of indexing is to develop strategies by which candidate matches can first be obtained at low computational cost, following which more discriminating but expensive matching procedures can be employed.

Inherent dimension The dimensionality of the subspace of feature space needed to encompass all entities in the population being described. More practically, all but a subpopulation of appropriately low probability needs to be described to within an appropriate tolerance.

Initialization When optimizing an objective function of some argument tuple by an iterative optimization technique, initialization chooses a value of the tuple at which to begin the iteration.

Integral over a medially implied region An integral of some function over a portion of the interior of a medially implied object or over a portion of the medially implied boundary. Also see skeletal integral.

Intensity profiles Intensity functions in an image along the boundary normals of an object.

Internal medial locus A Blum medial locus of the interior of an object, i.e., for which all of the bitangent spheres are within the object interior.

Inter-patient segmentation Segmentation of a new target patient trained by a population from various patients.

Interstitial voxels Voxels in an image space that are outside all of the objects being explicitly represented.

Intra-patient segmentation Segmentation of a given patient at a new time trained by the within-patient variations of the same or different patients.

Intrinsic coordinates A tuple accessing points in the interior and boundary of an object in a way determined by the object itself. Intrinsic coordinates are medial if they are determined using the medial locus of the object.

Irregularity penalty A term in an objective function in fitting an object model. The term penalizes irregular spacing of the model's primitives, for example, by a monotonic increasing function of the difference between each primitive and the average of its neighbors.

Iterative conditional modes A method of replacing every value in a grid by the mode (relative maximum) of the probability distribution given by its neighbors. If the probability distribution assumed is Gaussian, the mode is the mean of the neighbors.

Junction of figures See branch curve.

L_2 **norm** The Euclidean norm of a tuple or a continuous function, calculated as the square root of the sum or integral, respectively, of the squares of the tuple elements or the function values.

Label image A voxel array for which each voxel contains a label indicating which object or region it belongs to.

Lagrangian formulation A formulation of a variational problem, where the state of the system is described by points moving with velocities in a d-dimensional space. For a given Lagrangian, its integral gives a functional over a space of curves, and the solutions are curves which extremize this functional. For a given Lagrangian, several trajectories may pass through a point in the configuration space. Also see Hamiltonian formulation.

Lambda-medial axis In Voronoi-method-based skeletonization, under the assumption that the boundary has been sampled uniformly with distance between boundary samples $d(m)$, the parameter $\beta(m)$ at a medial point m is defined as the radius of the smallest enclosing ball of the intersection of the surface and the medial ball at m. The λ-medial axis is then defined as the subset of medial points with beta parameter greater than or equal to the constant λ.

Landmark A point that is identifiable by the local geometry. The term is also used for any point chosen to provide correspondence across instances of a geometric entity in a population.

Lateral ventricle A designated part of the fluid-filled ventricular system within a brain. Each human brain has two lateral ventricles.

Leave-one-out testing or experiment A means of validation of a segmentation method using a test set in which, successively over the test set elements, training is done on all but one of the elements and the segmentation is applied to the target case formed from the remaining element. The resulting segmentations are evaluated, and the evaluation results are described statistically. Such a means of evaluation gives somewhat unrealistic results, since the training set changes from target case to case.

Legal or legitimate shape See geometric illegality.

Legendre transformation The expression of a functional relationship $f(x)$ as a different function whose argument is the derivative of f. In Chapter 4 the Hamiltonian is expressed as the Legendre transform of the Lagrangian.

Level set, level surface A locus of points where some specified function takes on a fixed value.

Lie group A set of transformations that simultaneously has the following two properties. First, it forms a group, i.e., is closed under composition, has a unique identity, and provides a unique inverse in the group to each element of the group. Second, at each point (transformation) in the set, the set is differentiable.

Ligature This term literally refers to the glue that holds together the parts. In the context of the Blum skeleton, ligature regions are skeletal segments that can have significant length but contribute little to the reconstruction of the object's boundary. A typical example is the addition of a protrusion to an elongated part of an object, which leads to a ligature branch between their associated skeletal segments. Such segments cannot simply be pruned because this would result in alterations to the object's topology. Hence they are typically handled by being signaled as less salient during skeleton-based object matching. Also see significance measure.

Likelihood The likelihood function of a tuple of random parameters of interest is the probability or probability density of the data, given the value of that random parameter tuple.

Local form A local form is a classification of a generic point on the medial axis or the symmetry set, where the description is in terms of Arnold's A_k, D_k notation for the contact function. As described in Chapter 2, there are only a small number of possible local forms of the medial axis.

Local shape description A representation describing only a selected part of a geometric entity.

Log likelihood The logarithm of the probability or probability density of the data, given a random parameter tuple of interest.

Log posterior The logarithm of the probability or probability density of a random parameter tuple of interest, given the data.

Log prior The logarithm of the probability or probability density of a random parameter tuple of interest, given no data.

Magnetic resonance image, MRI A type of medical image produced in a high magnetic field using sequences of magnetic pulses. Typically MR images reflect at each voxel in 3D the local hydrogen chemistry there and thus the distribution of water, fat, etc.

Mahalanobis distance For a point in a feature space, for which the vector from the mean point to the target point is decomposed in terms of the principal directions, this distance is the square root of the sum of squares of the respective coordinate values expressed as z-values, i.e., each coordinate each has been normalized by the standard deviation in that direction.

Mallows distance A distance between two probability distributions. It is given by the L_p norm of the difference between the two quantile functions corresponding to the probability distributions. For $p = 2$ it is equal to the earth mover's distance between the probability densities.

Manifold An abstract mathematical space in which every point has a neighborhood that resembles Euclidean space.

Markov random field A grid or sequence of random variables such that each grid element has a set of neighbors and the joint probability of all of the random variables in the grid is entirely determined by each grid element's variable's conditional probability on its neighbors. In particular, at each grid position the conditional probability given values at its neighbors must be the same as the conditional probability given values at all other grid positions. Thus conditional probability on non-neighbors is transitively related to the conditional probabilities on neighbors.

MAT See medial axis transform.

Maximal common subgraph When considering two graphs to be matched, the maximal common subgraph problem is to find the largest subgraph in each for which the two subgraphs are isomorphic to one another.

Maximal/maximum clique A clique in a graph is a set of nodes such that there is an edge between each member of the set. A maximal clique is a clique that cannot be grown to form a clique with greater cardinality. A maximum clique is a maximal clique whose cardinality is greater than that of all other maximal cliques in the graph.

Maximal convexity core A height ridge of medial strength as a function of the parameters of a medial primitive, where the directions of analysis for maxima are eigenvectors of the Hessian matrix of medial strength with the largest eigenvalues.

Maximal inscribed ball (3D) or disk (2D) Any ball, i.e., filled sphere, that is a subset of the interior of an object and is not strictly contained within any other ball that is a subset of the object.

Maxwell set This is the set of locations within the boundary of an object which have more than one closest point on the boundary in the sense of Euclidean distance.

Measure on the skeletal sheet See medial measure.

Medial atom An order 0 medial atom consists of a maximal ball or disk indicated by its center point $\mathbf{p}$ and its radius r. An order 1 medial atom consists of a hub $\mathbf{p}$ and two spokes indicated by the two spoke direction vectors $\mathbf{U}^{+1}$ and $\mathbf{U}^{-1}$ and the common spoke length radius r.

Medial axis, stable subset Voronoi methods for skeletonization seek to develop approximations of stable subsets of the medial axis. These subsets have the property that they are less sensitive to small perturbations of the boundary points from which the Voronoi diagram has been calculated.

Medial axis transform The 1-to-1 relation between a boundary and its Blum medial axis.

Medial density For a region centered at a point on the medial surface (or curve), the limiting lower bound, as the region shrinks to zero volume (area), for the ratio of the area (perimeter) of the medial surface (curve) within the region to the area (perimeter) of the region.

Medial geometry See radial geometry.

Medial involute Given one point on an object boundary, its medial involute is that other point of tangency of the medial bitangent sphere that is tangent at the first point.

Medial labeling A digital classification of medial axis points on a cubic lattice. In 2D medial axis pixels can be end points, interior (curve) points or branch points.

In 3D medial axis voxels include, in addition, surface points, border points and different types of junction points.

Medial locus A collection of curves (in 2D) or sheets (in 3D) forming the centers of spheres bitangent to the boundary of an object and entirely contained within it. In 2D it is also called the medial axis.

Medial measure A differential measure on the medial surface that is proportional to the amount of object volume captured by the spokes (radial vectors) emanating from the surface point in question.

Medial point A center location of a maximal bitangent sphere for an object boundary. Also called a hub.

Medial radius The radius of a maximal bitangent sphere for an object boundary.

Medial scaffold A graph-based hierarchical representation of medial axis points in 3D, where the classification of the local form is used to define nodes, links and hyperlinks.

Medial sheet A surface, typically with boundary but not itself branching, made up of medial points. The sheet may have fins emanating from a fin curve on the sheet, where the normal to the sheet is discontinuous. Part of the end curve of the sheet may be a fin curve on another sheet.

Medial strength, medialness A function on medial atoms derived from an image giving the compatibility of each medial atom with the intensity patterns in the image data. Typically, it is greater as contrast is higher at and in the directions of the spoke ends of the medial atoms.

Meshing See tile.

Metric See Riemannian geometry.

Midpoint locus A type of medial surface formed by the locus of the midpoints of line segments between tangency points of each sphere bitangent to an object boundary.

Mixture of distributions A probability distribution made from a weighted sum of probability distributions. Multimodal distributions are typically represented as mixtures of unimodal distributions such as Gaussians.

Monge form An equation of a surface that expresses the surface in terms of its height above a plane. The expression takes the form $z = f(x, y)$, where x, y, z are Cartesian coordinates and $f = \partial f/\partial x = \partial f/\partial y = 0$ at $x = y = 0$, that is, the tangent plane to the surface at the origin is the plane $z = 0$.

M-rep A representation of one or more objects via a collection of curves or sheets of order 1 medial atoms.

Multifigure object An object whose medial representation has more than one curve (in 2D) or sheet (in 3D).

Multimodal probability distribution A probability distribution whose density function has multiple relative maxima.

Multiple comparison problem See correction for multiple comparisons.

Multiplicity A given condition holds with multiplicity n at a point if it holds n different times at that point. For example, a function has a root of multiplicity n at $x = x_0$ if $f(x)/(x - x_0)^n$ is finite at $x = x_0$.

Multiscale, multiple spatial scales An object representation or analysis sequence that operates sequentially at multiple levels of spatial scale. Typically the sequence is large scale to small. With a multi-object entity typically one of the scale levels treats an object in relation to its neighbor objects, a smaller scale level treats the object as a whole, and yet a smaller scale level treats sections of the objects corresponding to the primitives of the representation. Also see scale.

Narrowing rate The rate at which an object is changing width at a medial position. At a point on the medial curve in 2D the narrowing rate is the magnitude of the rate of change of the medial radius per unit step along the medial curve, i.e., $|dr/ds|$. At a point on the medial surface in 3D the narrowing rate is the magnitude of the maximum rate of change of the medial radius over all directions of step, i.e., $|\nabla r|$.

N-connectivity The type of connectivity of two adjacent digital points in 2D and 3D.

Negative curvature As used here, concave curvature in some direction on the tangent plane of a surface. However, in strict mathematical conventions, curvature that is convex is said to have negative sign.

Neighbor atoms For an atom (medial primitive or boundary primitive), those primitives in a grid defining an object that are immediately linked to the reference primitive. For an object, those primitives in other objects with which the reference object is most heavily correlated.

Nonlinear manifold A manifold that is not itself a Euclidean space, i.e., in which the Pythagorean theorem does not hold everywhere.

Null hypothesis See hypothesis testing.

Object angle The angle between a spoke in a skeletal or medial atom and the tangent to the skeletal or medial sheet from which the spoke emanates.

Object-relative coordinates A means of referring to space in terms centered on a specified set of points on the object or objects being represented, typically its (their) boundary, and in coordinate directions also determined by the set of points, e.g., normal and tangent directions.

Object representation A mathematical or computational means of capturing an object. There are three major types: sampled, parametrized, and implicit.

Objective function A function measuring the fitness to some objective. An optimization method is applied to find the optimum of the objective function over its arguments.

1-parameter family Transitions or changes to the local form of the medial axis or symmetry set in 2D are examined in this book under the assumption of a single parameter changing to define a family of curves. In 3D these transitions are examined under the assumption of a single parameter changing along a family of surfaces.

Optimal parameter core A height ridge of medial strength as a function of the parameters of a medial primitive, where the directions of analysis for the subdimensional maximum are a specified subset of the parameters or functions thereof.

Optimization Finding the values of the arguments of an objective function that maximizes or minimizes the function. Typically the desire is to find the global optimum, but frequently a local, i.e., non-global relative optimum is found.

Conjugate gradient optimization. A method of optimization that is quadratically convergent if the initialization is within a region about the optimum in which the objective function is nearly quadratically convex in its variables.

Order of contact An integer indicating up to how many derivatives are in agreement between two curves or two surfaces at a place where they intersect. First order contact means non-tangential intersection (only zeroth derivatives agree), second order means tangency, third order means osculation (tangency and agreement in curvatures), etc.

Orientation A property captured by a unit vector, giving a direction.

Osculation In 2D one curve is said to osculate another at a point if it agrees in both normal (and thus tangent) and curvature at that point. In 3D, a surface is said to osculate another at a point if it agrees in normal and, along some specified direction(s), also in curvature.

Pablo A computer program produced at the University of North Carolina at Chapel Hill, to segment an object or objects by posterior optimization of m-reps.

Binary Pablo. A computer program produced at the University of North Carolina at Chapel Hill to fit an m-rep to a binary image for one or more objects.

Parabolic curve A curve on a surface consisting of points that are locally cylindrical, i.e., have one principal curvature that is zero. Parabolic curves separate surface regions that are convex or concave from those that are hyperbolic, i.e., saddle shaped.

Parametrized representation, parametrization Also spelled parameterized, parameterization. A representation of a geometric entity that is given by a continuous, invertible function of parametric variables on a base object with the same topology. Frequently the function is a linear combination of basis functions.

Partial Blum condition A skeletal structure having the property that all of its spokes are orthogonal to the envelope of its spokes, i.e., the boundary implied by skeletal structure. A medial structure satisfies the partial Blum condition.

Partial test In a hypothesis test on multiple features, a partial test applies a hypothesis test on a subset of the features, often one. The results are then combined to form the overall test.

Permutation test A form of hypothesis testing on two classes in which the sample data from the two classes are pooled on the basis of the assumption of the null hypothesis that there is no difference in the probability distributions for the two classes. In the process, the pooled sample data is repeatedly divided via a random permutation into a purported class 1 and class 2 group. For each permutation, the test statistic, such as the difference between the means of the two classes, is computed, and the distribution of these statistics over the many permutations is used to compute the p-value for a target case.

Pixel An element of a discrete grid forming an image within a 2-space.

Posterior optimization Finding the value of a parameter tuple that has the largest probability or probability density, given the data. The probability distribution of the parameters given the data is called the posterior probability distribution.

Posterior probability The conditional probability of variables to be determined, e.g., describing an object, given data, e.g., an image.

Power crust/diagram/distance/shape These concepts are used to obtain Voronoi-diagram-based medial axis approximations of good quality, where the basic strategy is to guarantee that slivers cannot occur.

Power crust. The power crust is a representation of the surface passing through a set of boundary points. It is obtained by first computing the power shape and then considering the envelope of the union of the Voronoi balls associated with its poles.

Power diagram. The power diagram is a type of generalized Voronoi diagram which is obtained from the following sequence of operations. First the Voronoi diagram of a set of boundary points is computed and its poles are chosen. The medial balls associated with these poles are then selected. The Power diagram is then obtained as the spatial subdivision of the region into cells such that all points within a cell are closest in power distance to a particular ball.

Power distance. The power distance between a point $x \in \mathbb{R}^3$ and a ball of radius ρ centered at v is the square of the Euclidean distance from v to x, minus ρ^2.

Power shape. The power shape is a construct used to provide an approximation to the medial axis. It converges to the true medial axis when the boundary sampling density increases. It is obtained by first labeling the cells in the power diagram according to whether they arise from an interior or an exterior pole. A connection is then made between each pair of interior poles whose power diagram cells are connected. The set of connections between interior poles gives the power shape.

Precision versus recall Given a query to a database, precision is defined as the number of relevant instances retrieved, expressed as a fraction of the total number of instances retrieved. A score of 1.0 indicates that all the instances received were relevant. Recall is defined as the number of relevant instances retrieved expressed as a fraction of the total number of relevant instances in the database. The performance of a system in an information retrieval task is often characterized by plotting the precision at different levels of recall, resulting in a precision versus recall curve. Curves shifted upwards and to the right indicate better performance.

Pre-symmetry set The set of pairs of points along the boundary such that a circle (2D) or a sphere (3D) passes through them and is tangent to the boundary at these locations.

Primal medial atom An order 1 medial atom with its hub at the origin of the coordinate system and both of its spokes of unit length and in direction along the x-axis of the coordinate system.

Principal component Strictly, the coefficient of a principal direction in the expansion of a point in a feature space in a basis of principal directions, or the vector made from the coefficient times the principal direction. The term is sometimes used to mean the principal direction.

Principal component analysis A method of decomposing variation in a collection of samples in a Euclidean space by estimating the mean and covariance matrix and deriving principal directions that are statistically independent, orthogonal principal modes of variation as the eigenvectors of the covariance matrix and principal variances corresponding to the modes as the eigenvalues of the covariance matrix. This approach can be used to fit a Gaussian probability distribution to a collection of samples.

Principal direction A direction on a surface where the direction of the rate of change of the normal (i.e., the direction of swing of the normal) for translation along a tangential direction points in that tangential direction. Also, it is a direction in which the normal curvature has a relative maximum or minimum with respect to direction of motion. There are typically two orthogonal principal directions at a point, but at an umbilic (spherical) point, where both principal curvatures are the same, all directions are principal directions. This term is also used to describe an eigenvector of a covariance matrix, or equivalently a direction in a subspace of a Euclidean feature space that is orthogonal to all previously designated principal directions and over all directions of that subspace the line in that direction has minimal sum of squared distances from the training data points to the line.

Principal edge curvature The single variable generalized eigenvalue of the edge shape operator.

Principal geodesic In a feature space that is a Lie group or a symmetric space and with a sample set of points in that space, a principal geodesic manifold of dimension k is a submanifold which contains the the Fréchet mean of the sample set, in which all geodesics including that mean are also geodesics in the full feature space, and which of all such submanifolds of dimension k is closest in sum of squared geodesic distances to the sample set. The kth principal geodesic is a vector in the principal geodesic manifold of dimension k that is orthogonal at the Fréchet mean to all principal geodesics in the principal geodesic manifold of dimension $k - 1$. The term *principal geodesic* is also used for the projection onto the feature space of a principal direction in the tangent space to that space at the Fréchet mean. In a Euclidean space a principal direction is a principal geodesic, in either sense.

Principal line A line in a principal direction on a surface, i.e., a tangent to an integral curve of principal directions.

Principal modes of variation See principal geodesics.

Principal radial curvature An eigenvalue of the matrix representing the radial shape operator.

Principal radial curve An integral curve of principal radial direction, i.e., where each tangent to the curve is a principal radial direction.

Principal radial direction An eigenvector of the matrix representing the radial shape operator.

Procrustes alignment One geometric entity is Procrustes aligned to another by applying a translation such that the two entities share centers of gravity, a scaling such that the two entities have the same sum of squared distances of reference points in the representation (typically the boundary point samples) to the center of gravity, and a rotation such that after those normalizations the Procrustes distance between the corresponding reference points is minimized.

Procrustes distance between two sequences of points. The sum of squared distances between the corresponding points.

Procrustes initialization. Given the representation of an atlas object, a target object is Procrustes initialized by applying Procrustes alignment of the target object to the atlas object.

Projection Given a subspace of a feature space with a Riemannian metric, the projection of a point in the feature space onto the subspace produces the element of subspace that is closest according to the metric. With a Euclidean space and a Euclidean (flat) subspace with an orthonormal basis, projection of a point occurs by computing dot products between the point's representation and the subspace basis vectors, thereby computing coefficients of those basis vectors.

Proper geometry A representation of a geometric entity not having the property of geometric illegality.

Proper shape See geometric illegality.

Protrusion, protrusion figure A part of an object boundary corresponding to a branch in the medial axis or surface.

p-**value** The statistical significance threshold used in hypothesis testing.

p-**value statistic** A map of a random variable into the probability density value at which the null hypothesis on that random variable would be true at random.

Quantile function A function of cumulative probability giving the value of a scalar random variable for which the probability of that variable being less than the value has the specified argument value. The quantile function is the inverse of the cumulative distribution function.

  Regional intensity quantile function (RIQF). A quantile function on the distribution of intensity in a specified image region.

Quench point Upon erosion of an object boundary at a fixed speed (also called a grassfire), a quench point is where the eroded points emanating from two boundary points meet each other. Also see shock.

Quotient of Lie groups The quotient A/B of Lie groups A and B is the set of transformations in A modulo B, i.e., the set of nonequivalent transformations in A, where two transformations in A are equivalent if one can be obtained from the other by composition with some transformation in B. An example is the set of directions in 3D, which is formed as the set of frame rotations in 3D modulo the set of of rotations in the plane of the first two vectors in the frame.

Radial curvature See principal radial curvature.

Radial direction See principal radial direction.

Radial distance function A function on the medial curves or surfaces to a scalar giving the common distance to two or more nearest points (or a point of multiplicity 2) on the associated object boundary. For the associated medial atom, this distance is the common size of the spokes of the atom. For a skeletal curve or surface the function may give different values for the different spokes at the point. Also called the radius function.

Radial flow Flow outward from medial or skeletal surface points along their spokes at a speed proportional to the radial distance (spoke length). Thus at any point in time the front of the flow is at an equal proportion of every spoke's length, and conventionally at time 1 the flow is to the object boundary.

Radial geometry

  Edge shape operator. At end curves to a medial sheet, the edge shape operator S_E substitutes for the radial shape operator in describing spoke direction change. The substitution is necessary because of the tangency of radial vectors to the medial surface at these points.

  Radial shape operator. For a unit radial (spoke) vector field on one side of the medial sheet, the radial shape operator S_{rad} captures a the rate of change of the spoke direction unit vector in tangential directions on the medial sheet.

  Radial vector field. The field of vectors (also called spoke vectors) from points on the medial surface to the corresponding points on the object boundary where the maximal sphere at the medial point is tangent. A useful way to understand this field is as a function from positions on the double of M to spokes.

Radial line A line segment along a radial vector (spoke) of a medial atom.

Radial map The 1-up-to-4-valued function from points on a medial curve or surface to the corresponding points on the implied boundary, i.e., at the ends of the spokes in the medial atom there.

Radial parabolic curve A curve on the 2-sided skeletal surface at each point of which one of the principal radial curvatures is zero.

Radial scaling factor The multiplier by which a medial radius is lengthened or shortened under a medial deformation.

Radial umbilic point A point on the 2-sided skeletal surface (double of M) in 3D where the two principal radial curvatures, i.e., the eigenvalues of the radial shape operator, are equal.

Radius function See radial distance function.

Reaction-diffusion equation A partial differential equation which describes the evolution of the boundary of an object in the direction of its normal by a speed function that is a combination of a term that is independent of the local shape of the boundary (the reaction term) and one that is a function of the local curvature (the diffusion term). The reaction term when used on its own can give rise to singularities or shocks while the diffusion term smooths the boundary. A typical example is a linear combination of constant motion (grassfire flow) and motion proportional to the local curvature.

Regional intensity summary A means of representing the collection of image intensities in an image region. The three most common representations are a voxel by voxel list of intensities, a histogram of intensities, and a quantile function of intensities.

Regularity In a graph of geometric primitives irregularity is a measure of the differences of each primitive from the mean of its neighbors.

Relative geometry Geometric properties using groupings of points on an object boundary via their correspondence to points on the medial surface. Such geometry captures such notions as thickness of an object or regions of a local bulge of an object.

Relative critical set For a scalar function on a surface in 3D, the relative critical set is the set of points and curves where the function has a relative maximum or relative minimum in one of the eigendirections of the Hessian of the function.

Reparametrization Changing the function used in parametrization. The effect is to change the correspondence between locations on the target manifold and those on other instances of the manifold.

Residue The geodesic difference between a desired value of a geometric entity and its value computed to date, typically down to a present scale.

Ridge point A point of a surface where there is a sphere for which the contact function has type A_3. When a sphere has ordinary contact at two points with a surface and the sphere changes so that these two points come into coincidence, the point of coincidence is a ridge point.

Riemannian distance See Riemannian geometry.

Riemannian geometry The study of smooth manifolds with an inner product, and thus a metric, on the tangent space at each point which varies smoothly from point to point. Based on this metric, standard geometric concepts, such as area and length, are defined on the manifold.

Riemannian gradient The gradient in a space with a Riemannian metric.

r-regular shape An open set is said to be r-regular if it can be opened and closed using mathematical morphology, with a structuring element that is a disk of radius r.

Salience, saliency See significance measure.

Sample mean The estimated mean of a random variable from a sample set of values of the variable. Typically the sample mean is computed as the Fréchet mean or, for variables that are vectors in a linear feature space, as the average of the vectors.

Sampling In this book sampling refers to evaluating geometric features at discrete positions from a continuous locus of these features.

Scale An entity with spatial aspects is measured by features given not only at specified positions but also at specified scale values. There are two types of scale used in this book. The first describes the size of spatial details captured. The second refers to the aperture within which measurements are made, i.e., through a summary of nearby feature values via weights defined by the aperture centered at the measurement position.

Scale-level See multiscale.

Sculpting The process of removing a sub-part of an object.

Shape operator An operator, introduced in differential geometry, that given any vector in the tangent plane of a curve or surface yields certain components of the derivatives of the vector on that curve (in 2D) or surface (in 3D). Typically this term is used to describe the operator yielding derivatives of the normal to the curve or surface, but the radial shape operator generalizes this notion to spokes on a skeletal curve or surface.

Sheet of medial atoms A 2-manifold of medial atoms, with boundary and possibly nonsmoothnesses, at fin curves.

Second fundamental form A symmetric, bilinear function of two vectors at a point on a surface which describes the rate of change of the normal, or equivalently the tangent plane, at the point. One of the vector arguments gives a velocity vector in the tangent plane, and the other gives a direction along which the rate of change is measured. This form is equivalent to the shape operator.

Shape Sometimes used to mean any tuple describing the geometry of an entity. Sometimes used to mean such tuples that do not vary with translation, rotation, or uniform scaling of the entity.

Change independent of neighbors. Changes in a target geometric entity such that the conditional probability distribution on those changes given neighbor entities' changes is equal to the unconditional probability distribution on those target entity changes.

Interrelationship with neighbors. The residue of changes in a geometric entity after the changes independent of neighbors are removed.

Shape discrimination Determination of the existence of differences between two populations of geometric objects. Typically, not only whether there is any difference but also locations and geometric types of difference are of interest.

Shape distinctions Differences in geometric features between classes.

Shape space A subspace of the feature space of geometric entities. Frequently it is chosen as the space spanned by the principal directions (geodesics) with dominant eigenvalues.

Shock This term arises in gas dynamics when two gases of different densities meet to form a discontinuity of the density function at their interface. In this book the term is used to represent the collision of distinct boundary points under the action of the grassfire flow, where the gradient of the underlying Euclidean distance function (the level curves of the flow) has a discontinuity. The locus of shocks gives the medial axis.

Shock graph A graph-based hierarchical representation of medial axis points in 2D. There are two common interpretations in the literature. In the first the classification of local forms is used to define nodes consisting of isolated medial axis points with curves of A_1^2 acting as links between them. In the second the derivatives of the radius function along the medial axis is used to identify minima, maxima and monotonically varying segments to provide a further classification into types and links formed between connected medial axis points.

Segmentation Extraction of the region of or boundary in an image corresponding to a desired object or objects.

Significance map At each location the map gives the p-value for which the null hypothesis for the random variable at that position holds at random.

Significance measure This term has two meanings in this book. In the first use the term provides a measure of the importance or saliency of a branch or subtree of a computed medial locus. In the second use, a positive value less than 1.0, typically slightly greater than zero, determines the stringency of a hypothesis test. Also see significance, statistical.

Significance, statistical The threshold chosen for hypothesis testing. The null hypothesis is rejected if the probability of the differences observed happening at random under that hypothesis is less than the threshold.

Similarity transform A geometric transform composed of translations, rotations, and uniform scaling.

Simple point In digital topology this term is used to denote a pixel (2D) or a voxel (3D) in a cubic lattice that can be removed from a digital object without altering its homotopy type. In other words, the number of connected components, holes and cavities is not changed when such a point is removed.

Simple surface A surface comprised of simple points.

Singularity A singularity is a location where a function is not differentiable. In this book a typical use of this term is in reference to the singularities or shocks of the Euclidean distance function. Also see shock.

6-junction point A generic point on certain medial loci in 3D, in which 6 surfaces meet non-smoothly. Four branch curves meet at such a point.

Skeletal hierarchy A coarse-to-fine organization of skeletal segments with the more salient ones being higher up in the hierarchy.

Skeletal integral An integral of a function in a region in the interior or on a region on the boundary of a skeletally represented object, where the integral is expressed as integrals along the skeletal curve or surface. In the case of integrals over the interior, the integrand for any point on the skeletal curve or surface involves integrals along the corresponding spoke or spokes.

Skeletal structure A generalization of the medial structure. The medial axis is replaced by a collection of piecewise smooth skeletal sheets. The radial vectors (spokes) in the skeletal structure need not meet a variety of medial conditions, in particular, that the spokes with a common hub have the same length, that the skeletal sheet bisects those two spokes, and that the boundary formed by the envelope of the spokes is orthogonal to the spokes. An m-rep is a special case of a skeletal structure.

Skeleton Any curve or sheet from which non-crossing spokes to an object boundary emanate at each point. A medial locus is one form of skeleton.

Skeletonization Computing a skeleton from an object boundary.

Slab An object or figure in 3D shaped such that its medial locus is a sheet.

Slivers In Voronoi-method-based skeletonization in 3D the Delaunay tetrahedralization of a sample of points can contain tetrahedra that are very flat, resulting in Voronoi vertices that are far from the medial axis. These tetrahedra are called slivers;

suitable methods to remove them from consideration have to be developed. Also see power crust.

Spatial scale See scale.

Spectral graph theory The study of the topology of a graph by creating appropriate low dimensional representations. One such representation in computer science theory and applications is the eigenvalue decomposition of the adjacency matrix. Also see adjacency matrix.

Spherical harmonics A set of orthonormal basis functions from the sphere to any object with spherical topology. They are typically ordered by a frequency in each of the two dimensions of the argument of the spherical harmonics.

Spherical topology An object with a closed surface has spherical topology if it has no through holes or cavities.

Spline An approximating function or manifold made from typically polynomial patches which satisfy some smoothness constraint where the patches meet. A B-spline is based on a particular set of basis polynomials that, for each patch, are respectively multiplied by the control points associated with that patch.

Splines of medial atoms See continuous m-reps.

Spoke In a skeletal or medial atom a spoke is a vector going from the skeletal or medial locus to the corresponding object boundary.

Spoke end The terminus of a medial atom's spoke at the object boundary.

Spoke-length distance metric A distance along the spoke of a skeletal or medial atom given as a multiple of the length of the spoke.

Statistics on image intensities A learned probability distribution on a tuple describing image intensity patterns in one or more, typically object-relative image regions. The probability distribution may be parametrized, e.g., a Gaussian, or nonparametric, e.g., a histogram or quantile function. Examples are tuples of voxel intensities and tuples of quantile functions, each of which is a tuple of intensity values at the respective quantiles.

Stratified set See Whitney stratified set.

Subdivision surface method A method for evaluating a fine grid of sample locations of a surface, or in the limit, an infinitesimally fine grid, from an initial, coarse Cartesian grid of samples. The method operates by successively refining the grid spacing, typically by 2 in each dimension, and computing the sample values in the refined grid as weighted averages of the nearby sample values in the previous level of grid spacing.

Catmull-Clark subdivision. A particular subdivision surface method that converges to a B-spline whose control points are the original sample.

Subfigure A figure representing a protrusion adding to a host figure or an indentation subtracting from the host figure.

Subgraph isomorphism See maximal common subgraph.

Symmetric axis Another term for the medial axis.

Symmetric space A feature space with the abstract structure that it has a Riemannian metric and at each point A the mapping that reverses the geodesics from A to any point B is an isometry, i.e., does not change distances. Examples are the feature space of translations and rotations in Euclidean space and Cartesian products thereof.

Symmetry-curvature duality A theorem relating a section of a curve with exactly one curvature extremum with exactly one symmetry axis which terminates at the corresponding center of curvature.

Symmetry set A set of sphere centers and radii for the set of all bitangent spheres to an object boundary. These spheres need not be entirely or even partially interior to the object.

Symmetry, spherical or circular A function of position that has the property that points equidistant from a specified center point have the same function value.

Tangent hyperplane or plane The locally closest fitting linear manifold fit at a point to another, typically nonlinear manifold.

Target object An object being sought in an image, or an object on which a probability distribution is computed.

t-distribution A probability distribution on the difference of sample means between two Gaussian distributed random variables.

Template image An image to which a target image is compared in a registration or segmentation task, after the candidate geometric transformation has been applied.

Thinning Since some methods for computing a medial locus as a collection of voxels include regions where the locus is more than one voxel thick, thinning is a post-process that leaves the locus one voxel thick.
   Topology preserving thinning. Thinning such that the thinned object has the same homotopy type as the original object.

Thin-plate spline interpolation A method of producing a not necessarily diffeomorphic but smooth warp that matches landmarks and minimizes a bending energy summing the Frobenius norms of the deformation in x, of the deformation in y, and of the deformation in z.

Tile, mesh of tiles A tile is a surface interpolated from line segments connecting a cycle of vertices in 3-space. A mesh of such tiles, i.e., of connected vertices sampling an object boundary (or medial surface), forms an approximation to that object boundary (or medial surface).

Topological signature vector (TSV) A low dimensional characterization of the topology of the subgraph rooted at a particular node of a directed acyclic graph. It is obtained by considering all the children of that node, and for each computing the sum of the k largest eigenvalues of the adjacency matrix of the subgraph rooted at them. These sums are then sorted to form the TSV.

Topological skeleton In the context of Voronoi-based skeletonization this refers to that part of the skeleton that cannot be further reduced by removing elements without topological changes occurring to it.

Topology In mathematics the topology of an object refers to spatial aspects that are maintained under all smooth, invertible deformations (diffeomorphisms). The major property of interest in this book is the number of through holes (and cavities) in the object. An object with spherical topology has no through holes and no cavities.

Training population A collection of instances of an object or an image used to estimate a probability distribution on the object geometry or the image.

Transitions Changes or transformations from one local form to another that occur on the medial axis or symmetry set along a 1-parameter family of curves (in 2D) or surfaces (in 3D).

Tri-orthogonal display A form of display of a 3D image in which intensity values on 1-3 of three orthogonal planes are shown. Usually, the planes encompass cardinal directions and each plane's position along its normal is interactively chosen.

Trough A curve in the n-dimensional argument space of a function where the function is a relative minimum in a specified $n-1$ dimensional subspace. The trough depends on the definition of the subspace. Also called a valley.

Tube In most of this book this term refers to a class of objects in 3D made by sweeping a circle, of possibly continuously varying radius, along a curve, which is the medial axis. The circles are bases of cones whose axes are tangent to a medial curve. In chapter 3 the term is also used, as mathematicians have done, to refer to a generalized annulus of constant radius r with respect to a manifold.

Typicality, geometric A measure of how far geometric features of an entity differs from the expected features of that entity.

Umbilic point A point on a surface where both principal curvatures are the same, i.e., the point is infinitesimally like a sphere, either convex or concave.

Unimodal distribution A probability distribution whose density function has a single relative maximum.

Valley A form of trough in which the directions of minimization of the function are orthogonal to its gradient.

Velocity field A function on a stratified set that assigns a smoothly varying velocity to each point in the set.

Voronoi diagram Given a set of elements made from points and/or curves or patches, the Voronoi diagram assigns to each position in space the element or elements that it is closest to.

Voronoi edge. In 2D a Voronoi edge is a curve segment made of the set of elements of the Voronoi diagram that are equidistant to some specific pair of elements of the set.

Voronoi face. In 3D a Voronoi edge is a surface patch made of the set of elements of the Voronoi diagram that are equidistant to some specific pair of elements of the set.

Voronoi region. The set of points closer to one member of the set of elements than any other.

Voronoi skeleton. A skeleton computed as a subset of the Voronoi edges or faces.

Voronoi vertex. A point which lies at the intersection of two Voronoi edges.

Voxel An element of a discrete grid forming an image within a 3-space.

Warp A piecewise smooth deformation of a physical space, typically 2-space or 3-space. Typically used in registration, some warps are restricted to be diffeomorphisms.

Whitney stratified set A space formed from a collection of manifolds, possibly of varying dimensions, which are joined so that certain regularity conditions are satisfied. An example is the Blum medial axis for a generic region, which in 3D consists of surfaces, branch and end curves, and curve endpoints and junction points.

Y-junction curve The intersection of 3 medial sheets in 3D to form what resembles a Y-shaped collection of curves.

Y-junction point A branchpoint on the 2D medial axis, where 3 smooth curves meet to form a structure resembling a Y.

Amenta, N., T. J. Peters, and A. Russell: 2003, 'Computational Topology: Ambient Isotopic Approximation of 2-Manifolds'. *Theoretical Computer Science* **305**(1), 3–15. [233]

Anderson, T. W.: 1958, *An Introduction to Multivariate Statistical Analysis*. New York: Wiley. [295]

Angelidis, A., P. Jepp, and M.-P. Cani: 2002, 'Implicit Modeling with Skeleton Curves: Controlled Blending in Contact Situations'. In: *International Conference on Shape Modeling and Applications (SMI'02)*. Banff, Alberta, Canada, pp. 137–144. [337]

Ankerst, M., G. Kastenmüller, H. Kriegel, and T. Seidl: 1999, '3-D Shape Histograms for Similarity Search and Classification in Spatial Databases'. In: *Advances in Spatial Databases, 6th International Symposium*, Vol. 18, pp. 700–711. [312, 319, 320]

Arcelli, C. and G. Sanniti di Baja: 1986, 'Computing Voronoi Diagrams in Digital Pictures'. *Pattern Recognition Letters* **4**(5), 383–389. [209]

Arcelli, C. and G. Sanniti di Baja: 1988a, 'Finding Local Maxima in a Pseudo-Euclidean Distance Transform'. *Computer Vision, Graphics, and Image Processing* **43**, 361–367. [164, 166]

Arcelli, C. and G. Sanniti di Baja: 1988b, 'Weighted Distance Transforms: A Characterization'. In: *Image Analysis and Processing*, Vol. II. Plenum Press, New York, pp. 205–211. [164]

Arcelli, C. and G. Sanniti di Baja: 1992, 'Ridge Points in Euclidean Distance Maps'. *Pattern Recognition Letters* **13**(4), 237–243. [128]

Armstrong, C. G., S. Bridgett, R. Donaghy, R. McCune, R. McKeag, and D. Robinson: 1998, 'Techniques for Interactive and Automatic Idealisation of CAD Models'. In: *Numerical Grid Generation in Computational Field Simulations*. pp. 643–662. [341]

Arnheim, R.: 1974, *Art and Visual Perception: A Psychology of the Creative Eye*. Berkeley, CA: University of California Press. New version; expanded and revised edition of the 1954 original. [336, 337, 349]

Arnold, V. I.: 1989, *Mathematical Methods of Classical Mechanics*. Berlin: Springer. [130, 132]

Asada, H. and M. Brady: 1986, 'The Curvature Primal Sketch'. *IEEE Transactions on Pattern Analysis and Machine Intelligence* **8**(1), 2–14. [5, 194]

Attali, D.: 1995, 'Squelettes et Graphes de Voronoi 2D et 3D'. Ph.D. thesis, Université Joseph Fourier, Grenoble. [212, 214]

Attali, D.: 1997, 'r-Regular shape reconstruction from unorganized points'. *Computational Geometry: Theory and Applications* **10**(4), 248–273. [226]

Attali, D. and J.-D. Boissonnat: 2003, 'Complexity of the Delaunay Triangulation of Point on Polyhedral Surfaces'. *Discrete and Computational Geometry* **30**, 437–452. [232]

Attali, D. and A. Montanvert: 1994, 'Semicontinuous Skeletons of 2D and 3D Shapes'. In: C. Arcelli, L. Cordella, and G. Sanniti di Baja (eds.): *Aspects of Visual Form Processing*. World Scientific, Singapore, pp. 32–41. [201, 208, 209, 210, 212]

Attali, D. and A. Montanvert: 1996, 'Modeling Noise For a Better Simplification of Skeletons'. In: *International Conference on Image Processing*, Vol. 3. pp. 13–16. [226]

Attali, D. and A. Montanvert: 1997, 'Computing and Simplifying 2D and 3D Continuous Skeletons'. *Computer Vision and Image Understanding* **67**(3), 261–273. [234, 346]

Attali, D., G. Sanniti di Baja, and E. Thiel: 1995, 'Pruning Discrete and Semi-continuous Skeletons'. In: C. Braccini, L. DeFloriani, and G. Vernazza (eds.): *Image Analysis and Processing (Proc. ICIAP'95)*, Vol. 974 of *Lecture Notes in Computer Science*. Springer: Sanremo, Italy, pp. 488–493. [174]

Attali, D., J.-D. Boissonnat, and H. Edelsbrunner: 2008, 'Stability and Computation of the Medial Axis: A State-of-the-Art Report'. In: T. Möller, B. Hamann, and B. Russell (eds.): *Mathematical Foundations of Scientific Visualization, Computer Graphics, and Massive Data Exploration*. Springer, http://www.springer.com/math/cse/book/978-3-540-25076-0. [224]

Attali, D., G. Sanniti di Baja, and E. Thiel: 1997, 'Skeleton Simplification Through Nonsignificant Branch Removal'. *Image Processing and Communications* **3**(3–4), 63–73. [32]

Attneave, F.: 1954, 'Some Informational Aspects of Visual Perception'. *Psychological Review* **61**, 183–193. [194]

August, J., K. Siddiqi, and S. W. Zucker: 1999, 'Ligature Instabilities in the Perceptual Organization of Shape'. *Computer Vision and Image Understanding* **76**(3), 231–243. [6]

Aylward, S. R. and E. Bullitt: 2002, 'Initialization, Noise, Singularities, and Scale in Height Ridge Traversal for Tubular Object Centerline Extraction'. *IEEE Transactions on Medical Imaging* **21**(2), 61–75. [34]

Bader, R. F. W.: 1990, *Atoms in Molecules — A Quantum Theory*, Vol. 22 of *International Series of Monographs on Chemistry*. Oxford: Clarendon Press. Oxford University Press; Reprint edition (June, 1994). [328, 347]

Bajaj, C.: 2007, 'Geometric Modeling and Quantitative Visualization of Virus Ultrastructure'. In: M. Laubichler and G. B. Müller (eds.): *Modeling Biology: Structures, Behaviors, Evolution*. MIT Press, Cambridge, MA, USA. [346]

Bajaj, C., Z. Yu, and M. Auer: 2003, 'Volumetric Feature Extraction and Visualization of Tomographic Molecular Imaging'. *Journal of Structural Biology* **144**(1–2), 132–143. [346, 347]

Ban, Y.-E. A., H. Edelsbrunner, and J. Rudolph: 2004, 'Interface Surfaces for Protein-Protein Complexes'. In: *Proceedings of the 8th International Conference on Research in Computational Molecular Biolog*. San Diego, CA, pp. 205–212. [348]

Batty, M. and S. Rana: 2004, 'The Automatic Definition and Generation of Axial Lines and Axial Maps'. *Environment and Planning B* **31**(4), 615–640. [333]

Beil, M., H. Braxmeier, F. Fleischer, V. Schmidt, and P. Walther: 2005, 'Quantitative Analysis of Keratin Filament Networks in Scanning Electron Microscopy Images of Cancer Cells'. Submitted for publication. [346]

Betelu, S., G. Sapiro, A. Tannenbaum, and P. Giblin: 2000, 'Noise Resistant Affine Skeletons of Planar Curves'. In: *Proceedings of the European Conference on Computer Vision*. Dublin, Ireland, pp. 742–754. [6]

Biederman, I.: 1987, 'Recognition–by–components: A Theory of Human Image Understanding'. *Psychological Review* **94**(2), 115–147. [21, 313]

Binford, T. O.: 1971, 'Visual Perception by Computer'. In: *IEEE Conference on Systems and Control*. [313]

Bitter, I., A. E. Kaufman, and M. Sato: 2001, 'Penalized-Distance Columetric Skeleton Algorithm'. *IEEE Transactions on Visualization and Computer Graphics* **7**(3), 195–206. [343]

Björner, A.: 1995, 'Topological Methods'. In: R. Graham, M. Grötschel, and L. Lovász (eds.): *Handbook of Combinatorics*. North Holland, Amsterdam: Elsevier, pp. 1819–1872. [233]

Blanding, R., C. Brooking, M. Ganter, and D. Storti: 1999, 'A Skeletal-Based Solid Editor'. In: W. F. Bronsvoort and D. C. Anderson (eds.): *Proceedings of the Fifth Symposium on Solid Modeling and Applications (SSMA-99)*. ACM Press: New York, pp. 141–150. [7, 341]

Blanding, R., G. Turkiyyah, D. Storti, and M. Ganter: 2000, 'Skeleton-based Three-dimensional Geometric Morphing'. *Journal of Computational Geometry: Theory and Applications* **15**(1–3), 129–148. [341]

Bloomenthal, J. and K. Shoemake: 1991, 'Convolution Surfaces'. *Computer Graphics (SIGGRAPH '91 Proceedings)* **25**(4), 251–256. [7]

Blum, H.: 1967, 'A Transformation for Extracting New Descriptors of Shape'. In: W. Wathen-Dunn (ed.): *Models for the Perception of Speech and Visual Form*. Cambridge, MA: MIT Press. [1, 4, 7, 13, 224, 329]

Blum, H.: 1973, 'Biological Shape and Visual Science'. *Journal of Theoretical Biology* **38**, 205–287. [135, 200, 313, 315, 326, 327, 342]

Blum, H. and R. Nagel: 1978, 'Shape Description Using Weighted Symmetric Axis Features'. *Pattern Recognition* **10**(3), 167–180. [7, 13, 14, 86]

Bogaevsky, I. A.: 1990, 'Metamorphoses of Singularities of Minimum Functions, and Bifurcations of Shock Waves of the Burgers Equation with Vanishing Viscosity'. *St Petersburg (Leningrad) Mathematics Journal* **1**, 807–823. [38, 58]

Bogaevsky, I. A.: 2002, 'Perestroikas of Shocks and Singularities of Minimum Functions'. *Physica D* **173**, 1–28. [38, 58]

Boissonat, J. and M. Teillaud: 1989, 'On the Randomized Construction of the Delaunay Tree'. Technical Report TR-1140, INRIA. [208]

Boissonnat, J.-D. and F. Cazals: 2001, 'Natural Coordinates of Points on a Surface'. *Computational Geometry: Theory and Applications* **19**, 155–173. [233, 239]

Boissonnat, J.-D., O. Devillers, J. Duquesne, and M. Yvinec: 1994, 'Computing Connolly Surfaces'. *Journal of Molecular Graphics* **12**(1), 61–62. [348]

Bookstein, F. L.: 1991, *Morphometric Tools for Landmark Data*. Cambridge: Cambridge University Press. [2, 287]

Borgefors, G.: 1984, 'Distance Transformations in Arbitrary Dimensions'. *Computer Vision, Graphics, and Image Processing* **27**, 321–345. [136, 138, 146, 312]

Borgefors, G.: 1986, 'Distance Transformations in Digital Images'. *Computer Vision, Graphics, and Image Processing* **34**, 344–371. [161]

Borgefors, G.: 1991, 'Another Comment on "A Note on Distance Transformations in Digital Images"'. *CVGIP: Image Understanding* **54**(2), 301–306. [161]

Borgefors, G.: 1993, 'Centres of Maximal Discs in the 5-7-11 Distance Transform'. In: *Proceedings of 8th Scandinavian Conference on Image Analysis.* Tromsø, Norway, pp. 105–111. [164]

Borgefors, G.: 1994, 'Applications Using Distance Transforms'. In: C. Arcelli, L. P. Cordella, and G. Sanniti di Baja (eds.): *Aspects of Visual Form Processing.* Singapore: World Scientific, pp. 83–108. [161, 187]

Borgefors, G.: 1996, 'On Digital Distance Transforms in Three Dimensions'. *Computer Vision and Image Understanding* **64**(3), 368–376. [161]

Borgefors, G. and I. Nyström: 1997, 'Efficient Shape Representation by Minimizing the Set of Centres of Maximal Discs/Spheres'. *Pattern Recognition Letters* **18**, 465–472. [166]

Borgefors, G., T. Hartmann, and S. L. Tanimoto: 1990, 'Parallel Distance Transforms on Pyramid Machines: Theory and Implementation'. *Signal Processing* **21**, 61–86. [164]

Borgefors, G., I. Ragnemalm, and G. Sanniti di Baja: 1991, 'The Euclidean Distance Transform: Finding the Local Maxima and Reconstructing the Shape'. In: *Proceedings of 7th Scandinavian Conference on Image Analysis.* Aalborg, Denmark. pp. 974–981. [165, 166]

Borgefors, G., I. Nyström, and G. Sanniti di Baja: 1996, 'Surface Skeletonization of Volume Objects'. In: P. Perner, P. Wang, and A. Rosenfeld (eds.): *Proceedings of SSPR'96 - Leipzig: Advances in Structural and Syntactical Pattern Recognition,* Vol. 1121 of *Lecture Notes in Computer Science.* Berlin/Heidelberg: Springer, pp. 251–259. [177]

Borgefors, G., I. Nyström, and G. Sanniti di Baja: 1997, 'Connected Components in 3D Neighbourhoods'. In: *Proceedings of 10th Scandinavian Conference on Image Analysis.* Lappeenranta, Finland. pp. 567–572. [178]

Borgefors, G., I. Nyström, and G. Sanniti di Baja: 1999, 'Computing Skeletons in Three Dimensions'. *Pattern Recognition* **32**(7), 1225–1236. [177, 181, 337]

Borgefors, G., I. Nyström, G. Sanniti di Baja, and S. Svensson: 2000, 'Simplification of 3D Skeletons using Distance Information'. In: L. J. Latecki, R. A. Melter, D. M. Mount, and A. Y. Wu (eds.): *Vision Geometry IX.* pp. 300–309. [181, 182]

Borgefors, G., G. Ramella, and G. Sanniti di Baja: 2001, 'Hierarchical Decomposition of Multiscale Skeletons'. *IEEE Transactions on Pattern Analysis and Machine Intelligence* **23**(11), 1296–1312. [188]

Bouix, S., J. Pruessner, D. L. Collins, and K. Siddiqi: 2005a, 'Hippocampal Shape Analysis Using Medial Surfaces'. *Neuroimage* **25**(4), 1077–1089. [292]

Bouix, S., K. Siddiqi, and A. R. Tannenbaum: 2005b, 'Flux Driven Automatic Centerline Extraction'. *Medical Image Analysis* **9**(3), 209–221. [34]

Brand, S.: 2005, 'Environmental Heresies'. *TechnologyReview.com.* [331]

Brandt, J. W. and V. R. Algazi: 1992, 'Continuous Skeleton Computation by Voronoi Diagram'. *Computer Vision, Graphics, and Image Processing* **55**(3), 329–338. [196, 203, 207, 212, 225]

Braun, M.: 1992, *Picturing Time*. Chicago, IL: Chicago University Press. [338]

Braunstein, M. L., D. D. Hoffman, and A. Saidpour: 1989, 'Parts of Visual Objects: An Experimental Test of the Minima Rule'. *Perception* **18**, 817–826. [21]

Brechbühler, C., G. Gerig, and O. Kübler: 1995, 'Parameterization of Closed Surfaces for 3-D Shape Description'. *Computer Vision, Graphics, and Image Processing: Image Understanding* **61**, 195–170. [302]

Broadhurst, R., J. Stough, S. Pizer, and E. Chaney: 2006, 'A Statistical Appearance Model Based on Intensity Quantile Histograms'. In: *IEEE International Symposium on Biomedical Imaging (ISBI)*. pp. 422–425. [282, 289]

Broadhurst, R. E., J. Stough, S. M. Pizer, and E. L. Chaney: 2005, 'Histogram Statistics of Local Model-Relative Image Regions'. In: O. F. Olsen, L. Florack, and A. Kuijper (eds.): *International Workshop on Deep Structure, Singularities and Computer Vision*, Vol. LNCS. pp. 71–82. [276, 283]

Bruce, J. W. and P. J. Giblin: 1986, 'Growth, Motion and 1-Parameter Families of Symmetry Sets'. *Proceedings of the Royal Society of Edinburgh* **104A**, 179–204. [55]

Bruce, J. W. and P. J. Giblin: 1992, *Curves and Singularities*. Cambridge: Cambridge University Press, 2nd edition. [39, 328]

Bruce, J. W., P. J. Giblin, and C. G. Gibson: 1985, 'Symmetry Sets'. *Proceedings of the Royal Society of Edinburgh* **101**(A), 163–186. [5, 49]

Bruce, J. W., P. J. Giblin, and F. Tari: 1996, 'Ridges, Crests and Suparabolic Lines of Evolving Surfaces'. *International Journal of Computer Vision* **18**(3), 195–210. [88]

Bruck, J., J. Gao, and A. A. Jiang: 2005, 'MAP: Medial Axis based Geometric Routing in Sensor Networks'. In: *11th Annual International Conference on Mobile Computing and Networking (MobiCom'05)*. Cologne, Germany, pp. 88–102. [331, 332]

Bruzzone, E., L. De Floriani, and E. Puppo: 1990, 'Reconstructing Three-Dimensional Shapes through Euler Operators'. In: *Progress in Image Analysis and Processing*. World Scientific, Singapore. [208]

Burbeck, C. A. and S. M. Pizer: 1995, 'Object Representation by Cores: Identifying and Representing Primitive Spatial Regions'. *Vision Research* **35**(13), 1917–1930. [23]

Burbeck, C. A., S. M. Pizer, B. S. Morse, D. Ariely, G. S. Zauberman, and J. Rolland: 1996, 'Linking Object Boundaries at Scale: A Common Mechanism for Size and Shape Judgements'. *Vision Research* **36**(3), 361–372. [23, 24]

Burton, E.: 1995, 'Thoughtful Drawings: A Computational Model of the Cognitive Nature of Children's Drawing'. In: F. Post and M. Göbel (eds.): *EUROGRAPHICS'95*. Vol. 14. Oxford: Blackwell Publishers, pp. C–159–C–170. [337]

Cain, J. E., T. Nion, D. Jeulin, and J. F. Bertram: 2005, 'Exogenous BMP-4 Amplifies Asymmetric Ureteric Branching in the Developing Mouse Kidney in vitro'. *Kidney International* **67**(2), 420–431. [345]

Calabi, L.: 1965a, 'On the Shape of Plane Figures'. Technical Report PML No 60429 SR-1, Parke Mathematical Laboratories. [9]

Calabi, L.: 1965b, 'A Study of the Skeleton of Plane Figures'. Technical Report PML No 60429 SR-2, Parke Mathematical Laboratories. [9]

Calabi, L. and W. Hartnett: 1968, 'Shape Recognition, Prairie Fires, Convex Deficiencies and Skeletons'. *American Mathematical Monthly* **75**, 335–342. [9]

Car, A.: 1997, *Hierarchical Spatial Reasoning: Theoretical Consideration and its Application to Modeling Wayfinding*, Vol. 10 of *GeoInfo Series*. Austria: Department of Geoinformation, Technical University of Vienna. [332]

Catmull, E. and J. Clark: 1978, 'Recursively Generated B-spline Surfaces on Arbitrary Topological Meshes'. *Computer Aided Design* **10**, 183–188. [282]

Cayley, A.: 1859, 'On Contour and Slope Lines'. *The London, Edinburgh and Dublin Philosophical Magazine and Journal of Science* **18**(120), 264–268. [329]

CGAL, C.: 2004, 'CGAL 3.1-Computational Geometry Algorithms Library'. http://www.cgal.org. [209, 225]

Chang, M.-C., F. F. Leymarie, and B. B. Kimia: 2004, '3D Shape Registration using Regularized Medial Scaffolds'. In: *Proceedings of the Symposium on 3D Data Processing, Visualization and Transmission*. IEEE Computer Society Press, Thessaloniki, Greece, pp. 987–994. [51, 65, 342, 346]

Chapman, M. S. and M. L. Connolly: 2001, 'Molecular Surfaces: Calculations, Uses and Representations'. In: *International Tables for Crystallography*, Vol. F: Crystallography of Biological Macromolecules of *International Tables for Crystallography*. International Union of Crystallography, Chester, Kluwer. [348]

Chaturvedi, A. and Z. Lee: 2005, 'Three-Dimensional Segmentation and Skeletonization to Build an Airway Tree Data Structure for Small Animals'. *Physics in Medicine and Biology* **50**(7), 1405–1419. [345]

Chazal, F. and A. Lieutier: 2005a, 'The Lambda Medial Axis'. *Graphical Models* **67**(4), 304–331. [229]

Chazal, F. and A. Lieutier: 2005b, 'Weak feature size and Persistent Homology: Computing Homology of Solids in R^n from Noisy Data Samples'. In: *Proceedings of the 21st Annual ACM Symposium on Computational Geometry*. [230]

Christensen, G. E., S. C. Joshi, and M. I. Miller: 1997, 'Volumetric Transformation of Brain Anatomy'. *IEEE Transactions on Medical Imaging* **16**(6), 864–877. [287]

Cignoni, P., C. Montani, and R. Scopigno: 1992, 'A Merge-First Divide and Conquer Algorithm for E^d Delaunay Triangulation'. Technical Report, CNUCE. [209]

Clarkson, K.: 1992, 'Hull - A Program for Convex Hulls'. http://cm.bell-labs.com/netlib/voronoi/hull.html. [209, 225]

Coeurjolly, D.: 2003, 'd-Dimensional Reverse Euclidean Distance Transformation and Euclidean Medial Axis Extraction in Optimal Time'. In: I. Nyström, G. Sanniti di Baja, and S. Svensson (eds.): *Proceedings of Discrete Geometry for Computer Imagery (DGCI 2003)*, Vol. 2886 of *Lecture Notes in Computer Science*. Springer, Naples, Italy, pp. 327–337. [166]

Cohen, J. D., M. C. Lin, D. Manocha, and M. Ponamgi: 1995, 'I-COLLIDE: An Interactive and Exact Collision Detection System for Large-Scale Environments'.

In: *Proceedings of the ACM Symposium on Interactive 3D Graphics*. pp. 189–196. [340]

Connolly, M. L.: 1992, 'Shape Distributions of Protein Topography'. *Biopolymers* **32**(9), 1215–1236. [348]

Cooper, D. B. et al.: 2002, 'Bayesian Virtual Pot-Assembly from Fragments as Problems in Perceptual-Grouping and Geometric-Learning'. In: *Proceedings of the International Conference on Pattern Recognition (ICPR'02)*, Vol. 3. Quebec City, Canada, pp. 297–302. [334]

Cootes, T., C. Beeston, G. Edwards, and C. Taylor: 1999, 'A Unified Framework for Atlas Matching Using Active Appearance Models'. In: *Information Processing in Medical Imaging (IPMI)*, Vol. 1613. Springer: New York, pp. 322–333. [283]

Cootes, T. F., A. Hill, C. J. Taylor, and J. Haslam: 1993, 'The Use of Active Shape Models for Locating Structures in Medical Images'. In: *International Conference on Information Processing in Medical Imaging*. pp. 33–47. [273, 274, 282, 283]

Corbusier, L.: 1985, *Towards a New Architecture*. Dover. Translation by Frederick Etchells of "Vers une architecture," first published in French in 1923. [333]

Cordella, L. P. and G. Sanniti di Baja: 1989, 'Geometric Properties of the Union of Maximal Neighborhoods'. *IEEE Transactions on Pattern Analysis and Machine Intelligence* **11**(2), 214–217. [174]

Crisco, J. J., M. A. Upal, and D. C. Moore: 2003, 'Advances in Quantitative in vivo Imaging'. *Current Opinions in Orthopedics* **14**(5), 351–355. [343]

Crouch, J., S. M. Pizer, E. L. Chaney, and M. Zaider: 2003, 'Medially Based Meshing with Finite Element Analysis of Prostate Deformation'. In: R. E. Ellis and T. M. Peters (eds.): *Medical Image Computing and Computer-Assisted Intervention (MICCAI)*, Vol. 2878. Springer LNCS, Montréal, Canada. [256, 307]

Crouch, J., S. M. Pizer, E. L. Chaney, Y. Hu, G. S. Mageras, and M. Zaider: 2007, 'Automated Finite Element Analysis for Deformable Registration of Prostate Images'. *IEEE Transactions on Medical Imaging* **26**(10), 1379–1390. [34]

Cullen-McEwen, L. A., G. Fricout, I. S. Harper, D. Jeulin, and J. F. Bertram: 2002, 'Quantization of 3D Ureteric Branching Morphogenesis in Cultured Embryonic Mouse Kidney'. *International Journal of Developmental Biology* **46**(8), 1049–1055. [345]

Culver, T., J. Keyser, and D. Manocha: 1999, 'Accurate Computation of the Medial Axis of a Polyhedron'. In: *Proceedings of the ACM Symposium on Solid Modeling and Applications*. pp. 179–190. [225]

Czanner, S., R. Durikovivc, and H. Inoue: 2001, 'Growth Simulation of Human Embryo Brain'. In: *IEEE 17th Spring Conference on Computer Graphics (SCCG'01)*. Budmerice, Slovakia, pp. 139–146. [345]

Dam, E., P. T. Fletcher, S. M. Pizer, G. Tracton, and J. Rosenman: 2004, 'Prostate Shape Modeling Based on Principal Geodesic Analysis Bootstrapping'. In: C. Barillot, D. Haynor, and P. Hellier (eds.): *Medical Image Computing and Computer-Assisted Intervention (MICCAI) Conference*, Vol. 3217(2). Springer, Saint-Malo, France, pp. 1008–1016. [279]

Damon, J.: 1998, 'Generic Structure of Two Dimensional Images Under Gaussian Blurring'. *SIAM Journal of Applied Mathematics* **59**, 97–138. [93, 94]

Damon, J.: 1999, 'Properties of Ridges and Cores for Two-Dimensional Images'. *Journal of Mathematical Imaging and Vision* **10**, 163–174. [93]

Damon, J.: 2003, 'Smoothness and Geometry of Boundaries Associated to Skeletal Structures I: Sufficient Conditions for Smoothness'. *Annales de l'institut Fourier* **53**, 1941–1985. [7, 15, 71, 73, 74, 75, 77, 82, 85, 87, 89]

Damon, J.: 2004, 'Smoothness and Geometry of Boundaries Associated to Skeletal Structures II: Geometry in the Blum Case'. *Compositio Mathematica* **140**(6), 1657–1674. [71, 73, 77, 83, 87, 89, 91, 97, 98, 99]

Damon, J.: 2005, 'Determining the Geometry of Boundaries of Objects from Medial Data'. *International Journal of Computer Vision* **63**(1), 45–64. [71, 73, 91, 99]

Damon, J.: 2006, 'Global Medial Structure of Regions in $\mathbb{R}^3$'. *Geometry and Topology* **10**, 2385–2429. [71, 73, 114, 119, 120, 121, 122]

Damon, J.: 2007a, 'Generic Geometry of Functions'. *In preparation*. [95]

Damon, J.: 2007b, 'Global Geometry of Regions and Boundaries via Skeletal and Medial Integrals'. *Communications in Analysis and Geometry* **15**(2), 307–358. [71, 73, 101, 109]

Damon, J.: 2007c, 'Tree Structure for Contractible Regions of $\mathbb{R}^3$'. *International Journal of Computer Vision* **74**(2), 103–116. [71, 73, 114, 119, 120, 121]

Damon, J.: 2008, 'Swept Regions and Surfaces: Modeling and Volumetric Properties'. *Theoretical Computer Science* **392**, 66–91. [71]

Danielsson, P.-E.: 1980, 'Euclidean Distance Mapping'. *Computer Graphics and Image Processing* **14**, 227–248. [163]

Das, P. P. and B. N. Chatterji: 1990, 'Octagonal Distances for Digital Pictures'. *Information Sciences* **50**, 123–150. [161]

Davies, R., C. Twining, T. Cootes, J. Waterton, and C. Taylor: 2002, 'A Minimum Description Length Approach to Statistical Shape Modeling'. *IEEE Transactions on Medical Imaging* **21**(5), 525–537. [260, 281]

deBerg, M., M. van Kreveld, M. Overmars, and O. Schwarzkopf: 1997, *Computational Geometry, Algorithms and Applications*. Berlin: Springer. [224]

Delingette, H.: 1999, 'General Object Reconstruction Based On Simplex Meshes'. *International Journal of Computer Vision* **32**(2), 111–146. [2, 273]

Descartes, R.: 1664, *Le Monde ou Traité de la lumière*. Claude Clerselier. Written in 1633; published posthumously in 1664. [328]

Deschamps, T. and L. D. Cohen: 2001, 'Fast Extraction of Minimal Paths in 3D Images and Applications to Virtual Endoscopy'. *Medical Image Analysis* **5**(4), 281–299. [34]

Descoteaux, M., L. Collins, and K. Siddiqi: 2004, 'Geometric Flows for Segmenting Vasculature in MRI: Theory and Validation'. In: *International Conference On Medical Image Computing and Computer Assisted Intervention*, Vol. LNCS 3217(1). pp. 500–507. [34]

Deussen, O. and T. Strothotte: 2000, 'Computer-Generated Pen-and-Ink Illustration of Trees'. In: *Proceedings of the 27th ACM International Conference on Computer Graphics and Interactive Techniques (SIGGRAPH'00)*. pp. 13–18. [336]

Deussen, O., J. Hamel, A. Raab, S. Schlechtweg, and T. Strothotte: 1999, 'An Illustration Technique Using Hardware-Based Intersections and Skeletons'. In: *Proceedings of Graphics Interface*. Kingston, OT, Canada, pp. 175–182. [338]

Dey, T. K. and P. Kumar: 1999, 'A Simple Provable Algorithm for Curve Reconstruction'. In: *Proceedings of the 10th. ACM-SIAM Symposium on Discrete Algorithms*. pp. 893–894. [226]

Dey, T. K. and W. Zhao: 2003, 'Approximating the Medial Axis from the Voronoi Diagram with a Convergence Guarantee'. *Algorithmica* **38**(2), 179–200. [235, 238, 239]

Dey, T. K., H. Woo, and W. Zhao: 2003, 'Approximating Medial Axis for CAD Models'. In: *8th ACM Symposium on Solid Modeling and Applications*. pp. 280–285. [238]

Diatta, A. and P. J. Giblin: 2005, 'Pre-Symmetry Sets of 3D shapes'. In: O. F. Olsen, L. M. J. Florack, and A. Kuijper (eds.): *Proceedings of the First International Workshop on Deep Structure, Singularities and Computer Vision*, Vol. LNCS 3753. Maastricht, Denmark, Springer, pp. 36–48. [61, 62]

Dimitrov, P.: 2003, 'Flux Invariants For Shape'. M.Sc. thesis, School of Computer Science, McGill University. [28, 137, 153, 314]

Dimitrov, P., C. Phillips, and K. Siddiqi: 2000, 'Robust and Efficient Skeletal Graphs'. In: *Proceedings of the IEEE Conference on Computer Vision and Pattern Recognition*. Hilton Head, SC, pp. 417–423. [139, 144, 145]

Dimitrov, P., J. N. Damon, and K. Siddiqi: 2003, 'Flux Invariants for Shape'. In: *Proceedings of the IEEE Conference on Computer Vision and Pattern Recognition*, Vol. 1. Madison, WI, pp. 835–841. [27, 28, 113, 137, 138, 145, 146, 153, 314]

Dryden, I. L. and K. Mardia: 1998, *Statistical Shape Analysis*. Chichester: Wiley. [2]

Dubinski, J., L. N. da Costa, D. S. Goldwirth, M. Lecar, and T. Piran: 1993, 'Void Evolution and the Large-Scale Structure'. *Astrophysical Journal* **410**(2), 458–468. [328]

Duda, R., P. Hart, and D. Stork: 2001, *Pattern Classification*. New York: Wiley. [274]

Durikovic, R.: 2004, 'Growth Simulation of Digestive System Using Function Representation and Skeleton Dynamics'. *International Journal on Shape Modeling* **10**(1), 31–49. [345]

Durikovic, R. and S. Czanner: 2001, 'Growth Animation of Human Organs'. *Journal of Visualization and Computer Animation* **12**, 287–295. [345]

Duvernoy, H. M.: 1998, *The Human Hippocampus, Functional Anatomy, Vascularization and Serial Sections with MRI*. Berlin: Springer. [301]

Dwyer, R.: 1987, 'A Faster Divide-and-Conquer Algorithm for Constructing Delaunay Triangulations'. *Algorithmica* **2**, 137–151. [208]

Dwyer, R.: 1991, 'Higher-dimensional Voronoi Diagrams in Linear Expected Time'. *Discrete and Computational Geometry* **6**, 343–367. [208]

Eberly, D.: 1996, *Ridges in Image and Data Analysis*, Computational Imaging and Vision Series. Dordrecht, The Netherlands: Kluwer. [30, 55, 93, 94]

Eberly, D., R. Gardner, B. Morse, S. Pizer, and C. Scharlach: 1994, 'Ridges for Image Analysis'. *Journal of Mathematical Imaging and Vision* **4**, 351–371. [29]

Edgington, E. S. (ed.): 1995, *Randomization Tests*. London: Academic. [294]

Ehmann, S. and M. Lin: 2000, 'Accelerated Proximity Queries Between Convex Polyhedra By Multi-Level Voronoi Marching'. In: *Proceedings of the IEEE/RSJ International Conference on Intelligent Robots and Systems*. pp. 2101–2106. [340]

Elad, M., A. Tal, and S. Ar: 2001, 'Content Based Retrieval of VRML Objects - An Iterative and Interactive Approach'. In: *6th Europgraphics Workshop on Multimedia*. Manchester, pp. 107–118. [312, 319, 320]

Fang, T.-P. and L. Piegl: 1993, 'Delaunay Triangulation Using a Uniform Grid'. *IEEE Computer Graphics and Applications* **13**(3), 36–47. [209]

Ferley, E., M.-P. Gascuel, and D. Attali: 1997, 'Skeletal Reconstruction of Branching Shapes'. *Computer Graphics Forum* **16**(5), 283–293. [337, 346]

Ferraro, P., C. Godin, and P. Prusinkiewicz: 2005, 'Toward a Quantification of Self-Similarity in Plants'. *Fractals* **13**(2), 91–109. [335, 336]

Finn, P. W. and L. E. Kavraki: 1999, 'Computational Approaches to Drug Design'. *Algorithmica* **25**, 347–371. [348]

Fletcher, P., C. Lu, S. Pizer, and S. Joshi: 2004, 'Principal Geodesic Analysis for the Study of Nonlinear Statistics of Shape'. *IEEE Transactions on Medical Imaging* **23**(8), 995–1005. [253, 303]

Fletcher, P. T., S. M. Pizer, and S. C. Joshi: 2006, 'Shape Variation of Medial Axis Representations by a Principal Geodesic Analysis on Symmetric Spaces'. In: H. Krim and A. Yezzi (eds.): *Statistics and Analysis of Shapes*. Boston, MA: Birkhäuser. [250]

Foskey, M., M. Garber, M. Lin, and D. Manocha: 2001, 'V-Plan: A Voronoi-Based Hybrid Motion Planner'. In: *Proceedings of the IEEE/RSJ International Conference on Intelligent Robots and Systems*. pp. 55–60. [340]

Foskey, M., M. Lin, and D. Manocha: 2003, 'Efficient Computation of a Simplified Medial Axis'. In: *8th AMC Symposium on Solid Modeling and Applications*. [229]

Foskey, M., B. Davis, L. Goyal, S. Chang, E. Chaney, N. Strehl, S. Tomei, J. Rosenman, and S. Joshi: 2005, 'Large Deformation 3D Image Registration in Image-Guided Radiation Therapy'. *Physics in Medicine and Biology* **50**, 5869–5892. [290]

Fridman, Y.: 2004, 'Modeling Tubuar Structures Using Cores'. Ph.D. thesis, University of North Carolina at Chapel Hill, Chapel Hill, NC. [29]

Fridman, Y., S. M. Pizer, S. Aylward, and E. Bullitt: 2004, 'Extracting Branching Tubular Object Geometry via Cores'. *Medical Image Analysis* **8**(3), 169–176. [30, 34]

Fritsch, D., E. Chaney, A. Boxwala, M. McAuliffe, S. Raghavan, A. Thall, and E. J.R.D.: 1995a, 'Core-based Portal Image Registration for Automatic Radiotherapy Treatment Verification'. *International Journal of Radiation, Oncology, Biology, Physics (special issue on Conformal Therapy)* **33**(5), 1287–1300. [29]

Fritsch, D., D. Eberly, S. Pizer, and M. McAuliffe: 1995b, 'Stimulated Cores and their Applications in Medical Imaging'. In: *Information Processing in Medical Imaging 1995 (IPMI'95)*. pp. 365–368. [30]

Frome, F. S.: 1972, 'A Psychophysical Study of Shape Alignment'. Technical Report TR-198, University of Maryland, Computer Science Center. [22]

Funke, S.: 2005, 'Topological Hole Detection in Wireless Sensor Networks and its Applications'. In: *Proceedings of the Joint Workshop on Foundations of Mobile Computing*. Cologne, Germany, pp. 44–53. [331]

Funkhouser, T., P. Min, M. Kazhdan, J. Chen, A. Halderman, and D. Dobkin: 2003, 'A Search Engine for 3D Models'. *ACM Transactions on Graphics* **22**(1), 83–105. [312]

Furst, J. D. and S. M. Pizer: 1998, 'Marching Optimal Parameter Ridges: An Algorithm to Extract Shape Loci in 3D Images'. In: *International Conference on Medical Image Computing and Computer-Assisted Intervention*. pp. 780–787. [29, 30]

Garber, M. and M. Lin: 2002, 'Constraint-Based Motion Planning Using Voronoi Diagrams'. In: *Proceedings of the International Workshop on Algorithmic Foundations of Robotics*. pp. 514–530. [340]

Gauch, J. M. and S. M. Pizer: 1993, 'The Intensity Axis of Symmetry and its Application to Image Segmentation'. *IEEE Transactions on Pattern Analysis and Machine Intelligence* **15**(8), 753–770. [328, 330]

Geiger, B.: 1993, 'Three-dimensional Modelling of Human Organs and its Application to Diagnostic and Surgical Planning'. Technical Report RR-2105, INRIA. [209]

Gelston, S. M. and D. Dutta: 1995, 'Boundary Surface Recovery from Skeleton Curves and Surfaces'. *Computer Aided Geometric Design* **12**(1), 27–51. ISSN 0167-8396. [14]

Geraerts, R. and M. H. Overmars: 2004, 'Clearance Based Path Optimization for Motion Planning'. In: *EEE International Conference on Robotics and Automation (ICRA'04)*. pp. 2386–2392. [340]

Geraerts, R. and M. H. Overmars: 2005, 'On Improving the Clearance for Robots in High-Dimensional Configuration Spaces'. In: *IEEE/RSJ International Conference on Intelligent Robots and Systems (IROS'05)*. pp. 4074–4079. [340]

Gerig, G., M. Styner, D. Jones, D. R. Weinberger, and J. A. Lieberman: 2001, 'Shape Analysis of Brain Ventricles Using SPHARM'. In: *Proceedings of the IEEE Workshop on Mathematical Methods in Biomedical Image Analysis*. pp. 171–178. [302]

Giblin, P.: 2000, 'Symmetry Sets and Medial Axes in Two and Three Dimensions'. In: R. Cipolla and R. Martin (eds.): The Mathematics of Surfaces IX. Heidelberg: Springer, pp. 306–321. [55]

Giblin, P. J. and S. A. Brassett: 1985, 'Local symmetry of plane curves'. *American Mathematical Monthly* **92**, 689–707. [5, 6, 8]

Giblin, P. J. and B. B. Kimia: 2000, 'A formal Classification of 3D Medial Axis Points and their Local Geometry'. In: *Proceedings of the IEEE Conference on Computer Vision and Pattern Recognition*, Vol. 1. pp. 566–573. [9, 10, 18]

Giblin, P. J. and B. B. Kimia: 2002, 'Transitions of the 3D Medial Axis under a One-Parameter Family of Deformations'. In: *Proceedings of the European Conference on Computer Vision*, Vol. 2351 of *Lecture Notes in Computer Science*. pp. 718–734, Springer, Copenhagen. [7, 58]

Giblin, P. J. and B. B. Kimia: 2003a, 'On the Intrinsic Reconstruction of Shape from its Symmetries'. *IEEE Transactions on Pattern Analysis and Machine Intelligence* **25**(7), 895–911. [51, 52, 66, 67]

Giblin, P. J. and B. B. Kimia: 2003b, 'On the Local Form and Transitions of Symmetry Sets, Medial Axes, and Shocks'. *International Journal of Computer Vision* **54**(1–3), 143–157. [55]

Giblin, P. J. and B. B. Kimia: 2004, 'A Formal Classification of 3D Medial Axis Points and their Local Geometry'. *IEEE Transactions on Pattern Analysis and Machine Intelligence* **26**(2), 238–251. [7, 48, 54, 55, 314, 316]

Giblin, P. J., B. B. Kimia, and A. J. Pollitt: 2008, 'Transitions of the 3D Medial Axis Under a One-Parameter Family of Deformations'. *IEEE Transactions on Pattern Analysis and Machine Intelligence*, to appear. [58]

Gibson, C. G., K. Wirthmiller, A. A. du Plessis, and E. J. N. Looijenga: 1976, *Topological Stability of Smooth Mappings*, Vol. 552 of *Lecture Notes in Mathematics*. Berlin/Heidelberg: Springer. [71]

Gold, C. M.: 2000, 'Primal/Dual Spatial Relationships and Applications'. In: *9th International Symposium on Spatial Data Handling (SDH)*, Vol. 4a. Beijing, China, pp. 15–27. [330]

Gold, C. M. and J. Snoeyink: 2001, 'A One-Step Crust and Skeleton Extraction Algorithm'. *Algorithmica* **30**(2), 144–163. [226]

Gold, C. M. and D. Thibault: 2001, 'Map Generalization by Skeleton Retraction'. In: *Proceedings of the 20th International Cartographic Conference (ICC)*. Beijing, China, pp. 2071–2081. [330]

Goldak, J. A., X. Yu, A. Knight, and L. Dong: 1991, 'Constructing Discrete Medial Axis of 3-D Objects'. *International Journal of Computational Geometry and its Applications*. pp. 327–339. [239]

Golland, P., W. Grimson, and R. Kikinis: 1999, 'Statistical Shape Analysis Using Fixed Topology skeletons: Corpus Callosum Study'. In: *International Conference on Information Processing in Medical Imaging*. pp. 382–388. [292]

Golubitsky, M. and V. Guillemin: 1974, *Stable Mappings and their Singularities*, Graduate Texts in Mathematics. Berlin/Heidelberg: Springer. [73]

Gomes, J. and O. Faugeras: 1999, 'Reconciling Distance Functions and Level Sets'. Technical Report TR3666, INRIA. [128]

Gooch, B., G. Coombe, and P. Shirley: 2002, 'Artistic Vision: Painterly Rendering Using Computer Vision Techniques'. In: *Proceedings of the 2nd International Symposium on Non Photorealistic Rendering (NPAR'02)*. Annecy, France, pp. 83–91. [337]

Gopal, K., T. D. Romo, E. Mckee, K. C. Childs, L. Kanbi, R. Pai, J. Smith, J. C. Sacchettini, and T. R. Ioerger: 2005, 'TEXTAL: Automated Crystallographic Protein Structure Determination'. In: *Proceedings of the Innovative Applications of Artificial Intelligence Conference*. pp. 1483–1490. [347]

Greer, J.: 1974, 'Three-Dimensional Pattern Recognition: An Approach to Automated Interpretation of Electron Density Maps of Proteins'. *Journal of Molecular Biology* **82**, 279–301. [347]

Greer, J.: 1985, 'Computer Skeletonization and Automatic Electron Density Map Analysis'. *Methods in Enzymology* **115**, 206–224. [347]

Grenander, U.: 1976, *Pattern Synthesis: Lectures in Pattern Theory*, Vol. I. New York: Springer. [33]

Grenander, U.: 1978, *Pattern Synthesis: Lectures in Pattern Theory*, Vol. II. New York: Springer. [33]

Grenander, U.: 1981, *Regular Structures: Lectures in Pattern Theory*, Vol. III. New York: Springer. [33]

Grenander, U.: 1996, *Elements of Pattern Theory*. Baltimore, MD: John Hopkins University Press. [242]

Grenander, U. and M. I. Miller: 1998, 'Computational Anatomy: An Emerging Discipline'. *Quarterly of Applied Mathematics* **56**, 617–694. [3]

Guibas, L. J., C. Holleman, and L. E. Kavraki: 1999, 'A Probabilistic Roadmap Planner for Flexible Objects with a Workspace Medial Axis based Sampling Approach'. In: *Proceedings of Intl. Conf. Intelligent Robots and Systems.* Kyongju, Korea, pp. 254–260. [340]

Hallinan, P. L., G. G. Gordon, A. L. Yuille, P. J. Giblin, and D. Mumford: 1999, *Two and Three Dimensional Patterns of the Face*. Natick, MA: A. K. Peters. [41, 43]

Hamza, A. B. and H. Krim: 2003, 'Geodesic Object Representation and Recognition'. In: *Proceedings of DGCI*, Vol. LNCS 2886. pp. 378–387. [313]

Han, Q., C. Lu, G. Liu, S. Pizer, S. Joshi, and A. Thall: 2004, 'Representing Multi-Figure Anatomical Objects'. In: *International Symposium on Biomedical Imaging (ISBI)*, Vol. Catalog Number 04EX821C. pp. 1251–1254. [246]

Han, Q., S. M. Pizer, and J. N. Damon: 2005a, 'Medial Atom Interpolation Via the Radial Shape Operator'. Technical Report in preparation, MIDAG, Department of Computer Science, UNC Chapel Hill. [252]

Han, Q., S. M. Pizer, D. Merck, S. Joshi, and J.-Y. Jeong: 2005b, 'Multi-Figure Anatomical Objects for Shape Statistics'. In: G. Christensen and M. Sonka (eds.): *International Conference on Information Processing in Medical Imaging*, Vol. 3565. pp. 701–712. [288]

Hatcher, B., K. Aspelund, A. Willis, J. Speicher, D. B. Cooper, and F. F. Leymarie: 2005, 'Computational Schemes for Biomimetic Sculpture'. In: *Proceedings of the 5th International Conference on Creativity and Cognition*. London, pp. 22–31. [338, 339]

Held, M.: 1991, *On the Computational Geometry of Pocket Machining*, No. 500 in Lecture Notes in Computer Science. Berlin: Springer. [341]

Held, M.: 2001, 'VRONI: An Engineering Approach to the Reliable and Efficient Computation of Voronoi Diagrams of Points and Line Segments'. *Computational Geometry: Theory and Applications* **18**(2), 95–123. [225, 341]

Hilaga, M., Y. Shinagawa, T. Komura, and T. L. Kunii: 2001, 'Topology Matching for Full Automatic Similarity Estimation of 3D Shape'. In: *Proceedings of SIGGRAPH 2001*. Los Angeles, CA, pp. 203–212. [313, 346]

Hilditch, C. J.: 1969, 'Linear Skeletons from Square Cupboards'. In: B. Meltzer and D. Michie (eds.): *Machine Intelligence 4*. Edinburgh: Edinburgh University Press, pp. 403–420. [168]

Hillier, B.: 1996, *Space is the Machine — A Configurational Theory of Architecture*. Cambridge: Cambridge University Press. [333]

Hillier, B., J. Hanson, and J. Peponis: 1987, 'Syntactic Analysis of Settlements'. *Architecture et Comportment/Architecture and Behavior* 3(3), 217–231. [333]

Ho, S.: 2004, 'Profile Scale Spaces for Statistical Image Match in Bayesian Segmentation'. Ph.D. thesis, University of North Carolina, Chapel Hill, NC. [282]

Hoffman, D. D. and W. A. Richards: 1984, 'Parts of Recognition'. *Cognition* **18**, 65–96. [21]

Hoffmann, C. M.: 1990, 'How to Construct the Skeleton of CSG Objects'. In: *Proceedings of the Fourth IMA Conference on the Mathematics of Surfaces*. [225]

Hoffmann, C. M.: 1995, 'Geometric Approaches to Mesh Generation'. In: I. Babuska et al. (eds.): *Modeling, Mesh Generation, and Adaptive Numerical Methods for Partial Differential Equations*, Vol. 75 of *IMA Volumes in Mathematics and its Applications*. Springer, Minnesota, USA, pp. 31–52. [341]

Hoffmann, C. M. and P. J. Vermeer: 1996, 'Validity Determination for MAT Surface Representation'. In: G. Mullineux (ed.): *Proceedings of the 6th IMA Conference on the Mathematics of Surfaces (IMA-94)*, Vol. VI of *Mathematics of Surfaces*. Oxford: Clarendon Press, pp. 249–266. [14]

Holleman, C. and L. E. Kavraki: 2000, 'A Framework for Using the Workspace Medial Axis in PRM Planners'. In: *Proceedings of the International Conference on Robotics and Automation*. San Fransisco, CA, pp. 1408–1413. [340]

Huan, J., D. Bandyopadhyay, W. Wang, J. Snoeyink, J. Prins, and A. Tropsha: 2005, 'Comparing Graph Representations of Protein Structure for Mining Family-Specific Residue-Based Packing Motifs'. *Journal of Computational Biology* **12**(6), 657–671. [348]

Hubbard, P. M.: 1996, 'Approximating polyhedra with spheres for time-critical collision detection'. *ACM Transactions On Graphics* **15**(3), 179–210. [340]

Icke, V. and R. v. d. Weygaert: 1987, 'Fragmenting the Universe. I. Statistics of Two-Dimensional Voronoi Foams'. *Astronomy and Astrophysics* **184**, 16–32. [328]

Icke, V. and R. v. d. Weygaert: 1991, 'The Galaxy Distribution as a Voronoi Foam'. *Quaterly Journal of the Royal Astronomical Society* **32**(2), 85–112. [328]

Igarashi, T., S. Matsuoka, and H. Tanaka: 1999, 'Teddy: A Sketching Interface for 3D Freeform Design'. In: *Proceedings of the 26th annual conference on Computer graphics and interactive techniques (SIGGRAPH'99)*. pp. 409–416. [7, 337]

Ioerger, T. R. and J. C. Sacchettini: 2002, 'Automatic Modeling of Protein Backbones in Electron Density Maps via Prediction of C-alpha Coordinates'. *Acta Crystallographica* **D58**(12), 2043–2054. [347]

Jeong, J., S. Pizer, and S. Ray: 2006, 'Statistics on Anatomic Objects Reflecting Inter-Object Relations'. In: *International Workshop on Mathematical Foundations of Computational Anatomy (MFCA-2006)*. pp. 136–145. [259, 291]

Jeong, J.-Y.: 2008, 'Estimation of Probability Distribution on Multiple Anatomical Object Complex'. Ph.D. thesis, Department of Computer Science, University of North Carolina-Chapel Hill, Chapel Hill, NC. [256]

Joe, B.: 1991, 'Construction of Three-Dimensional Delaunay Triangulations using Local Transformations'. *Computer Aided Geometric Design* **8**, 123–142. [208]

Johnson, C. K., M. N. Burnett, and W. D. Dunbar: 1996, 'Crystallographic Topology and its Applications'. In: *Crystallographic Computing 7: Macromolecular Crystallographic Data*. Oxford University Press, Oxford. http://www.ornl.gov/sci/ortep/topology.html. [328]

Jonker, P. and A. Vossepoel: 1995, 'On Skeletonization Algorithms for 2, 3 .. N Dimensional Images'. In: D. Dori and A. Bruckstein (eds.): *Shape, Structure and Pattern Recognition*. Singapore: World Scientific, pp. 71–80. [26]

Joshi, S. and M. I. Miller: 2000, 'Landmark Matching Via Large Deformation Diffeomorphisms'. *IEEE Transactions on Medical Imaging* **9**, 1357–1370. [2]

Kanisza, G.: 1979, *Organization in Vision: Essays on Gestalt Perception*. New York: Praeger. [350]

Katz, R.: 2002, 'Form Metrics for Interactive Rendering via Figural Models of Perception'. Ph.D. thesis, University of North Carolina at Chapel Hill, Chapel Hill, NC. [6]

Katz, R. and S. M. Pizer: 2003, 'Untangling the Blum Medial Axis Transform'. *International Journal of Computer Vision* **55**(3), 139–153. [32, 203]

Kazhdan, M., B. Chazelle, D. Dobkin, T. Funkhouser, and S. Rusinkiewicz: 2003a, 'A Reflective Symmetry Descriptor for 3-D Models'. *Algorithmica* **38**(1), 201–225. [312, 319, 320]

Kazhdan, M., T. Funkhouser, and S. Rusinkiewicz: 2003b, 'Rotation Invariant Spherical Harmonic Representation of 3D Shape Descriptors'. In: *Symposium on Geometry Processing*. [309, 314, 321]

Keim, D. A., C. Panse, and S. C. North: 2005, 'Medial-Axis-Based Cartograms'. *IEEE Computer Graphics and Applications* **25**(3), 60–68. [331]

Kelemen, A., G. Székely, and G. Gerig: 1999, 'Elastic Model-Based Segmentation of 3D Neuroradiological Data Sets'. *IEEE Transactions on Medical Imaging* **18**(10), 828–839. [3, 274]

Keller, R.: 1999, 'Generic Transitions of Relative Critical Sets in Parametrized Families with Applications to Image Analysis'. Ph.D. thesis, Department of Mathematics, University of North Carolina, Chapel Hill, NC. [93, 94]

Kendall, D. G.: 1989, 'A Survey of the Statistical Theory of Shape'. *Statistical Science* **4**, 87–120. [2]

Kikinis, R., M. Shenton, G. Gerig, H. Hokama, J. Haimson, B. O'Donnel, C. Woble, R. McCarley, and F. Jolesz: 1994, 'Temporal Lobe Sulco-Gyral Pattern Abnormalities in Schizophrenia: an MR Three-Dimensional Surface Rendering Study'. *Neuroscience Letters* **182**, 7–12. [219]

Kimia, B. B.: 2003, 'On the Role of Medial Geometry in Human Vision'. *Journal of Physiology* **97**(2–3), 155–190. [349, 350]

Kimia, B. B., A. Tannenbaum, and S. W. Zucker: 1995, 'Shape, Shocks, and Deformations I: The Components of Two-Dimensional Shape and the Reaction-

Diffusion Space'. *International Journal of Computer Vision* **15**, 189–224. [5, 9, 28, 47]

Kimia, B. B., A. R. Tannenbaum, and S. W. Zucker: 1990, 'Toward a Computational Theory of Shape: An Overview'. In: *Proceedings of the First European Conference on Computer Vision*. Antibes, France, Springer, pp. 402–407. [47, 71, 76]

Kimmel, R., D. Shaked, and N. Kiryati: 1995, 'Skeletonization Via Distance Maps and Level Sets'. *Computer Vision and Image Understanding* **62**(3), 382–391. [128, 129]

Klein, F.: 1987a, 'Euclidean Skeletons'. In: *Proceedings of 5th Scandinavian Conference on Image Analysis*. pp. 443–450. [191]

Klein, F.: 1987b, 'Vollständige Mittelachsenbeschreibung binärer Bildstrukturen mit euklidischer Metrik und korrekter Topologie'. Ph.D. thesis, ETH Nr. 8441. [191, 203]

Klein, P. N., T. B. Sebastian, and B. B. Kimia: 2001, 'Shape Matching using Edit-distance: An Implementation'. In: *Proceedings of the Twelfth Annual ACM-SIAM Symposium on Discrete Algorithms (SODA-01)*. ACM Press, New York, pp. 781–790. [310]

Koehler, W.: 1947, *Gestalt Psychology: An Introduction to New Concepts in Modern Psychology*. New York: Liveright Publication. [348]

Koenderink, J. J.: 1990, *Solid Shape*. Cambridge, MA: MIT Press. [247]

Koenderink, J. J. and A. J. van Doorn: 1998, 'The Structure of Relief'. *Advances in Imaging and Electron Physics* **103**, 65–150. [329]

Koffka, K.: 1935, *Principles of Gestalt Psychology*. New York: Brace Harcourt. [348]

Kong, T. Y. and A. Rosenfeld: 1989, 'Digital Topology: Introduction and Survey'. *Computer Vision, Graphics, and Image Processing* **48**(3), 357–393. [139]

Kovács, I. and B. Julesz: 1993, 'A Closed Curve is Much More Than an Incomplete One: Effect of Closure in Figure-Ground Segmentation'. *Proceedings of the National Academy of Science of the USA* **90**(16), 7495–7497. [22, 349]

Kovács, I. and B. Julesz: 1994, 'Perceptual Sensitivity Maps within Globally Defined Visual Shapes'. *Nature* **370**, 644–646. [22, 349]

Kovács, I., A. Feher, and B. Julesz: 1998, 'Medial-Point Description of Shape: A Representation for Action Coding and its Psychophysical Correlates'. *Vision Research* **38**, 2323–2333. [22, 23, 349]

Kovalevsky, V.: 1987, 'The Topology of Cellular Complexes as Applied to Image Processing'. In: *Proceedings of 2nd International Conference on Computer Analysis of Images and Patterns*. pp. 162–173. [211]

Kovalevsky, V.: 1989, 'Finite Topology as Applied to Image Analysis'. *Computer Vision, Graphics, and Image Processing* **46**, 141–161. [211]

Kuijper, A. and O. F. Olsen: 2004, 'Transitions of the Pre-Symmetry Set'. In: *17th International Conference on Pattern Recognition*, Vol. 3. pp. 190–193. [44]

Kurdyka, K. and G. Raby: 1989, 'Densité des Esembles Sous–Analytiques'. *Annales de l'Institut Fourier* **39**, 753–771. [112]

Kwan, M.-P. and J. Lee: 2005, 'Emergency Response After 9/11: The Potential of Real-Time 3D GIS for Quick Emergency Response in Micro-Spatial Environments'. *Computers, Environment and Urban Systems* **29**(2), 93–113. [332, 333]

Lazarus, F. and A. Verroust: 1999, 'Level Set Diagrams of Polyhedral Objects'. In: *Proceedings of the Fifth ACM Symposium on Solid Modeling and Applications.* pp. 130–140. [339]

Lazarus, F., S. Coquillart, and P. Jancene: 1994, 'Axial Deformations: An Intuitive Deformation Technique'. *Computer-Aided Design* **26**(8), 607–613. [337]

Lee, J.: 2001, '3D Data Model for Representing Topological Relations of Urban Features'. In: *Proceedings of 21st Annual ESRI International User Conference.* San Diego, CA. [332]

Lee, J.: 2004, 'A Spatial Access-Oriented Implementation of a 3-D GIS Topological Data Model for Urban Entities'. *GeoInformatica* **8**(3), 237–264. [332]

Lee, T. S.: 1995, 'Neurophysiological Evidence for Image Segmentation and Medial Axis Computation in Primate V1'. In: J. Bower (ed.): *Fourth Annual Computational Neuroscience Meeting.* London: Academic, pp. 373–378. [24]

Lee, T. S., D. Mumford, R. Romero, and V. A. F. Lamme: 1998, 'The Role of the Primary Visual Cortex in Higher Level Vision'. *Vision Research* **38**, 2429–2454. [24]

Leherte, L., J. Glasgow, K. Baxter, E. Steeg, and S. Fortier: 1997, 'Analysis of Three-Dimensional Protein Images'. *Journal of Artificial Intelligence Research (JAIR)* **7**, 125–159. [347]

Leito, H. C. G. and J. Stolfi: 2000, 'Information Contents of Fracture Lines'. In: *8th International Conference in Central Europe on Computer Graphics, Visualization, and Interactive Digital Media*, Vol. 2. University of West Bohemia Press, Pilsen, pp. 389–395. [334]

Levina, E. and P. J. Bickel: 2001, 'The Earth Mover's Distance is the Mallows Distance: Some Insights from Statistics'. In: *Proceedings of the IEEE International Conference on Computer Vision.* pp. 251–256. [284]

Lewis, R. H. and S. Bridgett: 2003, 'Conic Tangency Equations and Apollonius Problems in Biochemistry and Pharmacology'. *Mathematics and Computers in Simulation* **61**(2), 101–114. [348]

Leymarie, F.: 1997, 'Exploitation of 3D Georeferenced Datasets in a GIS'. In: *Proceedings of International Workshop on HPCN Exploitation of Multimedia Databases.* Ispra, Italy, pp. 10–29, JRC, no. EUR 17349 EN. [332]

Leymarie, F. and M. Gruber: 1997, 'Applications of Virtual Reality: Cyber-Monument & CyberCity'. In: *Proceedings of the 3rd European Digital Cities Conference.* Berlin, Germany. http://www.lems.brown.edu/~leymarie/cybercity/. [332]

Leymarie, F. and M. D. Levine: 1992, 'Simulating the Grassfire Transform Using an Active Contour Model'. *IEEE Transactions on Pattern Analysis and Machine Intelligence* **14**(1), 56–75. [27, 128, 129, 202]

Leymarie, F. and M. D. Levine: 1993, 'Tracking Deformable Objects in the Plane Using an Active Contour Model'. *IEEE Transactions on Pattern Analysis and Machine Intelligence* **15**(6), 617–634. [345]

Leymarie, F., N. Boichis, S. Airault, and O. Jamet: 1996, 'Towards the Automation of Road Extraction Processes'. In: *SPIE European Symposium on Satellite and Remote Sensing III*, Vol. SPIE-2960. Taormina, Italy, pp. 84–95. [330]

Leymarie, F., D. Cooper, M. Joukowsky, B. Kimia, D. Laidlaw, D. Mumford, and E. Vote: 2001, 'The SHAPE Lab: New Technology and Software for Archaeologists'. In: *CAA 2000: Computing Archaeology for Understanding the Past*, Vol. 931 of *BAR International Series*. Archaeopress: Oxford, pp. 79–89. [334]

Leymarie, F. F.: 2003, '3D Shape Representation via Shock Flows'. Ph.D. thesis, Division of Engineering, Brown University, Providence, RI, 02912. [50, 51, 328]

Leymarie, F. F.: 2006, 'Aesthetic Computing and Shape'. In: Paul Fishwick (ed.): *Aesthetic Computing*. Cambridge, MA: MIT Press, Chapt. 14. [337]

Leymarie, F. F. and B. B. Kimia: 2001, 'The Shock Scaffold for Representing 3D Shapes'. In: C. Arcelli, L. Cordella, and G. S. di Baja (eds.): *Proceedings of the International Workshop on Visual Form*. Capri, Italy: Springer, pp. 216–228. [50]

Leymarie, F. F. and B. B. Kimia: 2003, 'Computation of the Shock Scaffold for Unorganized Point Clouds in 3D'. In: *Proceedings of the IEEE Conference on Computer Vision and Pattern Recognition*, Vol. 1. pp. 821–827. [313, 314, 326]

Leymarie, F. F., P. J. Giblin, and B. B. Kimia: 2004a, 'Towards Surface Regularization via Medial Axis Transitions'. In: *Proceedings of International Conference on Pattern Recognition*, Vol. 3. Computer Society Press, Cambridge, England, pp. 123–126. [65]

Leymarie, F. F., B. B. Kimia, and P. J. Giblin: 2004b, 'Towards Surface Regularization via Medial Axis Transitions'. In: *Proceedings of the International Conference on Pattern Recognition (ICPR)*, Vol. 3. Cambridge, pp. 123–126. [330, 334, 341, 342]

Leyton, M.: 1987, 'Symmetry-Curvature Duality'. *Computer Vision, Graphics, and Image Processing* **38**(3), 327–341. [5, 12, 57, 199]

Leyton, M.: 1988, 'A Process Grammar For Shape'. *Artificial Intelligence* **34**, 213–247. [5]

Leyton, M.: 1989, 'Inferring Causal History From Shape'. *Cognitive Science* **13**, 357–387. [5]

Leyton, M.: 1992, *Symmetry, Causality, Mind*. Cambridge, MA: MIT press. [5, 12, 328, 337]

Leyton, M.: 2001, *A Generative Theory of Shape*, No. LNCS 2145 in Lecture Notes in Computer Science. Springer, http://www.rci.rutgers.edu/m̃leyton/homepage.htm. [334, 337, 345]

Leyton, M.: 2006, 'The Foundations of Aesthetics'. In: Paul Fishwick, (ed.): *Aesthetic Computing,* Cambridge, MA: MIT Press, Chapt. 13. [337]

Li, H.: 2002, 'An Integrated Approach to Protein Backbone Modeling'. Master's thesis, Queen's University, Department of Computing and Information Science, Kingston, OT, Canada. [347]

Lin, S. L., R. Nussinov, D. Fischer, and H. J. Wolfson: 1994, 'Molecular Surface Representations by Sparse Critical Points'. *Proteins: Structure, Function, and Genetics* **18**, 94–101. [348]

Lindenmayer, A.: 1968, 'Mathematical Models for Cellular Interactions in Development: Parts I and II'. *Journal of Theoretical Biology* **18**, 280–315. [336]

Liu, X., J.-Y. Jeong, J. H. Levy, R. R. Saboo, E. L. Chaney, and S. M. Pizer: 2008, 'Local Residual Statistics for Segmentations Using Deformable Shape Models'. *submitted for conference review.* [256]

Locher, P. J.: 2003, 'An Empirical Investigation of the Visual Rightness Theory of Picture Perception'. *Acta Psychologica* **114**(2), 147–164. [337]

Lorensen, W. E. and H. E. Cline: 1987, 'Marching cubes: A High Resolution 3D Surface Construction Algorithm'. *Computer Graphics (SIGGRAPH '87 Proceedings)* **21**, 163–169. [30]

Lorigo, L. M., O. D. Faugeras, E. L. Grimson, R. Keriven, R. Kikinis, A. Nabavi, and C.-F. Westin: 2001, 'CURVES: Curve Evolution for Vessel Segmentation'. *Medical Image Analysis* **5**, 195–206. [34]

Lu, C., S. Pizer, and S. Joshi: 2003, 'A Markov Random Field Approach to Multi-scale Shape Analysis'. In: L. D. Griffin and M. Lillholm (eds.): *Fourth International Conference On Scale Space Methods in Computer Vision*, Vol. LNCS 2695. pp. 416–431. [279]

Luneburg, R. Y.: 1964, *Mathematical Theory of Optics*. Berkeley/Los Angeles, CA: University of California Press. [130, 133]

Macrini, D.: 2003, 'Indexing and Matching for View-Based 3-D Object Recognition Using Shock Graphs'. Ph.D. thesis, University of Toronto, Toronto. [319]

Malandain, G. and S. Fernandez-Vidal: 1998, 'Euclidean Skeletons'. *Image and Vision Computing* **16**, 317–327. [128, 150, 215]

Malandain, G., G. Bertrand, and N. Ayache: 1993, 'Topological Segmentation of Discrete Surfaces'. *International Journal of Computer Vision* **10**(2), 183–197. [141, 143, 177, 314, 315]

Mallows, C.: 1972, 'A Note on Asymptotic Joint Normality'. *Annals of Mathematical Statistics* **43**(2), 508–515. [284]

Mangin, J.-F., D. Rivire, A. Cachia, D. Papadopoulos-Orfanos, D. L. Collins, A. C. Evans, and J. Rgis: 2004, 'Object-Based Morphometry of the Cerebral Cortex'. *IEEE Transactions on Medical Imaging* **23**(8), 968–982. [343, 344, 345]

Marr, D. and H. K. Nishihara: 1978, 'Representation and Recognition of the Spatial Organisation of Three-Dimensional Shapes'. In: *Proceedings of the Royal Society of London. Series B, Biological Scences*, Vol. 200. pp. 269–294. [313]

Mather, J. N.: 1973, 'Stratifications and Mappings'. In: M. Peixoto (ed.): *Dynamical Systems*. New York: Academic Press. [71]

Mather, J. N.: 1983, 'Distance From a Submanifold in Euclidean Space'. *Proceedings of Symposia in Pure Mathematics* **40**(2), 199–216. [1, 5, 9, 73]

Matheron, G.: 1988, 'Examples of Topological Properties of Skeletons'. In: J. Serra (ed.): *Image analysis and Mathematical Morphology*, Vol. 2. Academic, London, pp. 217–238. [26, 136]

Maurer, C. R., R. Qi, and V. Raghavan: 2003, 'A Linear Time Algorithm for Computing Exact Euclidean Distance Transforms of Binary Images in Arbitrary Dimensions'. *IEEE Transactions on Pattern Analysis and Machine Intelligence* **25**(2), 265–270. [163]

Maxwell, J. C.: 1870, 'On Hills and Dales'. *The London, Edinburgh and Dublin Philosophical Magazine and Journal of Science* **40**(269), 421–425. [329]

McAllister, M. and J. Snoeyink: 1999, 'Extracting Consistent Watersheds from Digital River and Elevation Data'. In: *ASPRS Annual Conference Proceedings — From Image to Information*. Portland, OR. [329]

McAllister, M. and J. Snoeyink: 2000, 'Medial Axis Generalization of River Networks'. *CaGIS* **27**(2). Journal of the Cartography and Geographical Information Society. [330]

McCulloch, W. S.: 1965, *Embodiments of Mind*. Cambridge, MA: MIT Press. [22]

McInerny, T. and D. Terzopoulos: 1996, 'Deformable Models in Medical Image Analysis: A Survey'. *Medical Image Analysis* **1**(2), 91–108. [273, 274]

Merck, D., G. Tracton, S. Pizer, and S. Joshi: 2006, 'A Methodology for Constructing Geometric Priors and Likelihoods for Deformable Shape Models'. Technical Report, MIDAG, Department of Computer Science, UNC Chapel Hill. [279]

Meyer, F.: 1989, 'Skeletons and Perceptual Graphs'. *Signal Processing* **16**, 335–363. [201]

Meyer, F. and P. Maragos: 1999, 'Multiscale Morphological Segmentations Based on Watershed, Flooding, and Eikonal PDE'. In: M. Nielsen, P. Johansen, O. Olsen, and J. Weickert (eds.): *Scale-Space Theories in Computer Vision*, No. 1682 in Lecture Notes in Computer Science. Springer, Corfu, Greece, pp. 351–362. [329]

Milenkovic, V. J.: 1993, 'Robust Construction of the Voronoi Diagram of a Polyhedron'. In: *Proceedings of the 5th Canadian Conference on Computational Geometry*. pp. 473–478. [225]

Miller, J.: 1998, 'Relative Critical Sets in $\mathbb{R}^n$ and Applications to Image Analysis'. Ph.D. thesis, Department of Mathematics, University of North Carolina, Chapel Hill, NC. [93]

Miller, M. I., S. C. Joshi, and G. E. Christensen: 1999, 'Large Deformation Fluid Diffeomorphisms For Landmark and Image Matching'. In: A. W. Toga and J. Mazziotta (eds.): *Brain Warping*. Elsevier, http://www.elsevier.com/wps/find/bookdescription.cws_home/673350/description pp. 115–131. [287]

Millman, D.: 1980, 'The Central Function of the Boundary of a Domain and its Differential Properties'. *Journal of Geometry* **14**, 182–202. [1]

Modica, L. and S. Mortola: 1977, 'Il limite nella Γ-convergence di una familiglia di funzionali ellittici'. *Bollettino della Unione Matematica Italiana* **14-A**, 526–529. [28]

Morse, B., S. Pizer, D. Puff, and C. Gu: 1998, 'Zoom-Invariant Vision of Figural Shape: Effects on Cores of Image Disturbances'. *Computer Vision and Image Understanding* **69**, 72–86. [29]

Mumford, D.: 1996, 'Pattern theory: A Unifying Perspective'. In: D. C. Knill and W. Richards (eds.): *Perception as Bayesian Inference*. Cambridge University Press, pp. 25–62. [33]

Mumford, D. and J. Shah: 1989, 'Optimal Approximations by Piecewise Smooth Functions and Associated Variational Problems'. *Communications on Pure Applied Mathematics.* **42**(5), 577–684. [28]

Nackman, L. R.: 1982, 'Three-Dimensional Shape Description Using the Symetric Axis Transform'. Ph.D. thesis, University of North Carolina at Chapel Hill. Department of Computer Science, NC. [14, 86]

Nackman, L. R.: 1984, 'Two-Dimensional Critical Point Configuration Graphs'. *IEEE Transactions on Pattern Analysis and Machine Intelligence* **6**(4), 442–450. [329]

Nackman, L. R. and S. M. Pizer: 1985, 'Three-Dimensional Shape Description Using the Symmetric Axis Transform I: Theory'. *IEEE Transactions on Pattern Analysis and Machine Intelligence* **7**(2), 187–202. [86]

Naeve, A. and J.-O. Eklundh: 1994, 'Generalized Cylinders - What are they?'. In: C. Arcelli, L. Cordella, and G. S. di Baja (eds.): *Aspects of Visual Form Processing.* Singapore: World Scientific, pp. 384–409. Workshop held in Capri, Italy. [328]

Näf, M.: 1996, 'Voronoi Skeletons: A Semicontinous Implementation of the 'Symmetric Axis Transform' in 3D Space'. Ph.D. thesis, ETH Zürich , Commmunication Technology Institue, Image Analysis Group IKT/BIWI. [32, 208, 209, 211, 212, 214, 215, 216]

Näf, M., O. Kübler, R. Kikinis, M. Shenton, and G. Székely: 1996, 'Characterization and Recognition of 3D Organ Shape in Medical Image Analysis Using Skeletonization'. In: *Workshop on Mathematical Methods in Biomedical Image Analysis.* pp. 139–150. [202, 208, 209, 211, 215, 216]

Nichols, T. E. and A. P. Holmes: 2001, 'Nonparametric Permutation Tests for Functional Neuroimaging: A Primer with Examples'. *Human Brain Mapping* **15**, 1–25. [294]

Niethammer, M., S. Betelu, G. Sapiro, A. Tannenbaum, and P. J. Giblin: 2004, 'Area-Based Medial Axis of Planar Curves'. *International Journal of Computer Vision* **60**(3), 203–224. [6]

Nystrm, I. and Ö. Smedby: 2001, 'Skeletonization of Volumetric Vascular Images - Distance Information Utilized for Visualization'. *Journal of Combinatorial Optimization* **5**(1), 27–41. Special Issue on *Optimization Problems in Medical Applications.* [188]

O'Dunlaing, C., M. Sharir, and C. K. Yap: 1983, 'Retraction: A New Approach to Motion Planning'. In: *Proceedings of the Fifteenth Annual ACM Symposium on Theory of Computing.* pp. 207–220. [340]

Ogniewicz, R.: 1993, *Discrete Voronoi Skeletons.* Hartung-Gorre Verlag, Konstanz, Germany. [30, 32, 191, 203, 207]

Ogniewicz, R.: 1994, 'Skeleton-Space: A Multiscale Shape Description Combining Region and Boundary Information'. In: *Proceedings of Computer Vision and Pattern Recognition.* pp. 746–751. [226]

Ogniewicz, R. L. and O. Kübler: 1995, 'Hierarchic Voronoi skeletons'. *Pattern Recognition* **28**(3), 343–359. [31, 207]

Okabe, A., B. Boots, K. Sugihara, and S. N. Chiu: 2000, *Spatial Tessellations: Concepts and Applications of Voronoi Diagrams, Probability and Statistics*. New York: Wiley, 2nd edition. [327, 328]

Okamura, A. M.: 2000, 'Haptic Exploration of Unknown Objects'. Ph.D. thesis, Stanford University, Department of Mechanical Engineering, CA. [341]

Okamura, A. M.: 2003, 'Uniting Haptic Exploration and Display'. In: R. A. Jarvis and A. Zelinsky (eds.): *Proceedings of the 10th International Symposium on Robotics Research*, Vol. 6 of *Springer Tracts in Advanced Robotics*. Lorne, Victoria, Australia, pp. 225–238. [341]

O'Neill, B.: 1997, *Elementary Differential Geometry*. Orlando, FL: Academic Press, 2nd edition. 482 pages. [42, 79]

Osada, R., T. Funkhouser, B. Chazelle, and D. Dobkin: 2002, 'Shape Distributions'. *ACM Transactions on Graphics* **21**(4), 807–832. [309, 312, 314, 319, 320, 321]

Osher, S. and J. Sethian: 1988, 'Fronts Propagating with Curvature Dependent Speed: Algorithms Based on Hamilton-Jacobi Formaulation'. *Journal of Computational Physics* **79**, 12–49. [129]

Paik, D. S., C. F. Beaulieu, G. D. R. Brooke Jeffrey, Rubin, and S. Napel: 1998, 'Automated Path Planning for Virtual Endoscopy'. *Medical Physics* **25**(5), 629–637. [343, 344]

Pallini, A. and F. Pesarin: 1992, 'A Class of Combinations of Dependent Tests by a Resampling Procedure'. In: K. H. Jöckel, G. Rothe, and W. Sendler (eds.): *Bootstrapping and Related Techniques*, Vol. 376 of *Lecture Notes in Economics and Mathematical Systems*. Berlin, Springer, pp. 93–97. [299]

Pantazis, D., R. Leahy, T. Nichol, and M. Styner: 2004, 'Statistical Surface-Based Morphometry Using a Non-parametric Approach'. In: *International Symposium on Biomedical Imaging(ISBI)*. [294]

Papaioannou, G. and E.-A. Karabassi: 2003, 'On the Automatic Assemblage of Arbitrary Broken Solid Artefacts'. *Image and Vision Computing* **21**(5), 401–412. [334]

Parsons, D. and J. Canny: 1994, 'Geometric Problems in Molecular Biology and Robotics'. In: R. Altman, D. Brutlag, P. Karp, R. Lathrop, and D. Searls (eds.): *Proceedings of the Second International Conference on Intelligent Systems for Molecular Biology (ISMB-94)*. AAAI Press: Menlo Park CA. [348]

Patrikalakis, N. M. and T. Maekawa: 2002, *Shape Interrogation for Computer Aided Design and Manufacturing*. New York: Springer. [341]

Pelillo, M., K. Siddiqi, and S. Zucker: 1999, 'Matching Hierarchical Structures Using Association Graphs'. *IEEE Transactions on Pattern Analysis and Machine Intelligence* **21**(11), 1105–1120. [309, 310, 311, 314]

Penn, A.: 2003, 'Space Syntax And Spatial Cognition — Or Why the Axial Line?'. *Environment and Behavior* **35**(1), 30–65. [333]

Pentland, A.: 1986, 'Perceptual Organization and the Representation of Natural Form'. *Artificial Intelligence* **28**, 293–331. [313]

Pesarin, F.: 2001, *Multivariate Permutation Tests with Applications in Biostatistics*. Chirchester: Wiley. [297, 298]

Pilgram, R., P. T. Fletcher, S. M. Pizer, O. Pachinger, and R. Schubert: 2003, 'Common Shape Model and Inter-individual Variations of the Heart using Medial Representation: A Pilot Study'. Technical Report, Institute for Medical Knowledge Representation and Visualization, University for Health Informatics and Technology, Tyrol, Austria. [288]

Pizer, S., W. Oliver, and S. Bloomberg: 1987, 'Hierarchical Shape Description via the Multiresolution Symmetry Axis Transform'. *IEEE Transactions on Pattern Analysis and Machine Intelligence* **9**(4), 505–511. [207]

Pizer, S., D. Eberly, B. Morse, and D. Fritsch: 1998, 'Zoom-Invariant Figural Shape: The Mathematics of Cores'. *Computer Vision and Image Understanding (CVIU '98)* **69**, 55–71. [29, 93]

Pizer, S. M., J. M. Coggins, and C. A. Burbeck: 1991, 'Formation of Image Objects in Human Vision'. In: *Computer Assisted Radiology (Proceedings, CAR)*. pp. 535–542. [55]

Pizer, S. M., D. S. Fritsch, P. Yushkevich, V. Johnson, and E. Chaney: 1999, 'Segmentation, Registration and Measurement of Shape Variation via Image Object Shape'. *IEEE Transactions on Medical Imaging* **18**, 851–865. [19]

Pizer, S. M., P. T. Fletcher, S. Joshi, A. Thall, J. Z. Chen, Y. Fridman, D. S. Fritsch, A. G. Gash, J. M. Glotzer, M. R. Jiroutek, C. Lu, K. E. Muller, G. Tracton, P. Yushkevich, and E. L. Chaney: 2003a, 'Deformable M-reps for 3D Medical Image Segmentation'. *International Journal of Computer Vision* **55**(2–3), 85–106. [71]

Pizer, S. M., K. Siddiqi, G. Székeley, J. N. Damon, and S. W. Zucker: 2003b, 'Multiscale Medial Axes and Their Properties'. *International Journal of Computer Vision* **55**(3), 155–179. [25, 114]

Pizer, S. M., J. Jeong, R. Broadhurst, S. Ho, and J. Stough: 2005a, 'Deep Structure of Images in Populations via Geometric Models in Populations'. In: O. Olsen, L. Florack, and A. Kuijper (eds.): *International Workshop on Deep Structure, Singularities and Computer Vision (DSSCV)*, Vol. 3753. Springer LNCS, Maastricht, The Netherlands, pp. 48–58. [283]

Pizer, S. M., P. T. Fletcher, S. Joshi, A. G. Gash, J. Stough, A. Thall, G. Tracton, and E. L. Chaney: 2005b, 'A Method and Software for Segmentation of Anatomic Object Ensembles by Deformable M-Reps'. *Medical Physics* **32**(5), 1335–1345. [275]

Pizer, S. M., J. Y. Jeong, C. Lu, K. Muller, and S. Joshi: 2005c, 'Estimating the Statistics of Multi-Object Anatomic Geometry Using Inter-Object Relationships'. In: O. F. Olsen, L. Florack, and A. Kuijper (eds.): *International Workshop on Deep Structure, Singularities and Computer Vision*, Vol. LNCS. pp. 59–70. [259]

Pollitt, A. J.: 2004, 'Euclidean and Affine Symmetry Sets and Medial Axes'. Ph.D. thesis, University of Liverpool. http://www.liv.ac.uk/~pjgiblin/. [38, 61, 62, 66, 67, 68]

Pollitt, A. J., P. J. Giblin, and B. B. Kimia: 2004, 'Consistency Conditions on Medial Axis'. In: *Eigth European Conference on Computer Vision*, Vol. 2. Springer, Prague, Czech Republic, pp. 530–541. [66, 67, 68]

Pontier, S., B. Shariat, and D. Vandorpe: 1998, 'Weighted Skeleton for Implicit Object Reconstruction'. In: *Proceedings of the 8th ICECGDG*. Austin, TX, pp. 520–525. [337]

Potterton, L., S. McNicholas, E. Krissinel, J. Gruber, K. Cowtan, P. Emsley, G. N. Murshudov, S. Cohen, A. Perrakisd, and M. Noble: 2004, 'Developments in the CCP4 Molecular-Graphics Project'. *Biological Crystallography* **60**(Part 12, Number 1), 2288–2294. Acta Crystallographica, Section D. [347]

Pranovich, S.: 2004, 'Structural Sketcher: A Tool for Supporting Architects in Early Stages'. Ph.D. thesis, Computer Science Department, Technische Universiteit Eindhoven, The Netherlands. [333]

Pranovich, S., H. Achten, B. de Vries, and J. van Wijk: 2005, 'Structural Sketcher: Representing and Applying Well-Structured Graphic Representations in Early Design'. *International Journal of Architectural Computing* **3**(1), 75–92. [333]

Preparata, F. and M. Shamos: 1985, *Computational Geometry*. New York: Springer. [192]

Price, M. and C. G. Armstrong: 1995, 'Hexahedral Mesh Generation by Medial Surface Subdivision: Part I. Solids with Convex Edges'. *International Journal for Numerical Methods in Engineering* **38**(19), 3335–3359. [341]

Price, M. and C. G. Armstrong: 1997, 'Hexahedral Mesh Generation by Medial Surface Subdivision: Part II. Solids with Flat and Concave Edges'. *International Journal for Numerical Methods in Engineering* **40**, 111–136. [341]

Prusinkiewicz, P.: 2004, 'Modeling Plant Growth and Development'. *Current Opinion in Plant Biology* **7**(1), 79–83. [336]

Prusinkiewicz, P. and A. Lindenmayer: 1990, *The Algorithmic Beauty of Plants*. Springer-Verlag. http://algorithmicbotany.org/. [336]

Psotka, J.: 1978, 'Perceptual Processes That May Create Stick Figures and BalancE'. *Journal of Experimental Psychology: Human Perception and Peformance* **4**(1), 101–111. [22]

Pudney, C.: 1998, 'Distance-Ordered Homotopic Thinning: A Skeletonization Algorithm for 3D Digital Images'. *Computer Vision and Image Understanding* **72**(3), 404–413. [128, 142]

Ragnemalm, I.: 1993, 'The Euclidean Distance Transform in Arbitrary Dimensions'. *Pattern Recognition Letters* **14**, 883–888. [163]

Rao, M., J. Stough, Y.-Y. Chi, K. Muller, G. S. Tracton, S. M. Pizer, and E. L. Chaney: 2005, 'Comparison of Human and Automatic Segmentations of Kidneys from CT Images'. *International Journal of Radiation Oncology, Biology, Physics* **61**(3), 954–960. [288]

Reddy, J. and G. Turkiyyah: 1995, 'Computation of 3D Skeletons Using a Generalized Delaunay Triangulation Technique'. *Computer-Aided Design* **27**(9), 677–694. [208]

Reddy, M.: 1992, 'Skeletons: Generation, Representation and Applications to Finite Element Modeling'. Master's thesis, Carnegie Mellon University, Department of Civil Engineering. [207]

Remy, E. and E. Thiel: 2003, 'Look-Up Tables for Medial Axis on Squared Euclidean Distance Transform'. In: I. Nyström, G. Sanniti di Baja, and S.

Svensson (eds.): *International Conference on Discrete Geometry for Computer Imagery*, Vol. 2886 of *Lecture Notes in Computer Science*. Springer, Naples, Italy, pp. 224–235. [165]

Reyner, S. W.: 1977, 'An Analysis of a Good Algorithm for the Subtree Problem'. *SIAM Journal on Computing.* **6**, 730–732. [319]

Rice, S. V., G. Nagy, and T. A. Nartker: 1999, *Optical Character Recognition: An Illustrated Guide to the Frontier*. Amsterdam: Kluwer. [187]

Rock, I. and C. Linnett: 1993, 'Is a Perceived Shape Based on its Retinal Image?'. *Perception* **22**(1), 61–76. [21]

Rosenfeld, A. and J. L. Pfaltz: 1966, 'Sequential Operations in Digital Picture Processing'. *Journal of the Association for Computing Machinery* **13**(4), 471–494. [161, 166]

Rosenfeld, A. and J. L. Pfaltz: 1968, 'Distance Functions on Digital Pictures'. *Pattern Recognition* **1**, 33–61. [161]

Rosin, P. R. and G. A. W. West: 1995, 'Salience Distance Transforms'. *Graphical Models and Image Processing* **57**(6), 483–521. [161]

Rouy, E. and A. Tourin: 1992, 'A Viscosity Solutions Approach to Shape-from-Shading'. *SIAM Journal of Numerical Analysis* **29**(3), 867–884. [130]

Runions, A., M. Fuhrer, B. Lane, P. Federl, A.-G. Rolland-Lagan, and P. Prusinkiewicz: 2005, 'Modeling and Visualization of Leaf Venation Patterns'. *ACM Transactions on Graphics* **24**(3), 702–711. [336]

Rutovitz, D.: 1966, 'Pattern Recognition'. *Journal of the Royal Statistical Society* **129**, 504–530. [168]

Saha, P. K. and B. B. Chaudhuri: 1994, 'Detection of 3-D Simple Points for Topology Preserving Transformations with Application to Thinning'. *IEEE Transactions on Pattern Analysis and Machine Intelligence* **16**(10), 1028–1032. [177]

Sanniti di Baja, G.: 1994, 'Well-Shaped, Stable, and Reversible Skeletons from the (3,4)-Distance Transform'. *Journal of Visual Communication and Image Representation* **5**(1), 107–115. [171, 172, 174]

Sanniti di Baja, G. and S. Svensson: 2000, 'Surface Skeletons Detected on the D^6 Distance Transform'. In: F. J. Ferri, J. M. Inetsa, A. Amin, and P. Pudil (eds.): *Proceedings of SSSPR 2000 - Alicante: Advances in Pattern Recognition*, Vol. 1876 of *Lecture Notes in Computer Science*. Berlin/Heidelberg: Springer, pp. 387–396. [176, 181]

Sanniti di Baja, G. and E. Thiel: 1994, '(3,4)-Weighted Skeleton Decomposition for Pattern Representation and Description'. *Pattern Recognition* **27**(8), 1039–1049. [188]

Sanniti di Baja, G. and E. Thiel: 1996, 'Skeletonization Algorithm Running on Path-Based Distance Maps'. *Image and Vision Computing* **14**, 47–57. [168, 170]

Savadjiev, P., F. P. Ferrie, and K. Siddiqi: 2003, 'Surface Recovery from 3D Point Data Using a Combined Parametric and Geometric Flow Approach'. In: *International Workshop on Energy Minimization Methods in Computer Vision and Pattern Recognition*, Vol. LNCS 2683. Springer, Lisbon, Portugal, pp. 325–340. [147]

Schmitt, M.: 1989, 'Some Examples of Algorithms Analysis in Computational Geometry by Means of Mathematical Morphology Techniques'. In: J. Boissonnat and J. Laumond (eds.): *Geometry and Robotics*, Vol. LNCS 391. Springer, Toulouse, France, pp. 225–246. [31, 193, 208]

Schobel, S., M. Chakos, G. Gerig, H. Bridges, H. Gu, H. Charles, and J. Lieberman: 2001, 'Duration and Severity of Illness and Hippocampal Volume in Schizophrenia as Assessed by 3D-Manual Segmentation'. *Schizophrenia Research* **49**(1–2), 165. [301]

Schröder, G., S. Ramsden, A. Christy, and S. Hyde: 2003, 'Medial surfaces of Hyperbolic Structures'. *The European Physical Journal B* **35**(4), 551–565. [238]

Sebastian, T. B., P. N. Klein, and B. B. Kimia: 2001, 'Recognition of Shapes by Editing Shock Graphs'. In: *Proceedings of the IEEE International Conference on Computer Vision*. Vancouver, Canada, pp. 755–762. [309, 310, 314]

Sebastian, T. B., H. Tek, J. J. Crisco, S. W. Wolfe, and B. B. Kimia: 2003, 'Segmentation of Carpal Bones from 3D CT Images using Skeletally Coupled Deformable Models'. *Medical Image Analysis* **7**(1), 21–45. [343]

Sebastian, T., P. Klein, and B. Kimia: 2004, 'Recognition of Shapes by Editing their Shock Graphs'. *PAMI* **26**, 551–571. [57, 60, 310, 311, 328]

Seidel, R.: 1982, 'The complexity of Voronoi diagrams in higher dimension'. In: *20th Allerton Conference on Communication, Control and Computing*. [208]

Serra, J.: 1982, *Image Analysis and Mathematical Morphology*, Vol. 1. Academic Press. [26, 156, 173, 344]

Sethian, J. A.: 1996, 'A Fast Marching Level Set Method for Monotonically Advancing Fronts'. *Proceedings of the National Academy of Sciences, USA* **93**, 1591–1595. [129, 130, 136, 161]

Shah, J.: 1996, 'A Common Framework for Curve Evolution, Segmentation and Anisotropic Diffusion'. In: *Proceedings of the IEEE Conference on Computer Vision and Pattern Recognition*. pp. 136–142. [28]

Shah, J.: 2001, 'Segmentation of Shapes'. In: *Workshop on Scale-Space and Morphology*. [29]

Shah, J.: 2005a, 'Grayscale Skeletons and Segmentation of Shapes'. *Computer Vision and Image Understanding* **99**(1), 96–109. [29, 148, 150]

Shah, J.: 2005b, 'Skeletons of 3D Shapes'. In: *The Fifth International Conference on Scale-Space and PDE Methods in Computer Vision*. [29]

Shaked, D. and A. M. Bruckstein: 1998, 'Pruning Medial Axes'. *Computer Vision and Image Understanding* **69**(2), 156–169. [173]

Shankar, R.: 1994, *Principles of Quantum Mechanics*. Plenum Press. [130, 132]

Sharvit, D., J. Chan, H. Tek, and B. B. Kimia: 1998, 'Symmetry-based Indexing of Image Databases'. *Journal of Visual Communication and Image Representation* **9**(4), 366–380. [47]

Sheehy, D.: 1994, 'Medial Surface Computation Using a Domain Delaunay triangulation'. Ph.D. thesis, The Queen's University of Belfast. [238]

Sheehy, D., C. Armstrong, and D. Robinson: 1995, 'Computing the Medial Surface of a Solid From a Domain Delaunay triangulation'. In: *Proceedings of the Third ACM Symposium on Solid Modeling and Applications*. pp. 201–212. [238]

Sheehy, D. J., C. G. Armstrong, and D. J. Robinson: 1996, 'Shape Description By Medial Surface Construction'. *IEEE Transactions on Visualization and Computer Graphics* **2**(1), 62–72. [341]

Sherbrooke, E. C., N. M. Patrikalakis, and E. Brisson: 1996, 'An Algorithm for the Medial Axis Transform of 3D Polyhedral Solids'. *IEEE Transactions on Visualization and Computer Graphics* **2**(1), 44–61. [225]

Sherstyuk, A.: 1999, 'Shape Design Using Convolution Surfaces'. In: *Proceedings of Shape Modeling International '99*. [7]

Shilane, P., P. Min, M. Kazhdan, and T. Funkhouser: 2004, 'The Princeton Shape Benchmark'. In: *Shape Modeling International*. Genova, Italy. [320]

Shinagawa, Y., T. L. Kunii, and Y. L. Kergosien: 1991, 'Surface Coding Based on Morse Theory'. *IEEE Transactions On Computer Graphics and Applications* **11**(5), 66–78. [313]

Shlyakhter, I., M. Rozenoer, J. Dorsey, and S. Teller: 2001, 'Reconstructing 3D Tree Models from Instrumented Photographs'. *IEEE Computer Graphics and Applications* **21**(3), 53–61. [336]

Shokoufandeh, A., S. J. Dickinson, K. Siddiqi, and S. W. Zucker: 1999, 'Indexing Using a Spectral Encoding of Topological Structure'. In: *IEEE Conference on Computer Vision and Pattern Recognition*. Fort Collins, CO, pp. 491–497. [312, 314]

Shokoufandeh, A., D. Macrini, S. Dickinson, K. Siddiqi, and S. W. Zucker: 2005, 'Indexing Hierarchical Structures Using Graph Spectra'. *IEEE Transactions on Pattern Analysis and Machine Intelligence* **27**(7), 1125–1140. [312, 313, 316, 317, 318]

Siddiqi, K. and B. B. Kimia: 1996, 'A Shock Grammar for Recognition'. In: *Proceedings of the Conference on Computer Vision and Pattern Recognition*. pp. 507–513. [47]

Siddiqi, K., B. B. Kimia, and K. J. Tresness: 1996, 'Parts of Visual Form: Psychophysical Aspects'. *Perception* **25**(4), 399–424. [21]

Siddiqi, K., S. Bouix, A. Tannenbaum, and S. W. Zucker: 1999a, 'The Hamilton-Jacobi Skeleton'. In: *Proceedings of the IEEE International Conference on Computer Vision*. Kerkyra, Greece, pp. 828–834. [27, 138]

Siddiqi, K., A. Shokoufandeh, S. J. Dickinson, and S. W. Zucker: 1999b, 'Shock Graphs and Shape Matching'. *International Journal of Computer Vision* **35**(1), 13–32. [47, 309, 310, 311, 314, 316, 317, 319, 328]

Siddiqi, K., B. B. Kimia, A. R. Tannenbaum, and S. W. Zucker: 2001, 'On the Psychophysics of the Shape Triangle'. *Vision Research* **41**, 1153–1178. [24]

Siddiqi, K., S. Bouix, A. Tannenbaum, and S. W. Zucker: 2002, 'Hamilton-Jacobi Skeletons'. *International Journal of Computer Vision* **48**(3), 215–231. [27, 72, 76, 137, 142, 313, 314]

Siersma, D.: 1999, 'Properties of Conflict Sets in the Plane'. *Geometry and Topology of Caustics – Caustics 1998, Banach Center Publications*. **50**, 267–276. [86]

Singh, M. and D. D. Hoffman: 1997, 'Salience of Visual Parts'. *Cognition* **63**, 29–78. [21]

Soille, P.: 2004, *Morphological Image Analysis: Principles and Applications*. New York: Springer, corr. 2nd printing of the 2nd edition. http://ams.jrc.it/soille/book2ndprint/. [328, 344]

Sotomayor, J., D. Siersma, and R. Garcia: 1999, 'Curvatures of Conflict Surfaces in Euclidean 3-space'. *Geometry and Topology of Caustics – Caustics 1998, Banach Center Publications.* **50**, 277–285. [86]

Staib, L. H. and J. S. Duncan: 1996, 'Model based Deformable Surface Finding for Medical Images'. *IEEE Transactions on Medical Imaging* **15**(5), 1–13. [273]

Stavroudis, O. N.: 1972, *The Optics of Rays, Wavefronts and Caustics.* New York: Academic. [130, 133]

Stolpner, S. and K. Siddiqi: 2006, 'Revealing Significant Medial Structure in Polyhedral Meshes'. In: *3DPVT '06: Proceedings of the Third International Symposium on 3D Data Processing, Visualization, and Transmission (3DPVT'06).* IEEE Computer Society, Washington, DC, pp. 365–372. [136, 153, 326]

Storti, D. W., G. M. Turkiyyah, M. A. Ganter, C. T. Lim, and D. M. Stat: 1997, 'Skeleton-Based Modeling Operations on Solids'. In: *SMA '97: Proceedings of the Fourth Symposium on Solid Modeling and Applications.* pp. 141–154. Held May 14–16, 1997 in Atlanta, Georgia. [7]

Stough, J., S. M. Pizer, E. L. Chaney, and M. Rao: 2004, 'Clustering on Image Boundary Regions for Deformable Model Segmentation'. In: *International Symposium on Biomedical Imaging (ISBI)*, Vol. Catalog Number 04EX821C. pp. 436–439, IEEE. [282]

Styner, M.: 2001, 'Combined Boundary-Medial Shape Description of Variable Biological Objects'. Ph.D. thesis, University of North Carolina at Chapel Hill, Chapel Hill, NC. [32]

Styner, M., G. Gerig, S. C. Joshi, and S. M. Pizer: 2003a, 'Automatic and Robust Computation of 3-D Medial Models Incorporating Object Variability'. *International Journal of Computer Vision* **55**, 107–122. [279, 303]

Styner, M., G. Gerig, J. Lieberman, D. Jones, and D. Weinberger: 2003b, 'Statistical Shape Analysis of Neuroanatomical Structures Based on Medial Models'. *Medical Image Analysis* **7**(3), 207–220. [293]

Styner, M., G. Gerig, S. Pizer, and S. Joshi: 2003c, 'Automatic and Robust Computation of 3D Medial Models Incorporating Object Variability'. *International Journal of Computer Vision* **55**(2/3), 107–122. [249]

Styner, M., J. Lieberman, D. Pantazis, and G. Gerig: 2004, 'Boundary and Medial Shape Analysis of the Hippocampus in Schizophrenia'. *Medical Image Analysis* **8**(3), 197–203. [293]

Styner, M., J. A. Lieberman, R. K. McClure, D. R. Weinberger, D. W. Jones, and G. Gerig: 2005, 'Morphometric Analysis of Lateral Ventricles in Schizophrenia and Healthy Controls Regarding Genetic and Disease-Specific Factors'. *Proceedings of the National Academy of Science* **102**(12), 4872–4877. [303, 304]

Sundar, H., D. Silver, N. Gagvani, and S. Dickinson: 2003, 'Skeleton Based Shape Matching and Retrieval'. In: *International Conference On Shape Modeling International and Applications.* Seoul, Korea, pp. 130–142. [314]

Sussman, M., P. Smereka, and S. Osher: 1994, 'A Level Set Approach for Computing Solutions to Incompressible Two-Phase Flow'. *Journal of Computational Physics* **114**, 146–154. [130]

Svensson, S.: 2002, 'Reversible Surface Skeletons of 3D Objects by Iterative Thinning of Distance Transforms'. In: G. Bertrand, A. Imiya, and R. Klette (eds.): *Digital and Image Geometry*, Vol. 2243 of *Lecture Notes in Computer Science*. Springer, Dagstuhl, Germany, pp. 395–406. [177]

Svensson, S. and M. Aronsson: 2003, 'Using Distance Transform based Algorithms for Extracting Measures of the Fibre Network in Volume Images of Paper'. *IEEE Transactions on Systems, Man, and Cybernetics. Part B: Cybernetics* **33**(4), 562–571. Special issue on 3-D Image Analysis and Modeling. [188]

Svensson, S. and G. Borgefors: 2002a, 'Digital Distance Transforms in 3D Images using Information from Neighbourhoods up to $5 \times 5 \times 5$'. *Computer Vision and Image Understanding* **88**(1), 24–53. [161]

Svensson, S. and G. Borgefors: 2002b, 'Distance Transforms in 3D using Four Different Weights'. *Pattern Recognition Letters* **23**(12), 1407–1418. [161]

Svensson, S. and G. Sanniti di Baja: 2002, 'Using Distance Transforms to Decompose 3D Discrete Objects'. *Image and Vision Computing* **20**(8), 529–540. [188]

Svensson, S. and G. Sanniti di Baja: 2003, 'Simplifying Curve Skeletons in Volume Images'. *Computer Vision and Image Understanding* **90**, 242–257. [184, 185, 186, 187]

Svensson, S., I. Nyström, and G. Borgefors: 1999, 'Fully Reversible Skeletonization for Volume Images Based on Anchor-Points from the D^{26} Distance Transform'. In: *Proceedings of 11th Scandinavian Conference on Image Analysis*. Kangerlussuaq, Greenland. pp. 601–608. [177]

Svensson, S., I. Nyström, C. Arcelli, and G. Sanniti di Baja: 2002a, 'Using Grey-Level and Distance Information for Medial Surface Representation of Volume Images'. In: R. Kasturi, D. Laurendeau, and C. Suen (eds.): *Proceedings of 16th ICPR, Québec City*, Vol. II. IEEE Computer Society, Canada. pp. 324–327. [189]

Svensson, S., I. Nyström, and G. Sanniti di Baja: 2002b, 'Curve Skeletonization of Surface-Like Objects in 3D Images Guided by Voxel Classification'. *Pattern Recognition Letters* **23**(12), 1419–1426. [180, 182]

Székely, G.: 1996, 'Shape Characterization by Local Symmetries'. Habilitationsschrift, Institut fur Kommunikationstechnik, Fachgruppe Bildwissenschaft, ETH Zürich. [30, 31]

Székely, G., C. Brechbühler, O. Kübler, R. Ogniewicz, and T. Budinger: 1992, 'Mapping the Human Cerebral Cortex Using 3D Medial Manifolds'. In: R. Robb (ed.): *Proceedings of 2nd Int. Conf. Visualization in Biomed. Comp.*, Vol. SPIE 1808. pp. 130–144. [209, 212]

Székely, G., M. Näf, C. Brechbühler, and K. O.: 1994, 'Calculating 3D Voronoi Diagrams of Large Unrestricted Point Sets for Skeleton Generation of Complex 3D Shapes'. In: *Proceedings of 2nd International Workshop on Visual Form*. World Scientific, Singapore, pp. 532–541. [209]

Talbot, H. and L. Vincent: 1992, 'Euclidean Skeletons and Conditional Bisectors'. In: *Proceedings of SPIE Conf. Medical Imaging V: Image Processing*, Vol. SPIE Vol. 1818. pp. 862–876. [201]

Tamrakar, A. and B. B. Kimia: 2004, 'Medial Visual Fragments as an Intermediate Image Representation for Segmentation and Perceptual Grouping'. In: *Proceedings of IEEE Workshop on Perceptual Organization in Computer Vision, POCV.* p. 47. [52, 350, 351]

Tanemura, M., T. Ogawa, and N. Ogita: 1983, 'A New Algorithm for Three-Dimensional Voronoi Tesselation'. *Journal of Computational Physics* **51**, 191–207. [208]

Tari, S. and J. Shah: 1998, 'Local Symmetries of Shapes in Arbitrary Dimension'. In: *Proceedings of the IEEE International Conference on Computer Vision.* Bombay, India. [29]

Tari, S. and J. Shah: 2000, 'Nested Local Symmetry Set'. *Computer Vision and Image Understanding* **79**(2), 267–280. [29]

Tari, Z. S. G., J. Shah, and H. Pien: 1997, 'Extraction of Shape Skeletons from Grayscale Images'. *Computer Vision and Image Understanding* **66**, 133–146. [27, 28, 29]

Teichmann, M. and S. Teller: 1998, 'Assisted Articulation of Closed Polygonal Models'. In: *Proceedings of 9th Eurographics Workshop on Animation and Simulation.* Lisbon, Portugal. [340]

Teixeira, R. C.: 1998, 'Curvature Motions, Medial Axes and Distance Transforms'. Ph.D. thesis, Harvard University, Cambridge, MA. [14]

Tek, H. and B. B. Kimia: 1999, 'Symmetry Maps of Free-Form Curve Segments Via Wave Propagation'. In: *Proceedings of the IEEE International Conference on Computer Vision.* Kerkyra, Greece, pp. 362–369. [129]

Tek, H. and B. B. Kimia: 2001, 'Boundary Smoothing via Symmetry Transforms'. *Journal of Mathematical Imaging and Vision* **14**(3), 211–223. [47, 330, 342]

Tek, H. and B. B. Kimia: 2003, 'Symmetry Maps of Free-Form Curve Segments Via Wave Propagation'. *International Journal of Computer Vision* **54**(1–3), 35–81. [47, 330]

Terriberry, T. B.: 2006, 'Continuous Medial Models in Two-Sample Statistics of Shape'. Ph.D. thesis, Department of Computer Science, University of North Carolina, Chapel Hill, NC. [263, 264]

Thall, A.: 2004, 'Deformable Solid Modeling via Medial Sampling and Displacement Subdivision'. Ph.D. thesis, University of North Carolina, Chapel Hill. [259, 282]

Thibault, D. and C. M. Gold: 2000, 'Terrain Reconstruction from Contours by Skeleton Generation'. *GeoInformatica* **4**(4), 349–373. [330]

Thiel, E. and A. Montanvert: 1992, 'Chamfer Masks: Discrete Distance Functions, Geometrical Properties and Optimization'. In: *Proceedings of 11th ICPR, The Hague.* pp. 244–247. [161]

Thompson, D. C.: 2000, 'Feasibility of a Skeletal Modeler for Conceptual Mechanical Design'. Ph.D. thesis, University of Texas at Austin. [342]

Tikhonov, A. and V. Arsenin: 1977, *Solution of ill posed problems.* Winston. [200]

Toth, R. E.: 1988, 'Theory and Language in Landscape Analysis, Planning, and Evaluation'. *Landscape Ecology* **1**(4), 193–201. [336]

Tresset, P. and F. F. Leymarie: 2005, 'Generative Portrait Sketching'. In: H. Thwaites (ed.): *Proceedings of the 11th International Conference on Virtual Systems and Multimedia (VSMM)*. Ghent, Belgium, pp. 739–748. http://belgium.vsmm.org/. [337, 338]

Tschirren, J., G. McLennan, K. Palyagyi, E. Hoffman, and M. Sonka: 2005, 'Matching and Anatomic Labeling of Human Airway Tree'. *IEEE Transactions on Medical Imaging* **24**(12), 1540–1547. [34]

Turkiyyah, G., D. Storti, M. Ganter, H. Chen, and M. Vimawala: 1997, 'An Accelerated Triangulation Method for Computing Skeletons of Free-Form Solid Models'. *Computer Aided Design* **29**(1), 5–19. [237, 239]

Turner, A., A. Penn, and B. Hillier: 2005, 'An Algorithmic Definition of the Axial Map'. *Environment and Planning B: Planning and Design* **32**(3), 425–444. [333]

Ucoluk, G. and I. H. Toroslu: 1999, 'Automatic Reconstruction of Broken 3-D Surface'. *Computers and Graphics* **23**(4), 573–582. [334]

van Leemput, K., F. Maes, D. Vandermeulen, and P. Seutens: 1999, 'Automated Model-based Tissue Classification of MR Images of the Brain'. *IEEE Transactions on Medical Imaging* **18**, 897–908. [302]

van Manen, M.: 2003a, 'Curvature and Torsion Formulas for Conflict Sets'. *Caustics'02*. [86]

van Manen, M.: 2003b, 'The Geometry of Conflict Sets'. Ph.D. thesis, University of Utrecht. Department of Mathematics. [86]

van Overveld, C. W. and B. Wyvill: 1997, 'Polygon Inflation for Animated Models: A Method for the Extrusion of Arbitrary Polygon Meshes'. *Journal of Visualization and Computer Animation* **8**(1), 3–16. [337, 340]

van Tonder, G. J.: 2006a, 'Order and Complexity of Naturalistic Landscapes: On Creation, Depiction and Perception of Japanese Dry Rock Gardens'. In: *Visual Thought: The Depictive Space of the Mind*. Benjamins. [335]

van Tonder, G. J.: 2006b, 'Recovery of Visual Structure in Illustrated Japanese Gardens'. *Pattern Recognition Letters*. [335, 336]

van Tonder, G. J. and Y. Ejima: 2000a, 'From Image Segmentation to Anti-textons'. *Perception* **29**(10), 1231–1247. [349]

van Tonder, G. J. and Y. Ejima: 2000b, 'The Patchwork Engine: Image Segmentation from Image Symmetries'. *Neural Networks* **13**(3), 291–303. [349]

van Tonder, G. J. and Y. Ejima: 2003, 'Flexible Computation of Shape Symmetries Within the Maximal Disk Paradigm'. *IEEE Transactions on Systems, Man, and Cybernetics, Part B* **33**(3), 535–540. [335, 349]

van Tonder, G. J. and M. J. Lyons: 2005, 'Visual Perception in Japanese Rock Garden Design'. *Axiomathes* **15**(3), 353–371. [335]

van Tonder, G. J., M. J. Lyons, and Y. Ejima: 2002, 'Visual Structure of a Japanese Zen Garden'. *Nature* **419**, 359–360. [335, 336]

Vasilevskiy, A. and K. Siddiqi: 2002, 'Flux Maximizing Geometric Flows'. *IEEE Transactions on Pattern Analysis and Machine Intelligence* **24**(12). [34]

Vermeer, P. J.: 1994, 'Medial Axis Transform to Boundary Representation Conversion'. Ph.D. thesis, Purdue University. [14]

Verroust, A. and F. Lazarus: 2000, 'Extracting Skeletal Curves from 3D Scattered Data'. *The Visual Computer* **16**(1), 15–25. [337, 340]

Verwer, B. J. H.: 1991, 'Local Distances for Distance Transformations in Two and Three Dimensions'. *Pattern Recognition Letters* **12**(11), 671–682. [161]

Vincent, L. and P. Soille: 1991, 'Watersheds in Digital Spaces: An Efficient Algorithm Based on Immersion Simulations'. *IEEE Transactions on Pattern Analysis and Machine Intelligence* **13**(6), 583–598. [329]

Vossepoel, A. M.: 1988, 'A Note on 'Distance Transformations in Digital Images''. *Computer Vision, Graphics, and Image Processing* **43**, 88–97. [161]

Vranic, D. and D. Saupe: 2001, '3-D Model Retrieval with Spherical Harmonics and Moments'. In: *Proceedings of the DAGM*. pp. 392–397. [312, 319, 320]

Wade, L. and R. E. Parent: 2002, 'Automated Generation of Control Skeletons for Use in Animation'. *The Visual Computer* **18**(2), 97–110. [340]

Wan, M., F. Dachille, and A. Kaufman: 2001, 'Distance-Field Based Skeletons for Virtual Navigation'. In: *Proceedings of the conference on Visualization '01*. San Diego, CA, pp. 239–246. [343]

Wein, R., J. P. van den Berg, and D. Halperin: 2005, 'The Visibility-Voronoi Complex and its Applications'. In: *Proceedings of 21st ACM Symposium on Computational Geometry*. Pisa, Italy, pp. 63–72. [340]

Weinberger, D. R., M. F. Egan, A. Bertolino, J. H. Callicott, V. S. Mattay, B. K. Lipska, K. F. Berman, and T. E. Goldberg: 2001, 'Prefrontal Neurons and the Genetics of Schizophrenia'. *Biological Psychiatry* **50**, 825–844. [302]

Weiss, I.: 1986, 'Curve Fitting with Optimal Mesh Point Placement'. Technical Report CAR-TR-22, Comp. Vision Lab. University of Maryland. [194]

Weiss, I.: 1990, 'Shape Reconstruction on a Varying Mesh'. *IEEE Transactions on Pattern Analysis and Machine Intelligence* **12**(4), 345–362. [194]

Werner, S. and P. Long: 2003, 'Cognition Meets Le Corbusier — Cognitive Principles of Architectural Design'. In: C. Freksa et al. (eds.): *Spatial Cognition III*, Vol. 2685 of *Lecture Notes in Computer Science*. pp. 112–126. [333]

Wertheimer, M.: 1923, 'Laws of Organization in Perceptual Forms'. In: W. D. Ellis (ed.): *A Source Book of Gestalt Psychology*. New York: Brace Harcourt (1938). [348]

West, G. B., J. H. Brown, and B. J. Enquist: 1999, 'A General Model for the Structure and Allometry of Plant Vascular Systems'. *Nature* **400**, 664–667. [335]

Weygaert, R. v. d.: 2001, 'The Cosmic Foam: Stochastic Geometry and Spatial Clustering across the Universe'. In: E. Feigelson and G. Babu (eds.): *Statistical Challenges in Modern Astronomy III*. New York: Springer, pp. 175–196, (in 2003). [328, 329]

Weygaert, R. v. d. and V. Icke: 1989, 'Fragmenting the Universe. II. Voronoi vertices as Abell clusters'. *Astronomy and Astrophysics* **213**(1–2), 1–9. [328]

Wienstroeer, M. and H. Mathieu: 2002, 'Reverse Process Simulation for Forging Using the Medial Axis Transformation'. In: *Advanced Technology of Plasticity*.

Proceedings of the 7th ICTP, Oct. 2002, Vol. 2. Yokohama, Japan, pp. 973–978. [341]

Willis, A. and D. B. Cooper: 2004a, 'Alignment of Multiple Non-overlapping Axially Symmetric 3D Datasets'. In: *Proceedings of ICPR*, Vol. 4. Cambridge, pp. 96–99. [334]

Willis, A. and D. B. Cooper: 2004b, 'Bayesian Assembly of 3D Axially Symmetric Shapes from Fragments'. In: *Proceedings of CVPR*, Vol. 1. pp. 82–89. [334]

Wilmarth, S. A., N. M. Amato, and P. F. Stiller: 1999, 'Motion Planning for a Rigid Body Using Random Networks on the Medial Axis of the Free Space'. In: *Proceedings of the 15th Annual Symposium of Computational Geometry*. Miami, FL, pp. 173–180. [340]

Wolter, F.-E. and K.-I. Friese: 2000, 'Local and Global Geometric Methods for Analysis Interrogation, Reconstruction, Modification and Design of Shape'. In: *Proceedings of Computer Graphics International (CGI'00)*. IEEE Computer Society, Geneva, Switzerland, pp. 137–151. [341]

Wolter, F.-E., N. Peinecke, and M. Reuter: 2004, 'Geometric Modeling of Complex Shapes and Engineering Artifacts'. In: *Encyclopedia of Computational Mechanics*, Vol. 1 of *Encyclopedia of Computational Mechanics*. Wiley. [341]

Worsley, K. J., S. Marrett, P. Neelin, A. C. Vandal, K. J. Friston, and A. C. Evans: 1996, 'A Unified Statistical Approach for Determining Significant Signals in Images of Cerebral Activation'. *Human Brain Mapping* **4**, 58–73. [294]

Xia, Y.: 1989, 'Skeletonization via the Realization of the Fire Front's Propagation and Extinction in Digital Binary Shapes'. *IEEE Transactions on Pattern Analysis and Machine Intelligence* **11**(10), 1076–1086. [129]

Yamashita, M. and T. Ibaraki: 1986, 'Distance Defined by Neighborhood Sequences'. *Pattern Recognition* **19**, 237–246. [161]

Ye, Q.: 1988, 'The Signed Euclidean Distance Transform and its Applications'. In: *Proceedings of 9th ICPR*. Rome, pp. 495–499. [163]

Yokoi, S., J.-i. Toriwaki, and T. Fukumura: 1975, 'An Analysis of Topological Properties of Digitized Binary Pictures Using Local Features'. *Computer Graphics and Image Processing* **4**(1), 63–73. [168]

Yomdin, Y.: 1981, 'On the Local Structure of a Generic Central Set'. *Compositio Mathematica* **43**(2), 225–238. [1, 9]

Yu, X., J. Goldak, and L. Dong: 1991, 'Constructing 3-D Discrete Medial Axis'. In: *Symposium on Solid Modeling Foundations and CAD/CAM Applications*. pp. 481–489. [208]

Yu, Z.: 1989, 'Stabile Analyse von binärbildern'. Ph.D. thesis, Universität Hamburg, Inst. Angewandte Math. [200]

Yu, Z., C. Conrad, and U. Eckhardt: 1992, 'Regularization of the Medial Axis Transform'. In: R. Klette and W. Kropatsch (eds.): *Theoretical Foundations of Computer Vision*. pp. 13–24. [201]

Yuille, A. and M. Leyton: 1990, '3D Symmetry-Curvature Duality Theorems'. *Computer Vision, Graphics, and Image Processing* **52**, 124–140. [5, 12]

Yushkevich, P., P. T. Fletcher, S. Joshi, A. Thall, and S. M. Pizer: 2003, 'Continuous Medial Representations for Geometric Object Modeling in 2D and 3D'. *Image and Vision Computing* **21**(1), 17–28. [20, 262]

Yushkevich, P., H. Zhang, and J. Gee: 2005, 'Parametric Medial Shape Representation in 3-D via the Poisson Partial Differential Equation with Non-linear Boundary Conditions'. In: *International Conference on Information Processing in Medical Imaging*, Vol. LNCS 3565. pp. 162–173. [262, 263, 281]

Zeleznik, R. C., K. P. Herndon, and J. F. Hughes: 1996, 'SKETCH: An Interface for Sketching 3D Scenes'. In: *Computer Graphics Proceedings, SIGGRAPH'96.* New Orleans, LA. [337]

Zhao, H. K., S. Osher, and R. Fedkiw: 2001, 'Fast Surface Reconstruction Using the Level Set Method'. In: *IEEE Workshop on Variational and Level Set Methods.* Vancouver, BC, pp. 104–111. [147]

Zhou, Y., A. Kaufman, and A. W. Toga: 1998, 'Three-Dimensional Skeleton and Centerline Generation Based on an Approximate Minimum Distance Field'. *The Visual Computer* **14**(7), 303–314. [188, 343]

Zhu, S. C.: 1999, 'Stochastic Jump-Diffusion Process for Computing Medial Axes in Markov Random Fields'. *IEEE Transactions on Pattern Analysis and Machine Intelligence* **21**(11). [5]

Index